Anaerobic Fermentation—a Biological Route towards Achieving Net Neutrality

Editor

Sanjay Nagarajan

MDPI

Basel • Beijing • Wuhan • Barcelona • Belgrade • Novi Sad • Cluj • Manchester

Editor
Sanjay Nagarajan
Department of Chemical
Engineering
University of Bath
Bath
UK

Editorial Office
MDPI AG
Grosspeteranlage 5
4052 Basel, Switzerland

This is a reprint of articles from the Special Issue published online in the open access journal *Fermentation* (ISSN 2311-5637) (available at: https://www.mdpi.com/journal/fermentation/special_issues/anaerobic_fermentation).

For citation purposes, cite each article independently as indicated on the article page online and as indicated below:

Lastname, A.A.; Lastname, B.B. Article Title. *Journal Name* **Year**, *Volume Number*, Page Range.

ISBN 978-3-7258-2405-2 (Hbk)
ISBN 978-3-7258-2406-9 (PDF)
doi.org/10.3390/books978-3-7258-2406-9

Contents

About the Editor

Sanjay Nagarajan

Sanjay Nagarajan is a lecturer in Chemical Engineering and the Deputy Director of the Research Centre for Sustainable Energy Systems at the University of Bath. He obtained his BTech in Biotechnology (2009), MSc in Environmental Engineering (2011), and PhD in Chemical Engineering (2017) from Anna University, India, National University of Singapore, and Queens University Belfast, UK, respectively. He has more than 10 years of research experience in the field of advanced oxidation processes for biorefinery applications. His expertise in advanced oxidation processes mainly revolves around hydrodynamic cavitation and photocatalysis. His extensive experience in the area of hydrodynamic cavitation of lignocellulosic biomass for anaerobic digestion has been well received across the scientific community, as evidenced by his publications. He was also the 2018 Marie Curie Career FIT Plus Fellowship awardee. He has won notable national and international accolades, the most significant being the "Young Investigator Award" in chemical reaction engineering from the Italian Association of Chemical Engineering (2018). He has a broad academic and industrial network across the UK, the EU, and India. He has published more than 35 peer-reviewed papers in reputed international journals, written two book chapters, and authored a book on hydrodynamic cavitation. His scientific contributions have been recognized by the research community (¿1200 citations with an h-index of 20). He is involved in various national/international networks and consortia. The most notable of these was the position of management board member at the UK High Value Biorenewables Network, which focuses on establishing and enriching the critical mass of early-career researchers in the field of biorefineries.

fermentation

MDPI

Editorial

Anaerobic Fermentation—A Biological Route towards Achieving Net Neutrality

Sanjay Nagarajan [1,2,3]

[1] Department of Chemical Engineering, University of Bath, Claverton Down, Bath BA2 7AY, UK; sn908@bath.ac.uk
[2] Centre for Sustainable Energy Systems, University of Bath, Claverton Down, Bath BA2 7AY, UK
[3] School of Chemistry and Chemical Engineering, Queen's University Belfast, Belfast BT7 9AG, UK

Increasing greenhouse gas levels have led to the international community pledging to curb the mean global temperature increase to less than 1.5 °C. While the commitment to such stringent targets gained an increased support by the COP26 (November 2021) community, actions to achieving this have not been effective. The end of COP26 coincided with the launch of this Special Issue targeted towards "Net Neutrality" via "Anaerobic Fermentation" to understand the current status of research in this compelling area. With increased interest amongst the research community to contribute to this Special Issue, the final submission deadline was extended to November 2022, which coincided with COP27. The lack of stakeholder commitment seen in COP26 initiated the traction towards climate financing, leading to the 'Adaptation Fund' and the 'Least Developed Countries Fund' to help support needy countries to meet short-, medium-, and long-term climate action plan targets. This also led to the conversations of achieving net zero emissions rapidly as opposed to revisiting transition targets. As the guest editor, I feel extremely proud that a key takeaway from COP27—'nature-based solutions'—was addressed in this Special Issue to an extent and, therefore, I would like to thank all the authors for their valuable contributions.

Achieving net neutrality has to follow a sustainable circular economy pathway, and anaerobic-fermentation-based biological routes have a significant role to play in this remit. The Special Issue was vastly successful in capturing this research, with studies focused on the fate of enzymes, microbiomes and metabolic pathways, new product streams, and intensifying fermentation using engineering optimisation, as well as identifying routes for sustainable biorefineries via anaerobic fermentation.

Anaerobic fermentation is well established at a commercial scale in wastewater treatment plants around the globe. However, the removal of key nutrients such as orthophosphates is not often possible with conventional anaerobic digestion (AD) systems. Khumalo et al. investigated an aerobic–anaerobic sequencing batch reactor to tackle this problem and improve orthophosphate removal rates [1]. Conventional AD often focuses on biogas production for energy recovery from waste. While this aspect is commonly exploited for recovering value out of a variety of waste, its intensification for enhanced value addition is still lacking. On this front, Miftah et al. reported the use of choline chloride monoethanolamine as the most effective deep eutectic solvent to recover a cellulose-rich residue from sugarcane leaves upon pre-treatment [2]. Intensified biomethane production was observed alongside biohydrogen, leading to maximum energy recovery from the waste feedstock. In contrast to pre-treatment, Shin et al. investigated the use of additives, especially food-waste-derived biochar, for enhancing the biomethane yield [3]. Adding enzyme cocktails to AD reactors is a promising new strategy to enhance hydrolysis and methane production rates. However, this may not always result in positive biomethane yield enhancements due to a number of reasons, with enzyme stability being the predominant factor. Küchler et al. investigated this phenomenon and reported that lignocellulose-degrading

Citation: Nagarajan, S. Anaerobic Fermentation—A Biological Route towards Achieving Net Neutrality. *Fermentation* **2023**, *9*, 404. https://doi.org/10.3390/fermentation9040404

Received: 10 April 2023
Accepted: 18 April 2023
Published: 21 April 2023

enzymes had a half-life of ~1.5 h when added to an AD reactor [4]. They established a workflow to monitor the stability of such enzymes, which is crucial in determining the efficiency of the process. Buriánková et al. investigated the microbial communities found in underground gas reservoirs, examining their unique metabolic pathways [5]. Rigorous qPCR- and sequencing-based methods were utilised to decipher the novel communities (from water samples taken over a two-year period). They concluded that such reservoirs could behave as natural fermenters for the bioconversion of CO_2 and H_2 to CH_4. However, engineered fermenters are still the state of the art. One of the limitations of commercial fermenters is achieving an appropriate mass transfer enabled via overcompensated mixing. Such cases would lead to excessive energy use for the fermenter's operation. Singh et al. investigated the significance of intermittent mixing by employing a helical ribbon impeller in the digester for the production of biogas [6]. They determined that volatile fatty acids (VFA) accumulation in addition to the specific power consumption by the digester considerably reduced, leading to an enhanced biogas yield as a result of intermittent mixing.

Beyond biogas, Robazza et al. investigated the anaerobic co-fermentation of the pyrolysis aqueous condensate and syngas to produce L-Malate, an important high-value biochemical [7]. Their work showed the potential of simultaneous detoxification as well as valorisation. Pinto et al., on the other hand, optimised an open microbiome towards biobutanol production as opposed to pure culture-based fermentation [8]. The problem with identifying multiple fermentation by-products was correlated to the highly diverse microbial community due to the use of an undefined microbial inocula (via 16S rRNA amplicon analysis).

VFAs are key intermediates in AD, which are of high value when obtained in high concentrations. Therefore, these products present a potential opportunity. Highlighting the importance of VFAs in a biorefinery, Nzeteu et al. presented a review on the potential valorisation routes of waste biomass to VFAs, followed by key products such as bioplastics and other high-value biochemicals [9]. The potential is truly unique, as many fossil-fuel-derived chemicals can be replaced via the VFA-based biorefinery routes. While Nzeteu emphasized this, Nagarajan et al.'s review on the production facets of VFA added another dimension to the Special Issue. The production of VFAs have to be intensified to be able to make the valorisation pathways economically viable. Nagarajan et al. discussed these perspectives in light of biohydrogen and VFA production by critically analysing the available pilot-scale state-of-the-art examples [10].

Overall, the Special Issue can be viewed as having three sections: (i) progressing the understanding of anaerobic digestion across disciplines; (ii) highlighting the potential for new products via anaerobic digestion; and (iii) identifying the suitable valorisation pathways for enabling a circular bioeconomy via VFAs. These 'nature-based solutions' inline with COP27 takeaways reassure us that the scientific community is progressing in the right direction towards achieving net neutrality. The leap to 'achieving net zero' from the current 'transition' mindset however still requires significant efforts from political, scientific, and commercial stakeholders to ensure financial viability.

Funding: This research received no external funding.

Acknowledgments: I would like to thank the authors, reviewers, and editorial team at *Fermentation* who made this Special Issue possible.

Conflicts of Interest: The author declares no conflict of interest.

References

1. Khumalo, S.M.; Bakare, B.F.; Tetteh, E.K.; Rathilal, S. Sequencing Batch Reactor Performance Evaluation on Orthophosphates and COD Removal from Brewery Wastewater. *Fermentation* **2022**, *8*, 296. [CrossRef]
2. Miftah, A.K.; Sittijunda, S.; Imai, T.; Salakkam, A.; Reungsang, A. Biohydrogen and Methane Production from Sugarcane Leaves Pretreated by Deep Eutectic Solvents and Enzymatic Hydrolysis by Cellulolytic Consortia. *Fermentation* **2022**, *8*, 396. [CrossRef]
3. Shin, D.-C.; Kim, I.-T.; Jung, J.; Jeong, Y.; Lee, Y.-E.; Ahn, K.-H. Increasing Anaerobic Digestion Efficiency Using Food-Waste-Based Biochar. *Fermentation* **2022**, *8*, 282. [CrossRef]

4. Küchler, J.; Willenbücher, K.; Reiß, E.; Nuß, L.; Conrady, M.; Ramm, P.; Schimpf, U.; Reichl, U.; Szewzyk, U.; Benndorf, D. Degradation Kinetics of Lignocellulolytic Enzymes in a Biogas Reactor Using Quantitative Mass Spectrometry. *Fermentation* **2023**, *9*, 67. [CrossRef]
5. Buriánková, I.; Molíková, A.; Vítězová, M.; Onderka, V.; Vítěz, T.; Urbanová, I.; Hanišáková, N.; Černý, M.; Novák, D.; Lochman, J.; et al. Microbial Communities in Underground Gas Reservoirs Offer Promising Biotechnological Potential. *Fermentation* **2022**, *8*, 251. [CrossRef]
6. Singh, B.; Kovács, K.L.; Bagi, Z.; Petrik, M.; Szepesi, G.L.; Siménfalvi, Z.; Szamosi, Z. Significance of Intermittent Mixing in Mesophilic Anaerobic Digester. *Fermentation* **2022**, *8*, 518. [CrossRef]
7. Robazza, A.; Welter, C.; Kubisch, C.; Baleeiro, F.C.F.; Ochsenreither, K.; Neumann, A. Co-Fermenting Pyrolysis Aqueous Condensate and Pyrolysis Syngas with Anaerobic Microbial Communities Enables L-Malate Production in a Secondary Fermentative Stage. *Fermentation* **2022**, *8*, 512. [CrossRef]
8. Pinto, T.; Grimalt-Alemany, A.; Flores-Alsina, X.; Gavala, H.N.; Gernaey, K.V.; Junicke, H. Shaping an Open Microbiome for Butanol Production through Process Control. *Fermentation* **2022**, *8*, 333. [CrossRef]
9. Nzeteu, C.; Coelho, F.; Davis, E.; Trego, A.; O'Flaherty, V. Current Trends in Biological Valorization of Waste-Derived Biomass: The Critical Role of VFAs to Fuel A Biorefinery. *Fermentation* **2022**, *8*, 445. [CrossRef]
10. Nagarajan, S.; Jones, R.J.; Oram, L.; Massanet-Nicolau, J.; Guwy, A. Intensification of Acidogenic Fermentation for the Production of Biohydrogen and Volatile Fatty Acids—A Perspective. *Fermentation* **2022**, *8*, 325. [CrossRef]

fermentation

MDPI

Article

Microbial Communities in Underground Gas Reservoirs Offer Promising Biotechnological Potential

Iva Buriánková [1,†], Anna Molíková [1,†], Monika Vítězová [1,*], Vladimír Onderka [2], Tomáš Vítěz [1,3,*], Iva Urbanová [2], Nikola Hanišáková [1], Martin Černý [1], David Novák [4], Jan Lochman [4], Josef Zeman [5], Jakub Javůrek [6], Markéta Machálková [6], Linda Dengler [7] and Harald Huber [7]

1 Laboratory of Anaerobic Microorganisms, Section of Microbiology, Department of Experimental Biology, Faculty of Science, Masaryk University, 625 00 Brno, Czech Republic; ivaburiankova@seznam.cz (I.B.); molikova.anna@gmail.com (A.M.); 446645@mail.muni.cz (N.H.); 451182@muni.cz (M.Č.)
2 Testlab Geo-Services, RWE Gas Storage CZ, s.r.o., 627 00 Brno, Czech Republic; vladimir.onderka@rwe.com (V.O.); iva.urbanova2@rwe.com (I.U.)
3 Department of Agricultural, Food and Environmental Engineering, Faculty of AgriSciences, Mendel University in Brno, 613 00 Brno, Czech Republic
4 Department of Biochemistry, Faculty of Science, Masaryk University, 625 00 Brno, Czech Republic; davidnovak@mail.muni.cz (D.N.); jlochman@seznam.cz (J.L.)
5 Department of Geological Sciences, Faculty of Science, Masaryk University, 602 00 Brno, Czech Republic; jzeman@sci.muni.cz
6 Tescan Orsay Holding, 623 00 Brno, Czech Republic; jakub.javurek@tescan.com (J.J.); marketa.machalkova@tescan.com (M.M.)
7 Department of Biology and Preclinical Medicine, Institute of Microbiology and Archaea Centre, University of Regensburg, 93053 Regensburg, Germany; linda.dengler@microbify.com (L.D.); harald.huber@biologie.uni-regensburg.de (H.H.)
* Correspondence: vitezova@sci.muni.cz (M.V.); vitez@sci.muni.cz (T.V.)
† These authors contributed equally to this work.

Citation: Buriánková, I.; Molíková, A.; Vítězová, M.; Onderka, V.; Vítěz, T.; Urbanová, I.; Hanišáková, N.; Černý, M.; Novák, D.; Lochman, J.; et al. Microbial Communities in Underground Gas Reservoirs Offer Promising Biotechnological Potential. *Fermentation* **2022**, *8*, 251. https://doi.org/10.3390/fermentation8060251

Academic Editor: Sanjay Nagarajan

Received: 1 May 2022
Accepted: 24 May 2022
Published: 26 May 2022

Abstract: Securing new sources of renewable energy and achieving national self-sufficiency in natural gas have become increasingly important in recent times. The study described in this paper focuses on three geologically diverse underground gas reservoirs (UGS) that are the natural habitat of methane-producing archaea, as well as other microorganisms with which methanogens have various ecological relationships. The objective of this research was to describe the microbial metabolism of methane in these specific anoxic environments during the year. DNA sequencing analyses revealed the presence of different methanogenic communities and their metabolic potential in all sites studied. Hydrogenotrophic *Methanobacterium* sp. prevailed in Lobodice UGS, members of the hydrogenotrophic order *Methanomicrobiales* predominated in Dolní Dunajovice UGS and thermophilic hydrogenotrophic members of the *Methanothermobacter* sp. were prevalent in Tvrdonice UGS. Gas composition and isotope analyses were performed simultaneously. The results suggest that the biotechnological potential of UGS for biomethane production cannot be neglected.

Keywords: methanogenic archaea; methanogenesis; underground gas storage; power to methane; green energy; CO_2 utilization

1. Introduction

Microbiological studies of underground gas storage (UGS) have demonstrated the presence of viable microorganisms [1,2] and their significant influence on biological and geochemical processes in these environments [3]. It is likely that Šmigáň and co-workers in 1990 were the first to observe the changes in the composition of the town gas in UGS caused by microorganisms. The recorded decrease in the amount of H_2 and CO_2 combined with the increase in the amount of CH_4 in the Lobodice UGS, Czechia, indicated the possibility that microbial communities inhabit the UGS. The presence of methanogenic archaea, which appeared to be responsible for the changes in gas quality [3,4], was also confirmed. Thirty years later, our stable isotope measurements confirmed biological methane production at the same site and at other sites. Power-to-gas technology with biological methane production in underground reservoirs seems to be one of the most promising options for carbon-neutral fuel production and storage [5]. The UGS can serve as a reactor where the biological conversion of CO_2 and H_2 to biomethane takes place [6]. Sources of CO_2 can be the thermal sector, various industries, or direct capture of CO_2 from the air [7]. Hydrogen can be produced by electrolysis of water or using surplus renewable energy [8], while hydrogen from waste treatment could also be used in the future [9]. It is also clear that methane will be treated as an energy carrier until safe hydrogen storage and distribution technologies are available [10]. The objective of this study was to determine whether a UGS natural sedimentary rock environment could be a potential target for biomethane production through power-to-methane technology. Although many scientific teams have mapped the microbiome of the subsurface environment (Table 1), the idea of using UGS as a natural bioreactor is new and topical. The enormous capacity of UG enables the production and storage of biological methane in quantities of millions of cubic meters. Three geologically different sites in the Czech Republic—Lobodice, Dolní Dunajovice, and Tvrdonice—serving as UGS were studied in terms of microbial communities using sequencing technology and quantitative PCR to assess their methanogenesis processes.

Table 1. Selected investigations performed in similar rock environment.

Author, Reference	Year	Country	Focus of Study
Šmigán et al. [4]	1989	Czech Republic	methanogenic archaea
Buzek et al. [4]	1994	Czech Republic	microbial methane production
Pedersen et al. [11]	1996	Sweden	microbial diversity
Kotelnikova et al. [12]	1997	Sweden	methanogenic archaea, homoacetogenic bacteria
Fry et al. [13]	1997	USA	microbial diversity
Shimizu et al. [14]	2006	Japan	microbial diversity
Ivanova et al. [2]	2007	Russia	microbiological
Basso et al. [1]	2009	France	microbial diversity
Kimura et al. [15]	2010	Japan	microbial methane study
Flynn et al. [16]	2013	USA	functional microbial diversity
Wu et al. [17]	2016	Sweden	microbial diversity, metabolism
Frank et al. [18]	2016	Russia	variability in microbial composition
Kadnikov et al. [19]	2017	Russia	microbial diversity
Vigneron et al. [20]	2017	USA	microbial methane study

Methanogenesis

Methanogenesis, the final step in the decomposition of organic matter, is carried out by methanogenic archaea, which play an important role in the global carbon cycle and are responsible for more than half of all methane produced on Earth per year [21]. In the last four decades, methanogenesis has also been described as the dominant metabolic pathway in very deep aquifers [22]. Methanogenic archaea are strictly anaerobic microorganisms and require an environment with low redox potentials of about −300 mV for their growth. Under these conditions, where all other favorable electron acceptors, such as oxygen, nitrate, sulphate, and iron compounds, are depleted or absent, methanogenesis can occur [23].

Methanogenesis is not a uniform process. Methane can be formed via three major pathways: hydrogenotrophic, methylotrophic, or acetoclastic. However, one major characteristic enzyme is present in all types of methanogenesis: methyl-coenzyme M reductase (MCR), which catalyzes the final step of methyl group reduction to methane [24].

The most widespread and probably the oldest form of methane production is hydrogenotrophic methanogenesis. This pathway is characterized by the conversion of CO_2 and H_2 to methane; molecular hydrogen serves as the electron donor and CO_2 as the electron acceptor. The reduction of formate also occurs via the hydrogenotrophic pathway. The only difference is in the first step, where formate is oxidized to CO_2 and then the CO_2

continues the pathway. The hydrogenotrophic pathway is widely used; for example, by the microorganisms of the orders *Methanobacteriales*, *Methanomicrobiales*, and *Methanococcales*.

The second known pathway is the methylotrophic pathway, in which methylated C_1 compounds (methanol, methylamines, or methyl sulfides) are first activated by specific methyltransferases [25]. Usually, four C_1 compounds are involved in this reaction. One of the methyl groups is oxidized to CO_2 and the remaining three methyl groups are reduced to methane. Methylotrophic methanogens include members of the orders *Methanobacteriales* and *Methanosarcinales* as well as the recently proposed *Bathyarchaeota*, *Verstraetearchaeota*, and *Methanomassiliicoccales*. In addition, some methanogens (members of *Verstraetearchaeota*) were found to be exclusively methylotrophic. Based on this, it was hypothesized that methylotrophic methanogenesis evolved as an independent ancient pathway [26].

Furthermore, two processes have been described for methanogenesis from acetate. The first is acetoclastic methanogenesis, where acetate is cleaved into carboxyl and methyl groups. The substrate for methanogenesis is acetate, which undergoes a disproportionate reaction. The carbon in the methyl group is reduced to methane, while the carbon in the carboxyl group is oxidized to CO_2. Two main genera of methanogens (*Methanosarcina*, *Methanothrix*) are able to use this pathway [27]. The second process is based on syntrophic mutualistic reactions. Less is known about this syntrophic metabolism, which catalyzes the oxidation of acetate to hydrogen and carbon dioxide (SAO) by SAO bacteria. It is possible that syntrophic acetate-oxidizing (SAO) bacteria facilitate acetate consumption and could be coupled with hydrogenotrophic methanogenesis (SAO-HM) [28].

2. Materials and Methods

2.1. Locality and Geological Preconditions

The Lobodice UGS is an aquifer type of UGS located in the central part of the Carpathian Foredeep and formed by an anticlinal structure at an average depth of 450 m. The basement is formed by Proterozoic and Paleozoic crystalline rocks, such as amphibolitic shales. The overlying reservoir sediments consist of lower Badenian clastic sediments, mainly conglomerates, and the caprock sealing the entire structure consists of lower Badenian clays.

The Dolní Dunajovice UGS is a depleted gas reservoir located in the southern part of the Carpathian Foredeep filled with Miocene sediments. Its structure consists of Carpathian sediments, mudstones, sandstones, and siltstones overlying the Eggenburgian sediments. The Eggenburgian mudstones form the caprock of the UGS and the Eggenburgian basal clastics glauconitic sandstones serve as the UGS horizon at an average depth of 1100 m. Jurassic carbonates (Kurdejov limestones) form the UGS basement.

The Tvrdonice UGS is a depleted set of gas and oil reservoirs located in the Czech part of Vienna basin. The sedimentary complex overlies the crystalline rocks of Brunovistulikum, and the thickness of the sediments reaches over 5 km in some parts of the Vienna basin. The hydrocarbon reservoirs, later converted to UGS, are located at a depth of 900 to 1600 m from the Sarmatian–Badenian boundary interval.

Gas and oil fields are generally structural traps tied to a fault system and subdivided into specific smaller segments and horizons. The reservoir horizons are formed by sandy strata interbedded with mudstones.

2.2. Sampling

Two types of water samples were collected from 16 wells over a two-year period. Water samples from overflow and groundwater were collected using a special rope technique and a sterile subsurface sampler—a stainless steel capsule with a volume of 0.75 L (Leutert GmbH, Adendorf, Germany).

Samples were collected from the overflow by passing water from the well through a 200 L barrel into which a sterile 1 L glass bottle was submerged. This sample was used for molecular biological analyses. For sample cultivation, a 5 L canister with a drain valve was filled in a similar manner. The water sample (30 mL) from this canister was then transferred to a sterile 150 mL culture vial fitted with a rubber stopper. A sterile tube was connected to the drain valve of the canister, at the end of which a sterile needle was inserted into a Luer-Lock thread to pierce the rubber stopper of the vial. The sterile culture vials were flushed with nitrogen several times before use. The nitrogen was then aspirated, and the vial gas phase was under vacuum so that the sample could be aspirated from the canister. For the subsurface sampler, we connected a sterile hose to the sampler valve after releasing the pressure in the sampler, which led to a 0.5 L sampling tube. As the water flowed through the tube, we punctured the wall of the tube with a sterile syringe and needle and collected water for culture. This water (30 mL) was transferred with a syringe to a sterile culture vessel that was under negative pressure. The water that flowed through the tube into the sterile 0.5 L sample tube was then used for molecular biology analyses. In both cases, culture was performed according to the procedures described in [29,30]. Water samples were transported to the laboratory under anaerobic conditions in a special cooling box. A total of 20 samples were collected, comprising 11 overflow water samples and 9 groundwater samples (Table S1).

2.3. Physical-Chemical Parameters and Groundwater Chemical Composition

Measurements of pH, redox potential (ORP), electrical conductivity, and temperature were conducted on site using the WTW Multi 350i (accuracy ±0.01 for pH, ±0.2 for ORP, $\pm0.5\%$ for electrical conductivity and ±0.1 °C for temperature). The SenTix 41 electrode was used to measure pH and temperature. The SenTix ORP electrode was used to measure redox potential. The ORP value measured against Ag/AgCl electrode was recalculated against standard hydrogen electrode (SHE) according to the operating manual. The WTW TetraCon 325 electrode was used to measure electrical conductivity. Prior to analysis, all groundwater samples were filtered through a 0.45 μm membrane filter (Millipore, HAWG047S6). Chemical analyses of the main groundwater components were performed in the chemical laboratory of the Department of Geological Sciences, Faculty of Science, Masaryk University, according to standard laboratory procedures. Basic data processing was performed in Microsoft® Excel spreadsheet, advanced data processing, geochemical calculations and geochemical modelling were performed in Geochemist's Workbench®, release 12.0.4. (Aqueous Solutions LLC, Champaign, IL, USA).

2.4. Degas Analysis

2.4.1. Isotopic Determination

Isotopic determination of $\delta^{13}C$ in CH_4 and δD in CH_4 of the gas samples dissolved in water was performed using a Picarro Cavity Ringdown Spectrometer (CRDS), a G2201-*i* Analyzer for Isotopic CO_2/CH_4, and a G2182-*i* Analyzer for δD & $\delta^{13}C$ in CH_4. Isotopic determination of $\delta^{18}O$, δD, and $\delta^{17}O$ in water was performed using the CRDS L2140-i Analyzer for Isotopic H_2O (all by Picarro, Inc., Santa Clara, CA, USA). Instrument setup and sample preparation in terms of appropriate concentration range, standards, etc. are described in the manufacturer's guidelines and in the standard operating procedures of Testlab Geo-Services (RWE Gas Storage CZ, Ltd., Brno, Czech Republic).

2.4.2. Gas Chromatography

Analysis of the gas samples was performed using the Agilent 7890B Gas Chromatograph (Agilent Technologies, Inc., Santa Clara, CA, USA), a three-channel system using TCD-TCD-FID detectors. The gas chromatograph was equipped with two precolumns HiSep Q, and three separation columns (HP-Plot Q, Molsieve 5a, and HP Molsieve), all of which were 0.53 mm in diameter. The mobile phase was Ar 5.0 and He 5.0 (SIAD Czech, Ltd., Rajhradice, Czech Republic). The thermal program and other parameters were set according to the standard operating procedure of Testlab Geo-Services (RWE Gas Storage CZ, Ltd., Brno, Czech Republic). (Results shown in Table S2).

2.5. Microscopy

Samples selected for analysis by Scanning Electron Microscopy (SEM) were prepared to withstand the high vacuum conditions in the SEM chamber. An appropriate volume (20 mL) of the water sample was filtered onto a polycarbonate membrane filter (0.2 μm, Merck Millipore, Guyancourt, France) using a vacuum filtration device (Merck Millipore, Guyancourt, France). The cells on filters were fixed with 2% glutardialdehyde in 0.1 M sodium cacodylate buffer (pH 7.4). After one hour at room temperature, followed by overnight fixation and drain off fixative, the samples were immediately transferred to 50% ethanol solution (EtOH). After fixation, small pieces of the filters were dehydrated by a graded EtOH series (70%, 85%, 95% and twice 100% EtOH), each step taking approximately 20 min at room temperature. The filter pieces saturated with 100% EtOH were dried with CO_2 (K850 Critical Point Dryer, Quorum Technologies, Lewes, UK) at the critical point and placed on a stub with conductive carbon tape. To increase the conductivity of the samples, the filters were sputter-coated with 5 nm of iridium (Q150T ES, Quorum Technologies, Lewes, UK). Samples were analyzed using a high-resolution field emission scanning electron microscope TESCAN CLARA (TESCAN ORSAY HOLDING, Brno, Czech Republic). All images were acquired at a low accelerating voltage of 1 keV using the in-column Axial detector.

2.6. Molecular Biological Methods

2.6.1. DNA Isolation

Well water samples (0.5 L from the subsampler, 1.5–2 L from the overflow) were filtered on 0.22 μm membrane filters (GTTP, Millipore, France) and subjected to DNA isolation using a kit according to the manufacturer's instructions (DNeasy Power Water Kit, Quiagen, Hilden, Germany). DNA quality was measured using NanoDrop 2000 UV-Vis spectrophotometer and Qubit TM fluorometer (Thermo Fisher Scientific, Waltham, MA, USA).

2.6.2. Quantitave PCR

The qPCR (quantitative polymerase chain reaction) method was used for absolute quantification of methanogens. A pure culture of *Methanobrevibacter smithii* was used as a template for the standard. DNA concentration was measured using a fluorometer Qubit (Thermo Fisher Scientific, Waltham, MA, USA) and the sample was diluted to the required concentrations (10^4–10^8 copies per μL). The reaction was performed on Light Cycler 480 (Roche, Basel, Switzerland) in triplicate for each sample. The reaction volume was 14 μL, including 4 μL of template DNA and 9 μL of Luna Master Mix (BioLabs, Ipswich, MA, USA) with two forward (0.25 μL per one) and one reverse primer (0.5 μL), each at a final concentration of 250 nM. The primer was targeted to the mcrA gene, which is supposed to be a single copy gene (one gene per methanogen cell). A combination of three primers was designed for this study. The reverse mcrA primer 5′-CGTTCATBGCGTAGTTVGGRTAGT-3′ was used in combination with equal volumes of the two forward primers mcrAF1 5′-ACTTCGGTGGATCDCARAGRGC-3′ and mcrAF2 5′-ACTTCGGCGGTTCDCARAGRGC-3′ [31,32]. The reaction conditions included an initial denaturation step at 95 °C for 3 min, followed by 35 cycles of denaturation at 95 °C for 30 s, annealing at 55 °C for 45 s, and extension at 72 °C for 30 s, with a ramp rate of 0.1 °C/s from the annealing to the extension temperature, followed by a final extension step at 72 °C for 10 min. Due to fluctuations in the signal during the first five cycles, fluorescence was read from the sixth cycle onward. The expected length of the amplicons was approximately 300 bp.

2.6.3. Illumina—Next Generation Sequencing Method

DNA extracted from groundwater was used as the template for PCRs with specific primers flanking the V4 region of the 16S rRNA gene sequence [33]. Amplification was performed using Platinum™ II Taq Hot-Start DNA Polymerase (ThermoFisher) at 0.8× according to the Earth Microbiome protocol [34]. After PCR, the amplification products were purified using Agencourt® AMPure XP Beads (Beckman Coulter, Brea, CA, USA).

Subsequently, the purified PCR samples were quantified and normalized using the Qubit fluorometer. The normalized PCR products were pooled, and their length and quality were checked using the DNF-474 HS NGS Fragment Kit for Fragment Analyzer (Agilent, Santa Clara, CA, USA). The final library was sequenced using an Illumina MiniSeq sequencer together with the Mid Output Kit (2 × 150 paired end sequencing) according to the manufacturer's instructions (Table S3).

Raw fastq reads were processed using the DADA2 package (version 1.16.0), [35] in R (version 4.0.0). Analysis was performed according to the standard operating procedure. Reads were first filtered, then trimmed, de-replicated, and de-noised. Then, forward and reverse reads were merged, chimaeras were removed, and the taxonomy was assigned by the RDP naive Bayesian classification [36] against the Silva database [37]. Multiple alignments were performed using the DECIPHER package and a phylogenetic tree was constructed using the phangorn package [38]. Phylogenetic and statistical analyses were then performed in R using the phyloseq package [39]. The datasets generated and analyzed during the current study are available in SRA under the project number BioProject ID: PRJNA759841.

To assess the metabolic potential of the communities based on the 16S rRNA gene amplicon data alone, functional annotation was performed using the FAPROTAX database [40]. The normalized and curated OTU abundances were assigned to a phylogenetically conserved functional group from information based on functional annotations of cultivated representatives. The program assumes that all cultivated and non-cultivated members can perform the functions verified in the database. The total DNA extracted from UGS water samples does not exclusively represent the metabolically active part of the community, as DNA from dormant and dead organisms is also extracted simultaneously [41]. For a deeper understanding of the metabolism of the microbial community metabolism, transcriptomic or proteomic data would be required and should be considered.

3. Results

3.1. Physicochemical Parameters and Water Chemistry

Characteristics of Groundwater

The chemical composition and physicochemical parameters of the studied UGS groundwaters are shown in the diagrams of Piper and Durov in Figure S1. Lobodice UGS groundwaters are slightly alkaline, pH is in the range of 7.36–8.70, and redox potential varies in the range of −238 to −399 mV, which means strongly reducing conditions. The groundwater has a relatively low mineralization of 2.4–3.1 g/L. Among the cations, sodium ions are predominant; among the anions, chlorides and bicarbonate are comparable, with bicarbonate slightly predominant. Water from caprock shows about twice as much mineralization. The ratio of major anion and cation concentrations is almost the same as for typical groundwaters from the Lobodice UGS. The groundwaters from the Tvrdonice UGS are also slightly alkaline (pH 8.10 and 8.45) with reducing conditions (−130 and −115 mV). Compared to the Lobodice UGS, they have about three times the mineral content (10.21 and 10.61 g/L). Their chemical composition is quite simple: of the main constituents, only sodium and chloride ions are present, while the concentrations of the other constituents are only in the two-digit milligram range. The groundwater of the Dunajovice UGS is almost identical in chemical composition to the groundwater of the Tvrdonice UGS, except that it has a three-times higher mineral content of about 30 g/L. The pH values are also in the characteristic range (7.65 and 8.07), and the redox potential indicates a typical anoxic environment (−138 and −142 mV).

All three types of UGS groundwaters are typical synsedimentary waters, and their properties reflect the sedimentary conditions and characteristics of the rock environment in which they are contained. The carbon, nitrogen, and sulfur components are major constituents that determine the oxidation–reduction processes. For the carbon component, the pH values are just around the groundwater saturation limit for calcite (Figure S2a), indicating that the groundwater is saturated relative to calcite. The solubility product of calcite determines the concentrations of calcium and bicarbonate ions, and conversely, pH values are buffered by the equilibrium between dissolved carbonate species and calcite.

The redox potential of the samples where groundwater was discharged from the wells for several hours is, with two exceptions, in the range where nitrogen species are stable as NH_4^+ (Figure S2b), and approach the limit of redox transformation of SO_4^{2-}/ HS^- manifested by pyrite precipitation under the given conditions (Figure S2c). Pyrite is present in the UGS sedimentary rocks and provides reducing conditions. As for carbon species, the pH and Eh conditions are close to the transition boundary between dissolved carbonate species (CO_2 and HCO_3^-) and methane (CH_4). The speciation diagram for carbon species (Figure S2d) shows that the conversion of carbon dioxide to methane begins at a redox potential of −225 mV. The physicochemical conditions and chemical composition of the studied UGS groundwaters are suitable for biologically assisted methanation, i.e., the conversion of carbon dioxide to methane by the action of molecular hydrogen. These conditions are supported by other redox-active components (NH_4^+ and HS^-).

3.2. Isotopic Determination

The results of the isotopic analysis of the gas and water samples indicate the origin of these samples. These results and an example of the typical isotopic composition of the injected gas are given in Table S4. The raw δD values measured in CH_4 must be corrected using the δD values measured in water, because there is an equilibrium between these values, and the values in methane are affected by the values in water. The data listed in Table S4 are also shown in Figure 1. The structure in the background [42] helps to immediately assign the samples to a specific type of origin. As can be seen in Figure 1, the samples from Lobodice and Dolní Dunajovice are clearly of microbial origin. The results for the samples from Tvrdonice indicate that these degassing samples are a mixture of microbial and thermogenic gas. For comparison, the values for injected gas have also been added to this figure to illustrate the difference between the thermogenic gas typically stored in the UGS and the gases that were influenced (at least to some degree) by microbial processes.

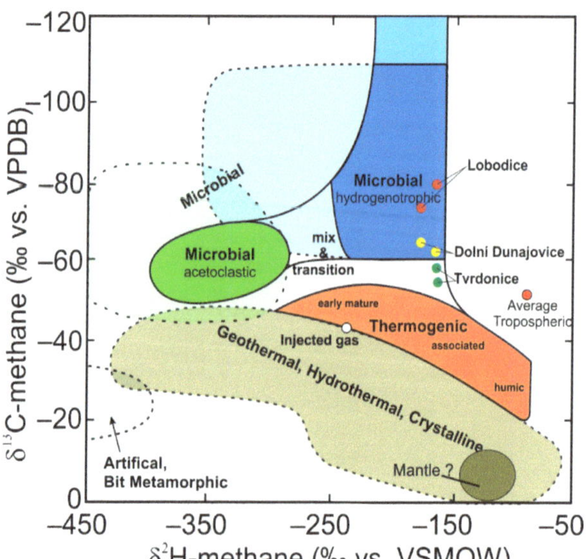

Figure 1. C–D plot [42] with measured data.

3.3. Microscopy

The SEM-micrograph shows a wide variety of morphological types of microorganisms, predominantly the accumulation of rod-shaped cells with a length of 1–4 μm and cocci with a diameter of about 0.3 μm (Figure 2).

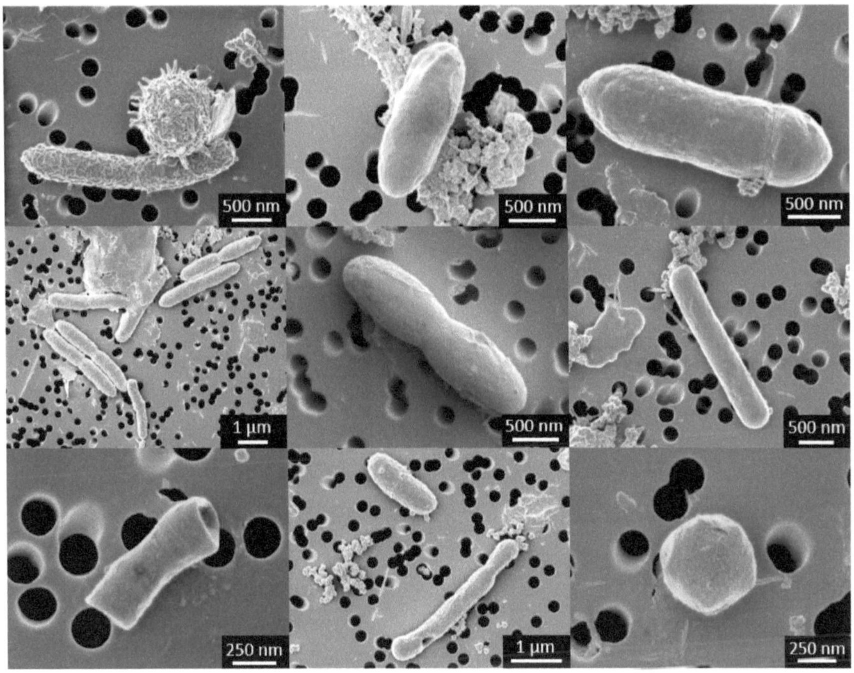

Figure 2. Examples of microbial morphotypes: rods and cocci found in well TVR-B captured by SEM. Inorganic nanoparticles which form conglomerates can be seen.

3.4. Quantitative Analyses of Methanogens (qPCR)

Of all sites, 14 wells were sampled, ten of which were sampled once and four of which were sampled repeatedly in different seasons, focusing on specific UGS regimes (injection–extraction gas periods) (Table S2). Of the total twenty samples from the dataset, DUN-B (December 2018), LO-H (November 2019), LO-H (October 2018), and LO-I (March 2019) were discarded due to low DNA concentration. The number of mcrA gene copies (Table S5) found in the Lobodice samples ranged from 7.20×10^1 to 2.40×10^7 in 1 μL of DNA isolated from 1 L of well water. The highest number of mcrA gene copies (2.40×10^7) was detected in the well LO-C (October 2018), while the lowest number (2.00×10^2) was found in LO-F (May 2018). In Dolní Dunajovice, the mcrA gene copies ranged from 1.7×10^5 to 1.03×10^6, and in Tvrdonice from 1.86×10^4 to 1.42×10^6. Certain wells in Lobodice were repeatedly sampled at different times of the year over a two-year period. When comparing the number of mcrA gene copies, the hypothesis about the influence of the UGS regime (injection/withdrawal periods for gas) on methanogen abundance cannot be confirmed.

3.5. Metagenomic Analyses of UGS Archaeal Community

3.5.1. Next-Generation Sequencing (NGS) Analysis

Next-generation sequencing (NGS) analysis targeting archaeal and bacterial 16S rRNA genes was performed to elucidate microbial community structures in anoxic groundwater from deep aquifers. The 24 samples were sequenced along with 16 independent samples on

Illumina MiniSeq using the MidOutput Kit (2 × 150 bp). The total run yield was 2.73 Gbp, with 88.39% of the reads passing the quality filter (>Q30), resulting in 8,689,550 reads passing the filter. The average error rate of the sequencing run was 0.98%. The DADA2 algorithm extracted 2864 unique ASVs from the 24 samples (Table S6). The sequencing depth obtained far exceeded the requirements, as shown by the rarefaction curve (Figure S3) with a minimum of 94,616 reads and a mean of 150,555 reads per sample.

For Illumina MiniSeq sequencing, the V4 region of the 16S rRNA gene was amplified [34,43]. The 16S rRNA gene is ubiquitous, occurs in several bacterial and archaeal species, and is highly conserved. Members of both domains were detected in varying unexpected proportions in all samples (archaea comprised 0.2%–75.7% of the microbial community in all sampled wells). Archaeal community composition differed slightly among the three sites, as did environmental and physicochemical conditions (Figure S4).

All samples collected were positive for the presence of methanogenic archaea. The 16S rRNA marker was used for sequence analyses and indicated the presence of an archaeal community in each well (Figure 3). The archaea were represented by five identified phyla, *Euryarchaeota* and *Crenarchaeota*, *Hydrothermarchaeota*, *Nanoarchaeaeota* and *Hadesarchaeaeota*, the last being a recently proposed phylum of thermophilic microorganisms found in deep mines, hot springs, marine sediments, and other subsurface environments [44]. This includes eight discovered classes, seven orders, thirteen families and fifteen genera. Sequencing of the 16S rRNA gene shows that the majority of the 140 archaeal OTUs belong to methanogens and consist of 10 genera and two recently discovered *Candidatus* species, whose abundance accounted for more than 5% of the total community.

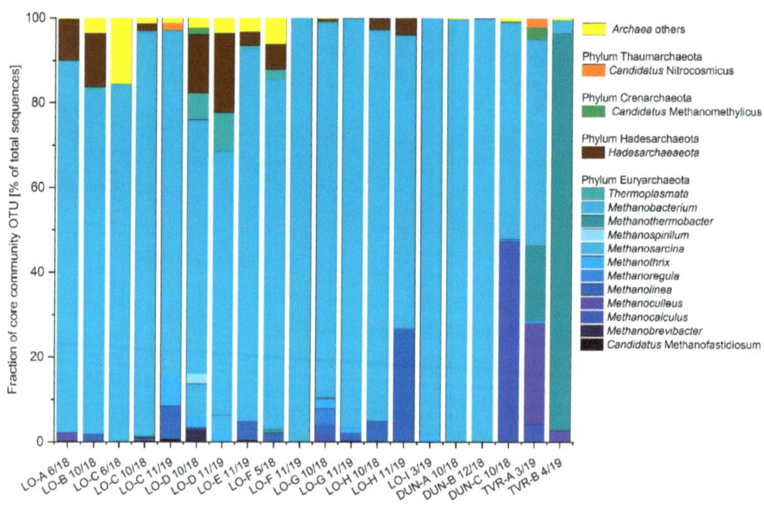

Figure 3. Archaeal community—16S rRNA gene sequence focused on methanogens detected in water samples from UGS's (5% cut).

The composition of the methanogenic archaea community detected in the UGS water samples shows that the major metabolic pathway in the UGS environment is hydrogenotrophic methanogenesis, with hydrogen and carbon dioxide as carbon and energy sources (Figure 4). In general, the genus *Methanobacterium* predominated in almost all samples, followed by the genus *Methanothermobacter* (present exclusively in Tvrdonice) and *Methanolinea*. The metabolism of these genera is exclusively hydrogenotrophic methanogenesis. The acetoclastic members of the genera *Methanothrix* and *Methanosarcina* were represented by large numbers of cells in the LO-A and LO-D wells.

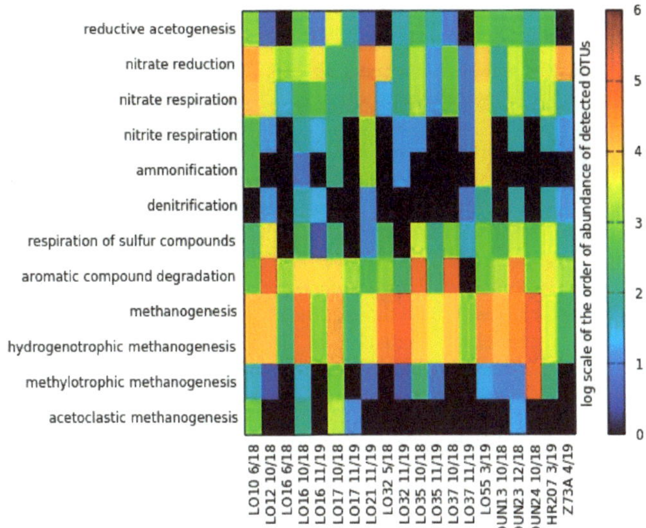

Figure 4. Selected metabolic groups predicted using the database of functional annotations of prokaryotic organisms, FAPROTAX.

Exceptionally high numbers of members of the genus *Methanocalculus* were found in the DUN-C well, where they make up half of the archaeal community. *Methanothermobacter* genera absolutely dominate in TVR-B and contribute to the archaeome in TVR-A, as do *Methanoculleus* members. In addition, new methanogens of interest were detected in interesting numbers in this well, namely the methylotrophs *Candidatus* Methanomethylicus and *Candidatus* Methanofastidiosum.

3.5.2. Biodiversity of Microbial Communities

Shannon and Unifrac indexes were chosen to describe the biodiversity of microbial communities. Alpha diversity refers to the diversity within a given ecosystem and is usually expressed by the number of species (or species richness). The diversity and richness of microbial communities inhabiting the different wells were determined using 16S rRNA gene analyses. Comparing the microbial diversity of each sample, Lobodice recorded the highest values with respect to all three sampling sites. Within all localities, the highest alpha diversity was found in the sample LO-D 11/19, while the opposite was found in LO-C 10/18. Relatively high values for the Shannon index were found in LO-C 11/19, LO-F 5/18, LO-F 11/19, and DUN-A 10/18 (Figure S5, Alpha biodiversity measure).

Beta diversity describes the structural complexity of the environment. It is a measure of the difference (or conversely, similarity) in species composition between communities along a given gradient of the environment, or between the community and its environment. Beta diversity is higher when a community contains uncommon species. Beta diversity shows the difference between microbial communities from different sampling sites. The obtained indices show no significant clustering among all samples. Only one distinct group formed from samples from Dolní Dunajovice and Tvrdonice, which originated from similar environments. In addition, some samples from Lobodice (LO-D 10/18, LO-E 11/19, LO-A 6/18) were clustered together with them. The most divergent samples, considering all sites, were LO-C 10/18 and LO-G 10/18 (Figures S5 and S6).

3.5.3. Metabolism Prediction—FAPROTAX

The potential metabolic functions of microorganisms were estimated using the database of functional annotations of prokaryotic organisms, FAPROTAX, which showed that the most frequent categories were related to carbon cycling under anaerobic conditions. Furthermore, metabolism involved in nitrate respiration and respiration of sulfur compounds was found to be marginal (Figure 4). During anaerobic decomposition of organic material, the hydrolysis products are decomposed into simple organic and inorganic substances (acids, alcohols, CO_2, H_2). Fermentation of these substances produces several reduced end products. In the next step—acetogenesis—the syntrophic acetogenic microorganisms produce hydrogen and decompose organic acids, alcohols, and some aromatic compounds. In the dataset, aromatic compound degraders, such as members of the genera *Pelotomaculum* or *Acinetobacter*, were detected in all samples, but significant levels were reached in the wells LO-B, LO-G, LO-H, and DUN-B. The presence of syntrophic acetogenic bacteria (*Sporomusa*, *Anoxynatronum*) was confirmed in the LO-D well. The dataset also reflects the composition of the methanogenic community in the well, showing a high abundance of acetoclastic methanogens. Acetogenic microorganisms are commonly found in syntrophy with methanogenic archaea, which consume the hydrogen they produce. Methanogens utilize substrates, such as carbonaceous substances (methanol, formic acid, methylamines, CO_2, CO) or acetic acid. The end products of their metabolism are methane and carbon dioxide. All types of methanogenic metabolism were recorded in the collected samples, but hydrogenotrophic methanogenesis, represented mainly by the genus *Methanobacterium*, was predominant. A significant potential of methylotrophic methanogenesis was observed only in DUN-C (*Methanocalculus*, *Methanospirillum*). Acetotrophic methanogens, represented exclusively by the genus *Methanosaeta*, were detected only in the LO-D well, corresponding to high acetogenic activity. The highest methanogenic activity was then found in DUN-C and LO-F. The intensity of predicted methanogenic metabolism varied by site and sampling, showing unexpected changes in metabolic activity. In general, it appears that this very particular type of ecosystem is more dynamic than should be expected.

4. Discussion

Over the course of four years of studying biomethanation in underground water reservoirs, we have built on the results summarized in the review paper [5].

The objective of the study was to determine part of the composition of the UGS microbiome, focusing on methanogenic archaea, biological methane production, and evaluation of the biotechnological potential for biomethane production from underground hydrocarbon reservoirs and aquifers. Although methanogens have been detected in aquifers [4,18,19] or in the sediments of UGS pipelines [45], to the best of our knowledge, this is the first thorough study addressing different types of reservoirs. We hypothesize that seasonal fluctuations of gas in UGS may have an impact on the abundance of methanogens in UGS [2].

To demonstrate the dynamics of their abundance during the year, some of the wells were sampled at different times of the year, focusing on specific UGS regimes (injection/withdrawal gas). The results show that the methanogenic community is dynamic throughout the year, but no significant trend was found with respect to the UGS regime (Figure 3). The results of the sequencing analyses confirmed that each well hosts a specific methanogenic community that accounts for 70%–100% of the total archaeal community. This means that methanogen abundance in UGS is generally greater than 10%.

Hydrogenotrophic *Methanobacterium* sp. prevailed in most samples from the Lobodice locality, as proven by NGS and cultivation. *Methanosarcina* sp. were also abundant, consuming acetate or methylated compounds as a substrate for methanogenesis. The second sampling site, Dolní Dunajovice, had a different reservoir environment and thus a different microbiome composition compared to Lobodice. Members of the hydrogenotrophic order *Methanomicrobiales* were most abundant. The third sampling site, Tvrdonice, was strongly influenced by the higher temperature in the UGS and therefore contained thermophilic members of the *Methanothermobacter* genus.

Methanogenic and fermentative microorganisms are often organized in mutualistic consortia to facilitate rapid electron exchange by diffusion of hydrogen or formate [46]. In addition, electron exchange by direct electron transfer between species has been discovered. *Geobacter metallireducens* transfers electrons directly to *Methanothrix harundinacea* during methanogenic degradation of ethanol, presumably through nanowires [47]. Members of the genera *Geobacter* and *Methanothrix* were found in samples from LO-A and DUN-A.

Anaerobic syntrophy is defined as a thermodynamically interdependent extreme lifestyle in which the degradation of an organic compound occurs only when the end products (usually hydrogen, formate, and acetate) are maintained at very low concentrations. Microbial syntrophy between *Bacteria* and methanogenic archaea enhances the methanogenic activity and methane yield. This type of syntrophy is related to the global carbon cycle in anaerobic environments, which is based on a complex community of metabolically coupled microorganisms that are highly adapted to the environment. This was clearly demonstrated in the publication [48].

The dynamics of the conversion of H_2 and CO_2 to methane varied in all samples depending on the UGS environment. The key parameters were temperature and chemical composition of the UGS groundwater. Biomethane production is also influenced by the composition of the microbial community in each UGS well [49]. Our study confirms the presence and natural activity of methanogenic archaea in underground gas storages. The function of underground gas storages as natural bioreactors is confirmed by the result of our field experiment under real reservoir conditions [50]. The strategic importance of this solution is quite clear. These principles can make an important contribution to reducing the impact of transport and energy on nature and decarbonizing the economy. The use of underground reservoirs for industrial production of biological methane is one of the ways by which the Czech Republic might achieve self-sufficiency, avoiding dependence on natural gas imports.

4.1. Lobodice

The low salinity and higher pH (close to eight) of the environment provide ideal conditions for the growth of members of the genus *Methanobacterium*. The high prevalence of this genus was confirmed in all but one of the sampled wells. This genus is hydrogenotrophic but can also metabolize formate. Several species have already been isolated from aquifers, so these genera seem to be widespread in this environment [3,15,51,52]. Moreover, this genus has even been isolated during previous experiments at the Lobodice UGS [4]. Members of the genus *Methanosarcina* were found in large numbers in LO-A and LO-D. The main substrates for growth are acetate, methanol, trimethylamines, or other methyl-containing compounds. Some of the species can form methane from H_2 and CO_2, or they use H_2 to reduce methanol to methane. Acetotrophic members of the genus *Methanothrix* were detected and were present in samples depending on morphological observation, especially in LO-C, LO-D, and LO-G samples. They are exclusively served by acetate as a substrate, and their affinity for this substrate is much higher than that of the genus *Methanosarcina*. Members of the genus *Methanoculleus* were found exclusively in the well LO-A. This genus usually lives in marine environments and brackish water, but is also widespread in other environments, such as bioreactors, landfills, or wastewater. Unlike other Archaea, *Methanoculleus* can use ethanol and some secondary alcohols as electron donors for final methane production [53]. Conditions at this site appear to support acetogenic bacteria of the genus *Acetobacterium*, which use H_2 and CO_2 to form acetate. The acetate formed does not remain in the liquid for long, being rapidly consumed by a narrow range of bacteria, or serving as a substrate for methanogenesis.

4.2. Dolní Dunajovice

The Dolní Dunajovice site is characterized by higher salinity and temperature compared to Lobodice. Due to these conditions, the composition of methanogens differs significantly from that in Lobodice. In the wells DUN-A and DUN-B, methanogens belonging to the order *Methanobacteriales* absolutely dominate. In the sample from the well DUN-C, two predominant archaeal taxons, *Methanobacterium* sp. and *Methanocalculus* sp., comprised around half of the organisms. Species of the genus *Methanocalculus* are very salt tolerant and can live at sodium chloride concentrations as high as 125 g/L [54]. Moreover, the higher temperature in the Dolní Dunajovice reservoir is ideal for them, as the optimal temperature of the species is 45 °C [55].

4.3. Tvrdonice

Tvrdonice offers unique conditions, with salinity levels somewhere between those of Lobodice and Dolní Dunajovice, at temperatures around 50 °C. The higher temperature fosters microbial communities distinct from those of mesophilic environments. The dominant methanogen, *Methanothermobacter* sp., was found in well TVR-B and was also detected in another deep aquifer [15]. Its predominance is likely to lead to faster reactions in methane formation. High temperatures are a key factor affecting microbial composition. The absolute prevalence of the thermophilic *Methanothermobacter* sp. (94%), which grows best at temperatures between 55 °C and 65 °C, was confirmed by 16S rRNA sequencing. Only two other genera were detected in well TVR-B by 16S rRNA sequencing: *Methanobacterium* (3.2%) and *Methanoculleus* (2.7%). These genera use carbon dioxide and hydrogen as substrates to produce methane for energy production.

Samples from TVR-A well water consisted of the genera *Methanobacterium* (approximately 50%), *Methanoculeus* (24%), *Methanothermobacter* (18%), and members of the recently proposed taxa *Candidatus* Methanomethylicus and *Candidatus* Methanofastidiosum [56]. *Candidatus* Methanofastidiosum is a unique methanogen that utilizes methylated thiol reduction and bridged carbon and sulfur cycles and may compete with CO_2-reducing methanogens and even sulfate reducers [56,57].

5. Conclusions

Based on our results, we can conclude that the underground gas storages assessed by us showed a microbiome composition suitable for biological methane production. It can be concluded that if underground gas storages offer suitable habitat, they can be used as fermenters for the biological conversion of CO_2 and H_2 into biomethane anywhere in the world. From a strategic perspective, underground storages can play an important role in the energy mix, as they can be used for long-term energy storage.

Supplementary Materials: The following supporting information can be downloaded at: https://www.mdpi.com/article/10.3390/fermentation8060251/s1, Table S1: Overview on sampling and on physiochemical parameters measured in sampled wells, Table S2: Table of gas composition measured in sampled wells, Table S3: Primer sequences for Illumina analysis, Table S4: Results of isotopic analysis of degas water and injected gas, Table S5: Absolute quantification of mcr-A gene via qPCR (gene copies per ml), Table S6: Table of Illumina reads for DADA2 analysis, Figure S1: Piper and Durov diagrams for groundwaters sampled at UGS Lobodice. UGS Tvrdonice. and UGS Dunajovice, Figure S2: Stability pH-Eh diagrams for carbon (a) nitrogen (b) and sulfur (c) dissolved components. Diagrams were prepared for UGS Lobodice groundwater conditions, conditions of UGS Tvrdonice and UGS Dunajovice are not significantly different. (d) Speciation of carbon dissolved species in dependence on redox potential, Figure S3: Rarefaction curve, Figure S4: Physical-chemistry well water properties in different sites and composition of archaeal community, Figure S5: The Alpha biodiversity index (Shannon) of sampled wells, Figure S6: The Beta biodiversity index (unweighted UniFrac) of sampled wells.

Author Contributions: I.B., M.V., and V.O. contributed to the conception and design of the study. I.B. and A.M. wrote the first draft of the manuscript. T.V., M.V., I.U., V.O. and J.Z. wrote sections of the manuscript. I.B., A.M., M.V., T.V., N.H. collected samples. I.B., A.M., and D.N. performed molecular biology analyses. I.B., D.N., M.Č., and J.L. performed bioinformatics analyses. I.U. performed degas analyses. J.Z. performed chemical analyses of samples. J.J. and M.M. performed SEM micrographs. H.H. and L.D. revised the manuscript. All authors have read and agreed to the published version of the manuscript.

Funding: The financial support was received from innogy Gas Storage, s.r.o.

Institutional Review Board Statement: Not applicable.

Informed Consent Statement: Not applicable.

Data Availability Statement: Not applicable.

Conflicts of Interest: The authors declare no conflict of interest.

References

1. Basso, O.; Lascourreges, J.F.; Le Borgne, F.; Le Goff, C.; Magot, M. Characterization by culture and molecular analysis of the microbial diversity of a deep subsurface gas storage aquifer. *Res. Microbiol.* **2009**, *160*, 107–116. [CrossRef] [PubMed]
2. Ivanova, A.E.; Borzenkov, I.A.; Tarasov, A.L.; Milekhina, E.I.; Belyaev, S.S. A microbiological study of an underground gas storage in the process of gas extraction. *Microbiology* **2007**, *76*, 461–468. [CrossRef]
3. Buzek, F.; Onderka, V.; Vančura, P.; Wolf, I. Carbon isotope study of methane production in a town gas storage reservoir. *Fuel* **1994**, *73*, 747–752. [CrossRef]
4. Šmigáň, P.; Greksák, M.; Kozánková, J.; Buzek, F.; Onderka, V.; Wolf, I. Methanogenic bacteria as a key factor involved in changes of town gas stored in an underground reservoir. *FEMS Microbiol. Lett.* **1990**, *73*, 221–224. [CrossRef]
5. Molíková, A.; Vítězová, M.; Vítěz, T.; Buriánková, I.; Huber, H.; Dengler, L.; Hanišáková, N.; Onderka, V.; Urbanová, I. Underground gas storage as a promising natural methane bioreactor and reservoir? *J. Energy Storage* **2022**, *47*, 103631. [CrossRef]
6. Rittmann, S.; Seifert, A.; Herwig, C. Essential prerequisites for successful bioprocess development of biological CH_4 production from CO_2 and H_2. *Crit. Rev. Biotechnol.* **2015**, *35*, 141–151. [CrossRef]
7. Koytsoumpa, E.I.; Bergins, C.; Kakaras, E. The CO_2 economy: Review of CO_2 capture and reuse technologies. *J. Supercrit. Fluids* **2018**, *132*, 3–16. [CrossRef]
8. Nazir, H.; Louis, C.; Jose, S.; Prakash, J.; Muthuswamy, N.; Buan, M.E.M.; Flox, C.; Chavan, S.; Shi, X.; Kauranen, P.; et al. Is the H_2 economy realizable in the foreseeable future ? Part I: H_2 production methods. *Int. J. Hydrogen Energy* **2020**, *45*, 13777–13788. [CrossRef]
9. Sampath, P.; Brijesh; Reddy, K.R.; Reddy, C.V.; Shetti, N.P.; Kulkarni, R.V.; Raghu, A.V. Biohydrogen Production from Organic Waste—A Review. *Chem. Eng. Technol.* **2020**, *43*, 1240–1248. [CrossRef]
10. Aguilera, R.F.; Aguilera, R. Revisiting the role of natural gas as a transition fuel. *Miner. Econ.* **2020**, *33*, 73–80. [CrossRef]
11. Pedersen, K.; Arlinger, J.; Ekendahl, S.; Hallbeck, L. 16S rRNA gene diversity of attached and unattached bacteria in boreholes along the access tunnel to the Äspö hard rock laboratory, Sweden. *FEMS Microbiol. Ecol.* **1996**, *19*, 249–262. [CrossRef]
12. Kotelnikova, S.; Pedersen, K. Evidence for methanogenic Archaea and homoacetogenic Bacteria in deep granitic rock aquifers. *FEMS Microbiol. Rev.* **1997**, *20*, 339–349. [CrossRef]
13. Fry, N.K.; Fredrickson, J.K.; Fishbain, S.; Wagner, M.; Stahl, D.A. Population structure of microbial communities associated with two deep, anaerobic, alkaline aquifers. *Appl. Environ. Microbiol.* **1997**, *63*, 1498–1504. Available online: https://aem.asm.org/content/63/4/1498.short (accessed on 21 May 2019). [CrossRef]
14. Shimizu, S.; Akiyama, M.; Ishijima, Y.; Hama, K.; Kunimaru, T.; Naganuma, T. Molecular characterization of microbial communities in fault-bordered aquifers in the Miocene formation of northernmost Japan. *Geobiology* **2006**, *4*, 203–213. [CrossRef]
15. Kimura, H.; Nashimoto, H.; Shimizu, M.; Hattori, S.; Yamada, K.; Koba, K.; Yoshida, N.; Kato, K. Microbial methane production in deep aquifer associated with the accretionary prism in Southwest Japan. *ISME J.* **2010**, *4*, 531–541. [CrossRef]
16. Flynn, T.M.; Sanford, R.A.; Ryu, H.; Bethke, C.M.; Levine, A.D.; Ashbolt, N.J.; Santo Domingo, J.W. Functional microbial diversity explains groundwater chemistry in a pristine aquifer. *BMC Microbiol.* **2013**, *13*, 146. [CrossRef]
17. Wu, X.; Holmfeldt, K.; Hubalek, V.; Lundin, D.; Åström, M.; Bertilsson, S.; Dopson, M. Microbial metagenomes from three aquifers in the Fennoscandian shield terrestrial deep biosphere reveal metabolic partitioning among populations. *ISME J.* **2016**, *10*, 1192–1203. [CrossRef]
18. Frank, Y.A.; Kadnikov, V.V.; Gavrilov, S.N.; Banks, D.; Gerasimchuk, A.L.; Podosokorskaya, O.A.; Merkel, A.Y.; Chernyh, N.A.; Mardanov, A.V.; Ravin, N.V.; et al. Stable and variable parts of microbial community in Siberian deep subsurface thermal aquifer system revealed in a long-term monitoring study. *Front. Microbiol.* **2016**, *7*, 2101. [CrossRef]
19. Kadnikov, V.V.; Frank, Y.A.; Mardanov, A.V.; Beletsky, A.V.; Ivasenko, D.A.; Pimenov, N.V.; Karnachuk, O.V.; Ravin, N.V. Variability of the composition of the microbial community of the deep subsurface thermal aquifer in Western Siberia. *Microbiology* **2017**, *86*, 765–772. [CrossRef]
20. Vigneron, A.; Bishop, A.; Alsop, E.B.; Hull, K.; Rhodes, I.; Hendricks, R.; Head, I.M.; Tsesmetzis, N. Microbial and isotopic evidence for methane cycling in hydrocarbon-containing groundwater from the Pennsylvania region. *Front. Microbiol.* **2017**, *8*, 593. [CrossRef]
21. Evans, P.N.; Boyd, J.A.; Leu, A.O.; Woodcroft, B.J.; Parks, D.H.; Hugenholtz, P.; Tyson, G.W. An evolving view of methane metabolism in the Archaea. *Nat. Rev. Microbiol.* **2019**, *17*, 219–232. [CrossRef]
22. Stevens, T.O.; McKinley, J.P.; Fredrickson, J.K. Bacteria Associated with Deep, Alkaline, Anaerobic Groundwaters in Southeast Washington T. *Microb. Ecol.* **1993**, *25*, 35–50. [CrossRef]
23. Thauer, R.K.; Kaster, A.K.; Seedorf, H.; Buckel, W.; Hedderich, R. Methanogenic archaea: Ecologically relevant differences in energy conservation. *Nat. Rev. Microbiol.* **2008**, *6*, 579–591. [CrossRef]

24. Hedderich, R.; Whitman, W.B. Physiology and biochemistry of the methane-producing Archaea. *Prokaryotes* **2006**, *2*, 1050–1079. Available online: http://www.academia.edu/download/63378241/Dworkin_The_Prokaryotes-A_Handbook_on_the_Biology_of_Bacteria_3rd_ed_Vol_220200520-24653-hqh6kx.pdf#page=1100 (accessed on 18 January 2021).
25. Lang, K.; Schuldes, J.; Klingl, A.; Poehlein, A.; Daniel, R.; Brune, A. New mode of energy metabolism in the seventh order of methanogens as revealed by comparative genome analysis of "Candidatus Methanoplasma termitum". *Appl. Environ. Microbiol.* **2015**, *81*, 1338–1352. [CrossRef]
26. Berghuis, B.A.; Yu, F.B.; Schulz, F.; Blainey, P.C.; Woyke, T.; Quake, S.R. Hydrogenotrophic methanogenesis in archaeal phylum Verstraetearchaeota reveals the shared ancestry of all methanogens. *Proc. Natl. Acad. Sci. USA* **2019**, *116*, 5037–5044. [CrossRef]
27. Hattori, S. Syntrophic acetate-oxidizing microbes in methanogenic environments. *Microbes Environ.* **2008**, *23*, 118–127. [CrossRef]
28. Mosbæk, F.; Kjeldal, H.; Mulat, D.G.; Albertsen, M.; Ward, A.J.; Feilberg, A.; Nielsen, J.L. Identification of syntrophic acetate-oxidizing bacteria in anaerobic digesters by combined protein-based stable isotope probing and metagenomics. *ISME J.* **2016**, *10*, 2405–2418. [CrossRef]
29. Hungate, R.E. Chapter IV A Roll Tube Method for Cultivation of Strict Anaerobes. *Methods Microbiol.* **1969**, *3*, 117–132. [CrossRef]
30. Miller, T.L.; Wolin, M.J. A Serum Bottle Modification of the Hungate Technique for Cultivating Obligate Anaerobes. *Appl. Microbiol.* **1974**, *27*, 985–987. [CrossRef]
31. Denman, S.E.; Tomkins, N.W.; McSweeney, C.S. Quantitation and diversity analysis of ruminal methanogenic populations in response to the antimethanogenic compound bromochloromethane. *FEMS Microbiol. Ecol.* **2007**, *62*, 313–322. [CrossRef] [PubMed]
32. Steinberg, L.M.; Regan, J.M. mcrA-targeted real-time quantitative PCR method to examine methanogen communities. *Appl. Environ. Microbiol.* **2009**, *75*, 4435–4442. [CrossRef] [PubMed]
33. Pichler, M.; Coskun, Ö.K.; Ortega-Arbulú, A.S.; Conci, N.; Wörheide, G.; Vargas, S.; Orsi, W.D. A 16S rRNA gene sequencing and analysis protocol for the Illumina MiniSeq platform. *Microbiologyopen* **2018**, *7*, e00611. [CrossRef] [PubMed]
34. Thompson, L.R.; Sanders, J.G.; McDonald, D.; Amir, A.; Ladau, J.; Locey, K.J.; Prill, R.J.; Tripathi, A.; Gibbons, S.M.; Ackermann, G.; et al. A communal catalogue reveals Earth's multiscale microbial diversity. *Nature* **2017**, *551*, 457–463. [CrossRef]
35. Callahan, B.J.; Sankaran, K.; Fukuyama, J.A.; McMurdie, P.J.; Holmes, S.P. Bioconductor workflow for microbiome data analysis: From raw reads to community analyses. *F1000Research* **2016**, *5*, 1496. [CrossRef]
36. Wang, Q.; Garrity, G.M.; Tiedje, J.M.; Cole, J.R. Naïve Bayesian classifier for rapid assignment of rRNA sequences into the new bacterial taxonomy. *Appl. Environ. Microbiol.* **2007**, *73*, 5261–5267. [CrossRef]
37. Quast, C.; Pruesse, E.; Yilmaz, P.; Gerken, J.; Schweer, T.; Glo, F.O.; Yarza, P. The SILVA ribosomal RNA gene database project: Improved data processing and web-based tools. *Nucleic Acids Res.* **2013**, *41*, 590–596. [CrossRef]
38. Schliep, K.P. Phangorn: Phylogenetic analysis in R. *Bioinformatics* **2011**, *27*, 592–593. [CrossRef]
39. McMurdie, P.J.; Holmes, S. Phyloseq: An R Package for Reproducible Interactive Analysis and Graphics of Microbiome Census Data. *PLoS ONE* **2013**, *8*, e61217. [CrossRef]
40. Louca, S.; Parfrey, L.W.; Doebeli, M. Decoupling function and taxonomy in the global ocean microbiome. *Science* **2016**, *353*, 1272–1277. [CrossRef]
41. Carini, P.; Marsden, P.J.; Leff, J.W.; Morgan, E.E.; Strickland, M.S.; Fierer, N. Relic DNA is abundant in soil and obscures estimates of soil microbial diversity. *Nat. Microbiol.* **2016**, *2*, 16242. [CrossRef]
42. Whiticar, M.J. Carbon and hydrogen isotope systematics of bacterial formation and oxidation of methane. *Chem. Geol.* **1999**, *161*, 291–314. [CrossRef]
43. Caporaso, J.G.; Lauber, C.L.; Walters, W.A.; Berg-Lyons, D.; Huntley, J.; Fierer, N.; Owens, S.M.; Betley, J.; Fraser, L.; Bauer, M.; et al. Ultra-high-throughput microbial community analysis on the Illumina HiSeq and MiSeq platforms. *ISME J.* **2012**, *6*, 1621–1624. [CrossRef]
44. Baker, B.J.; Saw, J.H.; Lind, A.E.; Lazar, C.S.; Hinrichs, K.U.; Teske, A.P.; Ettema, T.J.G. Genomic inference of the metabolism of cosmopolitan subsurface Archaea, Hadesarchaea. *Nat. Microbiol.* **2016**, *1*, 16002. [CrossRef]
45. Staniszewska, A.; Kunicka-Styczyńska, A.; Otlewska, A.; Gawor, J.; Gromadka, R.; Żuchniewicz, K.; Ziemiński, K. High-throughput sequencing approach in analysis of microbial communities colonizing natural gas pipelines. *Microbiologyopen* **2019**, *8*, e00806. [CrossRef]
46. Sieber, J.R.; McInerney, M.J.; Gunsalus, R.P. Genomic Insights into Syntrophy: The Paradigm for Anaerobic Metabolic Cooperation. *Annu. Rev. Microbiol.* **2012**, *66*, 429–452. [CrossRef]
47. Reguera, G.; McCarthy, K.D.; Mehta, T.; Nicoll, J.S.; Tuominen, M.T.; Lovley, D.R. Extracellular electron transfer via microbial nanowires. *Nature* **2005**, *435*, 1098–1101. [CrossRef]
48. Amend, J.P.; Shock, E.L. Energetics of overall metabolic reactions of thermophilic and hyperthermophilic Archaea and Bacteria. *FEMS Microbiol. Rev.* **2001**, *25*, 175–243. [CrossRef]
49. Yang, G.C.; Zhou, L.; Mbadinga, S.M.; Liu, J.F.; Yang, S.Z.; Gu, J.D.; Mu, B.Z. Formate-dependent microbial conversion of CO_2 and the dominant pathways of methanogenesis in production water of high-temperature oil reservoirs amended with bicarbonate. *Front. Microbiol.* **2016**, *7*, 365. [CrossRef]
50. Vítězová, M.; Urbanová, I.; Molíková, A.; Buriánková, I.; Hanišáková, N.; Onderka, V.; Vítěz, T.; Javůrek, J.; Machálková, M. In situ field experiment shows the potential of methanogenic archaea for biomethane production from underground gas storage. Unpublished work.

51. Godsy, E.M. Isolation of *Methanobacterium bryantii* from a deep aquifer by using a novel broth-antibiotic disk method. *Appl. Environ. Microbiol.* **1980**, *39*, 1074–1075. [CrossRef]
52. Kotelnikova, S.; Macario, A.J.L.; Pedersen, K. Methanobacterium subterraneum sp. nov., a new alkaliphilic, eurythermic and halotolerant methanogen isolated from deep granitic groundwater. *Int. J. Syst. Bacteriol.* **1998**, *48*, 357–367. [CrossRef]
53. Cheng, L.; Qiu, T.L.; Li, X.; Wang, W.D.; Deng, Y.; Yin, X.B.; Zhang, H. Isolation and characterization of Methanoculleus receptaculi sp. nov. from Shengli oil field, China. *FEMS Microbiol. Lett.* **2008**, *285*, 65–71. [CrossRef]
54. Ollivier, B.; Fardeau, M.; Cayol, J.; Magot, M.; Patel, B.K.C.; Prensiep, G.; Garcia, J. Methanocalculus halotolerans. *Int. J. Syst. Evol. Microbiol.* **1998**, *48*, 821–828.
55. Shlimon, A.G.; Friedrich, M.W.; Niemann, H.; Ramsing, N.B.; Finster, K. *Methanobacterium aarhusense* sp. nov., a novel methanogen isolated from a marine sediment (Aarhus Bay, Denmark). *Int. J. Syst. Evol. Microbiol.* **2004**, *54*, 759–763. [CrossRef]
56. Vanwonterghem, I.; Evans, P.N.; Parks, D.H.; Jensen, P.D.; Woodcroft, B.J.; Hugenholtz, P.; Tyson, G.W. Methylotrophic methanogenesis discovered in the archaeal phylum Verstraetearchaeota. *Nat. Microbiol.* **2016**, *1*, 16170. [CrossRef]
57. Nobu, M.K.; Narihiro, T.; Kuroda, K.; Mei, R.; Liu, W.T. Chasing the elusive Euryarchaeota class WSA2: Genomes reveal a uniquely fastidious methyl-reducing methanogen. *ISME J.* **2016**, *10*, 2478–2487. [CrossRef]

fermentation

MDPI

Article

Increasing Anaerobic Digestion Efficiency Using Food-Waste-Based Biochar

Dong-Chul Shin, I-Tae Kim, Jinhong Jung, Yoonah Jeong, Ye-Eun Lee and Kwang-Ho Ahn *

Department of Environmental Research, Korea Institute of Civil Engineering and Building Technology 283, Goyang-daero, Ilsanseo-gu, Goyang-si 10223, Gyeong-gi-do, Korea; dongchulshin@kict.re.kr (D.-C.S.); itkim@kict.re.kr (I.-T.K.); jinhong98@kict.re.kr (J.J.); yoonahjeong@kict.re.kr (Y.J.); yeeunlee@kict.re.kr (Y.-E.L.)
* Correspondence: khahn@kict.re.kr; Tel.: +82-31-910-0314

Abstract: The efficiency of methane production by anaerobic digestion (AD), during which energy is generated from organic waste, can be increased in various ways. Recent research developments have increased the volume of gas production during AD using biochar. Previous studies have used food waste itself in AD, or, added wood-biochar or sewage sludge charcoal as an accelerant of the AD process. The application of food-waste biochar in AD using activated sludge has not yet been studied and is considered a potential method of utilizing food waste. Therefore, this study investigated the use of biochar prepared by the thermal decomposition of food waste as an additive to AD tanks to increase methane production. The addition of food-waste biochar at 1% of the digestion tank volume increased the production of digestion gas by approximately 10% and methane by 4%. We found that food-waste biochar served as a medium with trace elements that promoted the proliferation of microorganisms and increased the efficiency of AD.

Keywords: anaerobic digestion; biochar; thermal decomposition; methane production; food waste

Citation: Shin, D.-C.; Kim, I.-T.; Jung, J.; Jeong, Y.; Lee, Y.-E.; Ahn, K.-H. Increasing Anaerobic Digestion Efficiency Using Food-Waste-Based Biochar. *Fermentation* **2022**, *8*, 282. https://doi.org/10.3390/fermentation8060282

Academic Editor: Sanjay Nagarajan

Received: 9 May 2022
Accepted: 14 June 2022
Published: 16 June 2022

1. Introduction

Anaerobic digestion (AD) is one of the best methods to utilize various organic waste materials for energy production [1]. Numerous methods have been applied to the process to increase bioenergy production from biomass waste [2].

The surplus sludge generated by sewage treatment produces biogas through AD. Therefore, energy can be recovered from the sludge, contributing to organic waste management—the production of biogas increases or decreases depending on the management of the AD tank. Accordingly, technologies are being developed to improve energy recovery and increase biogas production [3]. Most previous studies on increasing biogas production have focused on improving the microorganism activity or increasing the number of microorganisms [4].

Various AD studies have been conducted using food waste to produce biogas. When food waste is single-digested by AD, the accumulation of volatile fatty acids (VFAs) and the suppression of ammonia inhibit the digestion tank reactions [5,6]. The AD tank reactions are also inhibited by the salt concentrations of the food waste used in AD [7,8]. In addition, the high biodegradability of food waste inhibits reactions that produce methane [9].

Recent studies have focused on improving the yield of biogas by using various additives such as bio-based carbon materials including biochar, activated sludge, granulated activated carbon, and carbon cloth [2,10]. These additives improve the stability of active sludge in AD and increase biogas production [11,12]. Activated carbon increases sludge reduction and methane production [13], and biochar improves the rate of methane production [14,15].

Among these additives, biochar can negatively affect sludge production because of the composition of biomass in the AD, but positively affects methane production and can be used to recycle organic waste by biochar [16] Biochar, produced through pyrolysis of biomass, is an eco-friendly material with a high carbon content, porous structure, large specific surface area, and good biocompatibility. Biochar was reported to improve the efficiency of methane production by 32%, when added to the AD [17,18]. Wood-biochar or sewage sludge charcoal have been added to improve the methane production efficiency in AD [19,20].

However, in Korea, more than 50% of lignocellulosic waste is processed using paint, oil, and preservatives, and there is a limit in manufacturing biochar. In addition, sewage sludge contains heavy metals; therefore, continuous addition of sewage sludge to AD can lead to the accumulation of heavy metals [21].

To overcome these limitations, there is a need to use biochar using food waste. Because the generation of food waste is increasing, the production and use of biochar from food waste has been examined. The production of food-waste-based biochar has been studied to convert food waste into resources to be used as a soil conditioner or fuel [22,23]. The components of food-waste biochar can increase microorganism activity during the digestion process. Although the application of food-waste biochar in AD using activated sludge has not been evaluated, this method shows potential for utilizing food waste.

Therefore, in this study, we analyzed the physical properties of food-waste-based biochar to determine which trace elements are necessary for AD, and to evaluate whether it is useful as media in Brunauer, Emmett, Teller analysis. In addition, the biochemical methane potential (BMP) test confirmed that the amount of biogas generated was determined by the amount of biochar injected. The methane production potential confirmed the methane gas generation characteristics. We examined the effect of biochar produced by the thermal decomposition of food waste on the improvement of methane yield during AD.

2. Materials and Methods

2.1. Production of Biochar from Food Waste

The biochar used in the experiment was produced using primarily processed (screened, crushed, and dried) food waste at the resource recycling center in City Gimpo, Korea. The raw material contained 70.66% volatile matter, 9.35% ash, 10.74% fixed carbon, and 9.26% moisture.

Food waste usually has a high salt content. The food waste used in this study contained 1.2–1.8 $g{\cdot}L^{-1}$ salt (average 1.52 $g{\cdot}L^{-1}$). After thermal decomposition, the resulting biochar contained 0.3–1.0 $g{\cdot}L^{-1}$ salt (average 0.78 $g{\cdot}L^{-1}$). According to Gao et al. [24], the change in the digestion efficiency is insignificant when the salt content is below 5 $g{\cdot}L^{-1}$. Roberts et al. [25] reported that the effect of salt on microorganisms is not significant when the content is below 1 $g{\cdot}L^{-1}$. The salt tolerance of the microorganisms increased to 10 $g{\cdot}L^{-1}$ with adaption to salt. The current study did not consider the effect of salt on microorganisms in AD.

The biochar was prepared by slow pyrolysis by carbonizing the raw material at 500 °C for 10 min. The hopper of the pyrolizer was designed to minimize the inflow of external air. Nitrogen was used as a carrier gas to meet anoxic conditions and a "three-screw"-type internal system was created to facilitate heat transfer and mixing.

A proximate analysis was conducted to determine the composition of biochar, X-ray fluorescence spectrometry (XRF) was used to identify trace elements affecting digestion efficiency, and the Brunauer, Emmett, Teller (BET) analysis was used to examine porosity.

2.2. Operating Conditions for the BMP Test

The digestion efficiency was examined based on the volume of biogas produced and the methane content. For this, the BMP test was performed as suggested by Owen [26] and Shelton and Tiedje [27]. The microorganisms applied in the BMP test were seeded using sludge from the digestion tank operated in the field. The surplus and digested sludge used for the test were collected from the sewage treatment plant in City Ilsan, Korea. Table 1 shows the sludge properties.

Table 1. Sludge properties.

	COD (mg/L)	TSS (mg/L)	VSS (mg/L)	NH_4^+-N (mg/L)	VFA (mg/L)	Alkalinity (mg/L as $CaCO_3$)
Excess sludge	15,184	13,380	9170	8.1	22.5	110
Digested sludge	-	17,630	8870	-	198.5	4140

COD: chemical oxygen demand; TSS: total suspended solids; VSS: volatile suspended solids; VFA: volatile fatty acids.

Experiments were performed under six different conditions to examine the volume of gas produced based on the quantity of biochar used (0% control, 0.1%, 0.5%, 1.0%, 3.0%, and 5.0% of the volume ratio (v/v%), respectively). The ratio of conditions to microorganisms was the same in all conditions: 150 mL of surplus sludge and 200 mL of digested sludge (microorganisms). At this time, the organic material load for each case was 4.83 g COD/L, and the VS load was 3.21 g VS/L. To prevent a pH decrease due to acid production during AD, 1.2 g·L^{-1} sodium bicarbonate ($NaHCO_3$) was added. A 500-mL serum bottle was thoroughly purged with nitrogen gas to ensure that no oxygen remained in the headspace of the bottle and the bottle was sealed with a butyl rubber and aluminum cap. A BMP test was performed at 40 °C and 65 rpm until no gas was produced (60 days).

2.3. *Analysis of Gas Production and Composition*

The volume of digestion gas produced was measured simultaneously every day for the first seven days (24-h intervals), after which the time interval was adjusted according to the gas production rate. The volume of gas generated was measured using a gas syringe with capillary tube (capacity: 100 mL). The methane content was measured by analyzing the concentrations of N_2, CH_4, and CO_2 in the digestion gas. The gas chromatograph used in the gas analysis was a CTGC 1000 (Chemtekins Co. Ltd., Seoul, Korea). The maximum temperature of the oven inside the GC was 280 °C. A thermal conductivity detector with a minimum detection limit of 50 pg was used.

2.4. *Analysis of Biochar before and after the Biochemical Methane Potential (BMP) Test*

Surface and elemental analyses were performed to predict the change in biochar during the BMP test. A field emission scanning electron microscope (JEOL-7610F-Plus, JEOL Ltd., Tokyo, Japan) was used for surface analysis, and pre-treatment (lyophilization) was conducted before the analysis to consider the microorganisms that could be present in the biochar. Additionally, energy dispersive spectrometry (EDS) was used during the surface analysis to investigate changes in the biochar.

2.5. *Analysis of Methane Production Potential*

This study analyzed the effect of the addition of biochar on the methane generation potential using experimental data from the cumulative production curves of methane gas. The cumulative production curves obtained from the BMP test were applied to the modified Gompertz model [28], and the best-fit method by trial and error was used to estimate various parameters included in the model equation. Equation (1) is the modified Gompertz model [29].

$$M = P \times \exp\left[-\exp\left(\frac{R_m \times e}{P}(\lambda - t) + 1\right)\right] \quad (1)$$

where: M, cumulative methane production (mL); P, methane production potential (mL); R_m, methane production rate (mL·day^{-1}); e, exp(1); λ, lag growth phase time (days), and t, time (days).

2.6. Statistical Analyses

Experimental data were statistically analyzed using Microsoft Excel (2016). The means and standard deviations of the biogas production and methane content were calculated. The samples were analyzed thrice or more to ensure precision.

3. Results and Discussion

3.1. Biochar Composition

Proximate analysis revealed that the food-waste biochar contained 41.63% fixed carbon, 30.42% volatile matter, 26.82% ash, and 1.12% moisture. It consisted of 42.54% C, 7.28% H, and 3.46% N. Table 2 shows the component analysis results using XRF. Due to the characteristics of food waste, Ca was the highest, at approximately 60%, followed by Cl. Notably, it included trace elements such as Ti, Cu, Cr, and Zn, which are not found in wood-biochar [17].

Table 2. Food-waste-biochar composition.

Element	Ca	Cl	K	Fe	P	S	Sr
Fraction (%)	59.16	17.63	15.39	4.48	1.73	0.83	0.25
Element	Mn	Ti	Br	Cu	Cr	Rb	Zn
Fraction (%)	0.15	0.13	0.08	0.07	0.06	0.02	0.02

The trace elements in the biochar, including Ca, Fe, Cu, and Zn, exhibited the following characteristics with regard to AD microorganisms:

1. Ca is a key component for the growth of some methanogens and is critical for the formation of microbial aggregates [30];
2. Trace elements act as a cofactor for enzymes involved in methane formation [31];
3. Trace elements facilitate methane production [32];
4. Trace elements play an important role in the growth and metabolism of anaerobes [33];
5. Fe, Zn, and Ni are required for hydrogenase [34–37], and Fe is a key component for methane monooxygenase [38,39] and nitrogenase [40].

The BET analysis of the biochar indicated that the surface area was 1.2969 $m^2 \cdot g^{-1}$, the total pore volume was 0.004982 $cm^3 \cdot g^{-1}$, the adsorption average pore width was 153.6522 nm, and the adsorption average pore diameter was 179.996 nm. The graph of nitrogen isothermal adsorption (Figure 1) suggests Type II, according to the IUPAC classification [41]. According to Rouquerol et al. [42], biochar with such characteristics has mesopores, indicating that abundant pores were formed inside the biochar. As Yue [17] reported, the insides of the pores provide a beneficial environment for microbial growth [43,44].

3.2. Results of Digestion Gas Production

3.2.1. Trend in Gas Generation

Figure 2 shows the digestion gas generated with different quantities of biochar. The first two days were the stabilization phase, with little difference in the volume of gas produced in the control and experimental conditions. The 5% biochar condition produced 15 mL less digestion gas than the control condition due to increased trace elements and a high level of biochar, which suppressed microbial activity [45].

All conditions recorded the maximum volume of gas generation on day 4, which gradually decreased thereafter. The maximum volume of gas was higher in the added-biochar conditions than in the control condition. The less the biochar, the greater the maximum volume of gas generated on day 4 (133 mL at 0.1% biochar compared with 114 mL at 5.0%). In terms of the total amount of biogas generated during the period (Figure 3), however, increasing the amount of biochar increased the gas production (control condition: 1049 mL, 0.1%: 1084 mL, 0.5%: 1131 mL, 1.0%: 1155 mL, 3.0%: 1164 mL, and 5.0%: 1253 mL).

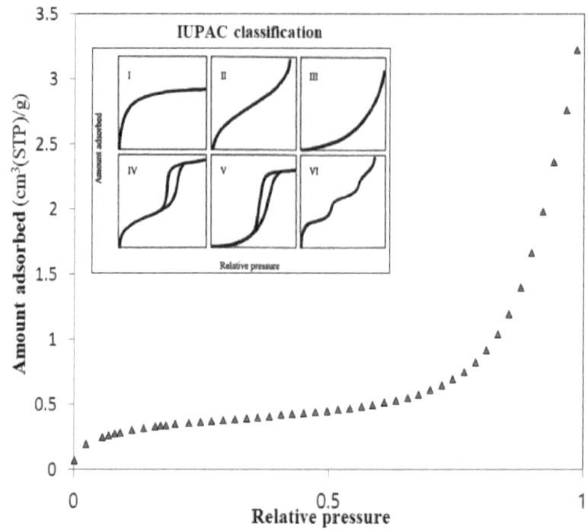

Figure 1. Adsorption-desorption isotherm curves of nitrogen.

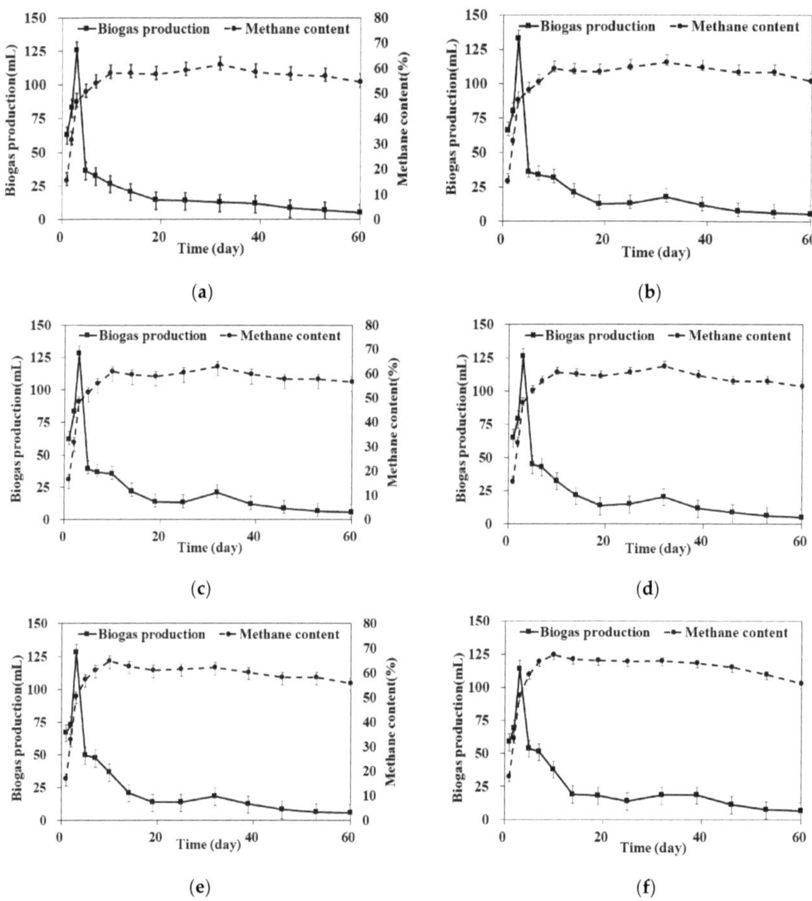

Figure 2. Methane content by condition (biochar content); (**a**) control 0%, (**b**) 0.1%, (**c**) 0.5%, (**d**) 1.0%, (**e**) 3.0%, (**f**) 5.0%.

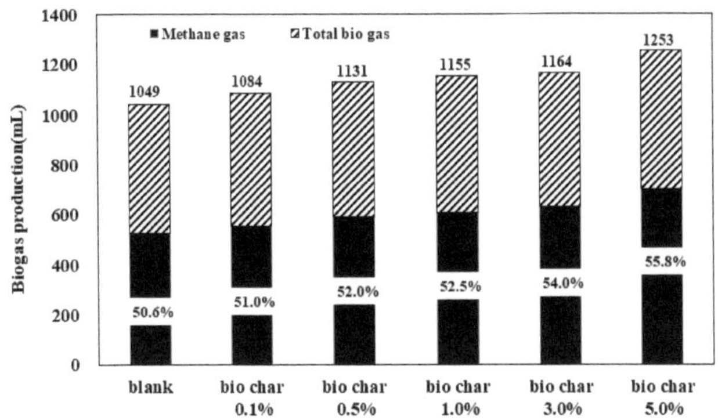

Figure 3. Comparison of total gas generation and methane content with biochar content.

The volume of digestion gas produced was converted into unit volume per initial volatile suspended solids (VSS): 116, 120, 125, 128, 129, and 139 mL $CH_4 \cdot g^{-1}$ $VSS_{in,}$ increasing from the control with increasing quantities of biochar, respectively. The unit amount increased with increasing biochar, with the largest increase at 0.5% biochar.

Similar to gas production, the methane content (Figure 2) was low at 16.5% for the first two days due to the stabilization of the reactor and the generation of N_2 by denitrification. However, from day 3, the methane content was maintained at 50% or higher. The average methane content of the control condition and the 0.1–5.0% biochar conditions (excluding the first two days of the stabilization phase) increased to 55.6%, 56.5%, 57.5%, 57.7%, 59.4%, and 61.1%, respectively (Figure 3).

The maximum daily gas production occurred on the third day with 0.1% biochar, while the methane content was highest at 5.0% on day 10. Zhang et al. [46] reported that the growth and activity of anaerobes could be increased by supplying trace elements. Therefore, it can be inferred that trace substances were eluted from the biochar, leading to the growth of methane-producing microorganisms and the continuous increase in the number of microorganisms.

3.2.2. Changes in Total Suspended Solids and Volatile Suspended Solids

Total suspended solids (TSS) and VSS before and after the BMP test were analyzed to examine sludge reduction in the digestion reaction. TSS and VSS after the BMP test were analyzed including biochar (Figure 4). Both TSS and VSS decreased in biochar in the various conditions up to 0.5% and increased rapidly when 3% biochar was added. The VSS was lower than the initial value of 9020 $mg \cdot L^{-1}$ when less than 3% biochar was added; however, it increased with more than 3% biochar. These results are opposite to the typical AD digestion reaction, in which the decomposition of organic matter increases with gas production, and thus the VSS decreases. Biochar is reported to serve as a medium when it is added to the digestion tank [47], and according to Montalvo [48], the population of microorganisms increases when zeolite, which plays the role of medium, is put into the digestion tank.

3.3. Changes in Biochar before and after the Biochemical Methane Potential Test

Figure 5 shows the results of the scanning electron microscopy (SEM) analysis performed to examine the changes in biochar before and after digestion (biochar was added to increase digestion efficiency.) As in the biochemical methane potential analysis, many pores were found in the biochar (Figure 5a,b), which appeared similar to those of the ceramic carrier suggested by Sun et al. [49]. Foreign substances were attached to the pores and surfaces after the digestion reaction (Figure 5c,d). The substances were similar to the images of methane-forming microorganisms reported by Yu et al. [50].

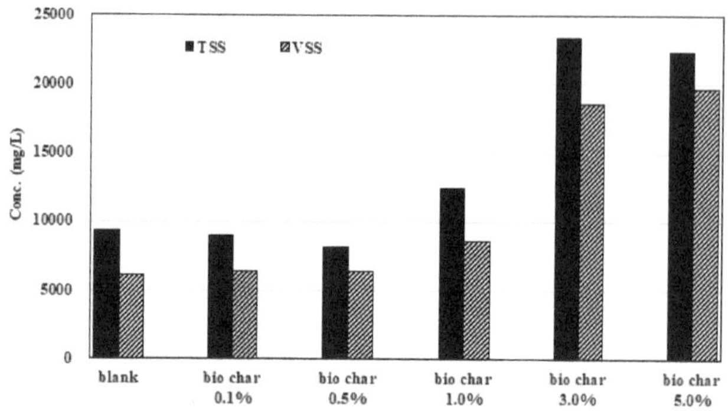

Figure 4. Total suspended solids (TSS) and volatile suspended solids (VSS) before and after the biochemical methane potential test.

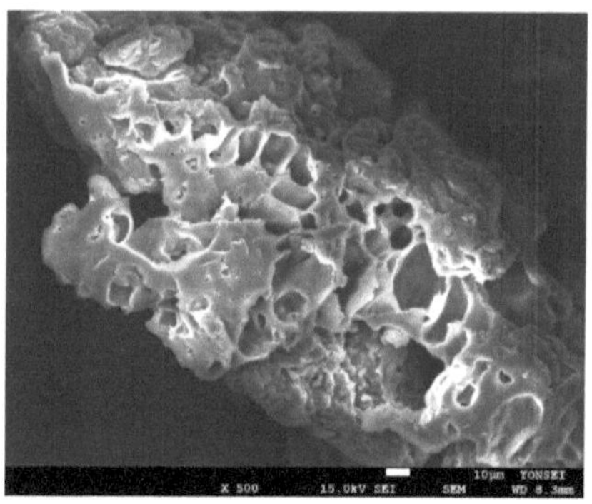

(**a**)

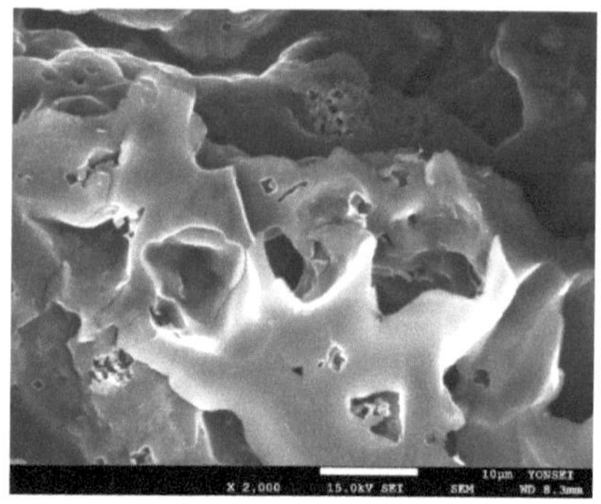

(**b**)

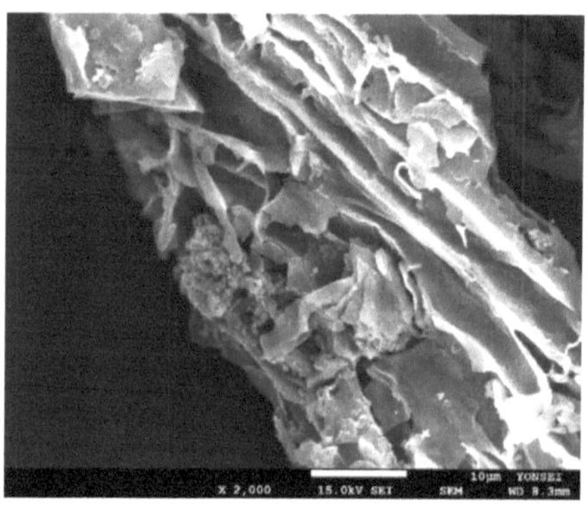

(**c**)

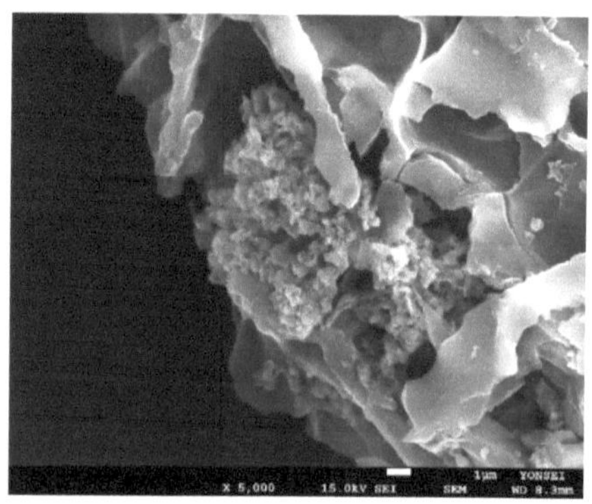

(**d**)

Figure 5. Scanning electron microscopy analysis before and after the biochemical methane potential test (**a**) before, ×500; (**b**) before, ×2000; (**c**) after, ×2000; and (**d**) after, ×5000.

The EDS analysis indicated that O and P content, constituting significant proportions of the microorganisms, increased after the BMP test, leading to a relative reduction in C content (Table 3). Other trace substances, including Na, Mg, Al, and K, were not found after the BMP test.

Table 3. Energy dispersive spectroscopy results before and after the biochemical methane potential test.

Element	C	O	P	Na	Mg	Al	S	Cl	K	Ca
Before	84.98	10.85	-	1.04	0.14	0.08	0.11	1.33	1.10	0.36
After	66.92	31.51	0.48	-	-	-	0.37	-	-	0.71

3.4. Results of Methane Production Potential

Applying the cumulative production curves of methane gas obtained from the BMP test to the modified Gompertz model for optimization showed that R^2 was 0.95 or higher (maximum 0.9673), indicating that the results of the BMP test on methane production with organic waste were correctly simulated (Figure 6). Accordingly, the methane potential (the volume of methane produced) increased as biochar was added (182.8, 191.2, 201.9, 209.1, 222.3, and 228.8 mL $CH_4 \cdot g^{-1}$ VS), and the highest methane production rates were 21.5, 21.8, 22.1, 22.8, 25.0, and 22.5 mL $CH_4 \cdot g^{-1}$ g VS per day. Based on the methane potential and the highest methane production rates, 3% biochar achieved the highest efficiency. The rates increased up to 3% biochar but decreased at 5%, which can be attributed to the decrease in microbial activity due to the increased addition of biochar [44].

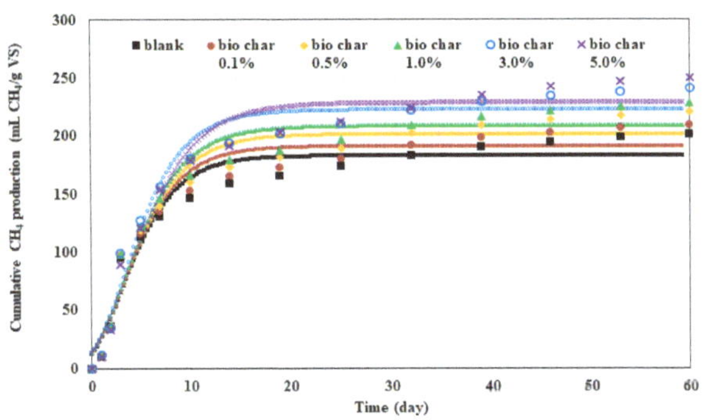

Figure 6. Comparison of the derived modified Gompertz equations using the results obtained in the current study.

Figure 6 describes the fitted results obtained using the modified Gompertz equation. The experimental value matches the model value at 32 days. The dots indicate the experimentally obtained values. The model shows that the maximum gas production volume was achieved near day 17. However, the dots show that gas was continuously generated for more than 60 days, indirectly indicating that methane is produced from the organic material of the biochar.

4. Conclusions

Comparing and examining changes in digestion efficiency by adding food-waste-based biochar to the digestion tank indicated that the efficiency increased in the biochar-added conditions compared with the control condition as follows:

1. Food-waste biochar added at a rate of 1% of the volume of the digestion tank increased the production of digestion gas by approximately 10% and methane by 4%;
2. Increasing the biochar increased the number of microorganisms in the biochar. The 3% biochar condition had a higher VSS after the reaction compared to the initial stage of the reaction;
3. The 3% biochar achieved the maximum methane production rate of 25.0 mL $CH_4 \cdot g^{-1}$ VS per day.

The results confirmed that food-waste-based biochar injected into digestion tanks enhanced digestion efficiency by serving as a medium contributing trace elements and increased the number of microorganisms. Therefore, using food-waste-based biochar to improve AD tank methane-production efficiency could be a practical and effective method for recycling food waste.

Author Contributions: Conceptualization, D.-C.S. and K.-H.A.; methodology, D.-C.S.; software, D.-C.S.; validation, D.-C.S., K.-H.A. and J.J.; formal analysis, Y.-E.L.; investigation, J.J. and Y.J.; resources, D.-C.S.; data curation, D.-C.S.; writing—original draft preparation, D.-C.S.; writing—review and editing, D.-C.S.; visualization, D.-C.S.; supervision, K.-H.A.; project administration, I.-T.K.; funding acquisition, I.-T.K. All authors have read and agreed to the published version of the manuscript.

Funding: Research for this paper was carried out under the KICT Research Program (project no. 20220079-001: Development of renewable energy source on power plant scale from food waste for a residential complex) funded by the Ministry of Science and ICT.

Institutional Review Board Statement: Not applicable.

Informed Consent Statement: Not applicable.

Data Availability Statement: Not applicable.

Conflicts of Interest: The authors declare no conflict of interest.

References

1. Luz, F.C.; Cordiner, S.; Manni, A.; Mulone, V.; Rocco, V. Biochar characteristics and early applications in anaerobic digestion: A review. *J. Environ. Chem. Eng.* **2018**, *6*, 2892–2909. [CrossRef]
2. Abbas, Y.; Yun, S.; Wang, Z.; Zhang, Y.; Zhang, X.; Wang, K. Recent advances in bio-based carbon materials for anaerobic digestion. *Renew. Sustain. Energy Rev.* **2021**, *135*, 110378. [CrossRef]
3. Ugwu, S.; Harding, K.; Enweremadu, C. Comparative life cycle assessment of enhanced anaerobic digestion of agro-industrial waste for biogas production. *J. Clean. Prod.* **2022**, *345*, 131178. [CrossRef]
4. Ming, G.; Shuang, Z.; Xinxin, M.; Weijie, G.; Na, S.; Qunhui, W.; Chuanfu, W. Effect of yeast addition on the biogas production performance of a food waste anaerobic digestion system. *R. Soc. Open Sci.* **2021**, *7*, 200443. [CrossRef]
5. Zhang, C.; Su, H.; Baeyens, J.; Tan, T. Reviewing the anaerobic digestion of food waste for biogas production. *Renew. Sustain. Energy Rev.* **2014**, *38*, 383–392. [CrossRef]
6. Rabii, A.; Aldin, S.; Dahman, Y.; Elbeshbishy, E. A Review on anaerobic co-digestion with a focus on the microbial populations and the effect of multi-stage digester configuration. *Energies* **2019**, *12*, 1106. [CrossRef]
7. Fijoo, G.; Soto, M.; Méndez, M.; Lema, J.M. Sodium inhibition in the anaerobic digestion process: Antagonism and adaptation phenomena. *Enzym. Microb. Technol.* **1995**, *17*, 180–188. [CrossRef]
8. Liu, Y.; Wang, W.; Wachem, A.; Zou, D. Effects of adding osmoprotectant on anaerobic digestion of kitchen waste with high level of salinity. *J. Biosci. Bioeng.* **2019**, *128*, 723–732. [CrossRef]
9. Bo, Z.; Pin-jing, H. Performance assessment of two-stage anaerobic digestion of kitchen wastes. *Environ. Technol.* **2014**, *34*, 1277–1285. [CrossRef] [PubMed]
10. Świechowski, K.; Matyjewicz, B.; Telega, P.; Białowiec, A. The influence of low-temperature food waste biochars on anaerobic digestion of food waste. *Materials* **2022**, *15*, 945. [CrossRef]
11. Yun, S.; Fang, W.; Du, T.; Hu, X.; Huang, X.; Li, X. Use of bio-based carbon materials for improving biogas yield and digestate stability. *Energy* **2018**, *164*, 898–909. [CrossRef]
12. Zhang, L.; Zhang, J.; Loh, K.-C. Activated carbon enhanced anaerobic digestion of food waste–Laboratory-scale and Pilot-scale operation. *Waste Manag.* **2018**, *75*, 270–279. [CrossRef]
13. Lee, J.; Lee, S.-H.; Park, H. Enrichment of specific electro-active microorganisms and enhancement of methane production by adding granular activated carbon in anaerobic reactors. *Bioresour. Technol.* **2016**, *205*, 205–212. [CrossRef] [PubMed]
14. Cai, J.; He, P.; Wang, Y.; Shao, L.; Lü, F. Effects and optimization of the use of biochar in anaerobic digestion of food wastes. *Waste Manag. Res.* **2016**, *34*, 409–416. [CrossRef] [PubMed]
15. González, J.; Sánchez, M.E.; Gómez, X. Enhancing anaerobic digestion: The effect of carbon conductive materials. *J. Carbon Res.* **2018**, *4*, 59. [CrossRef]
16. Indren, M.; Birzer, C.H.; Kidd, S.P.; Hall, T.; Medwell, P.R. Effects of biochar parent material and microbial pre-loading in biochar-amended high-solids anaerobic digestion. *Bioresour. Technol.* **2020**, *298*, 122457. [CrossRef]
17. Yue, X.; Arena, U.; Chen, D.; Lei, K.; Dai, Z. Anaerobic digestion disposal of sewage sludge pyrolysis liquid in cow dung matrix and the enhancing effect of sewage sludge char. *J. Clean. Prod.* **2019**, *235*, 801–811. [CrossRef]
18. Mumme, J.; Srocke, F.; Heeg, K.; Werner, M. Use of biochars in anaerobic digestion. *Bioresour. Technol.* **2014**, *164*, 189–197. [CrossRef]
19. Cimon, C.; Kadota, P.; Eskicioglu, C. Effect of biochar and wood ash amendment on biochemical methane production of wastewater sludge from a temperature phase anaerobic digestion process. *Bioresour. Technol.* **2020**, *297*, 122440. [CrossRef]

20. Qi, Q.; Sun, C.; Zhang, J.; He, Y.; Tong, Y.W. Internal enhancement mechanism of biochar with graphene structure in anaerobic digestion: The bioavailability of trace elements and potential direct interspecies electron transfer. *Chem. Eng. J.* **2021**, *406*, 126833. [CrossRef]
21. Jeong, Y.; Lee, Y.; Kim, I. Characterization of sewage sludge and food waste-based biochar for co-firing in a coal-fired power plant: A case study in Korea. *Sustainability* **2020**, *12*, 9411. [CrossRef]
22. Yrjälä, K.; Lopez-Echartea, E. Chapter Ten: Structure and function of biochar in remediation and as carrier of microbes. Advances in Chemical Pollution. *Adv. Chem. Pollut. Environ. Manag. Prot.* **2021**, *7*, 269–294. [CrossRef]
23. Ahn, K.; Shin, D.; Jung, J.; Jeong, Y.; Lee, Y.; Kim, I. Physicochemical properties of torrefied and pyrolyzed food waste biochars as fuel: A pilot-scale study. *Energies* **2022**, *15*, 333. [CrossRef]
24. Gao, Y.; Fang, Z.; Liang, P.; Zhang, X.; Qiu, Y.; Kimura, K.; Haung, X. Anaerobic digestion performance of concentrated municipal sewage by forward osmosis membrane: Focus on the impact of salt and ammonia. *Bioresour. Technol.* **2019**, *276*, 204–210. [CrossRef] [PubMed]
25. Roberts, K.P.; Heaven, S.; Banks, C.J. Quantification of methane losses from the acclimatisation of anaerobic digestion to marine salt concentrations. *Renew. Energy* **2016**, *86*, 497–506. [CrossRef]
26. Owen, W.F.; Stuckey, D.C.; Healy, J.B.; Young, L.Y.; McCarty, P.L. Bioassay for monitoring biochemical methane potential and anaerobic toxicity. *Water Res.* **1979**, *13*, 485–492. [CrossRef]
27. Shelton, D.R.; Tiedje, J.M. General method for determining anaerobic biodegradation potential. *Appl. Environ. Microbiol.* **1984**, *47*, 850–857. [CrossRef] [PubMed]
28. Lay, J.-J.; Li, Y.-Y.; Noike, T. Influences of pH and moisture content on the methane production in high-solids sludge digestion. *Water Res.* **1997**, *31*, 1518–1524. [CrossRef]
29. Lay, J.J.; Li, Y.Y.; Noike, T. Developments of bacterial population and methanogenic activity in a laboratory-scale landfill bioreactor. *Water Res.* **1998**, *32*, 3673–3679. [CrossRef]
30. Murray, P.A.; Zinder, S.H. Nutritional requirements of *Methanosarcina* sp. strain TM-1. *Appl. Environ. Microbiol.* **1985**, *50*, 49–55. [CrossRef]
31. Zandvoort, M.H.; van Hullebusch, E.D.; Gieteling, J.; Lens, P.N.L. Granular sludge in full-scale anaerobic bioreactors: Trace element content and deficiencies. *Enzym. Microb. Technol.* **2006**, *39*, 337–346. [CrossRef]
32. Zimmerman, A.R. Abiotic and microbial oxidation of laboratory-produced black carbon (biochar). *Environ. Sci. Technol.* **2010**, *44*, 1295–1301. [CrossRef] [PubMed]
33. Molaey, R.; Bayrakdar, A.; Sürmeli, R.O.; Çalli, B. Influence of trace element supplementation on anaerobic digestion of chicken manure: Linking process stability to methanogenic population dynamics. *J. Clean. Prod.* **2018**, *181*, 794–800. [CrossRef]
34. Albracht, S.P.J. Nickel hydrogenases: In search of the active site. *Biochim. Biophys. Acta Bioenerg.* **1994**, *1188*, 167–204. [CrossRef]
35. Fauque, G.; Peck, H.D., Jr.; Moura, J.J.G.; Huynh, B.H.; Berlier, Y.; DerVartanian, D.V.; Teixeira, M.; Przybyla, A.E.; Lespinat, P.A.; Moura, I.; et al. The three classes of hydrogenases from sulfate-reducing bacteria of the genus *Desulfovibrio*. *FEMS Microbiol.* **1988**, *4*, 299–344. [CrossRef]
36. Sawers, G. The hydrogenases and formate dehydrogenases of *Escherichia coli*. *Antonie Van Leeuwenhoek* **1994**, *66*, 57–88. [CrossRef]
37. Takashima, M.; Speece, R.E.; Parkin, G.F. Mineral requirements for methane fermentation. *Crit. Rev. Environ. Control* **1990**, *19*, 465–479. [CrossRef]
38. Lipscomb, J.B. Biochemistry of the soluble methane monooxygenase. *Annu. Rev. Microbiol.* **1994**, *48*, 371–399. [CrossRef]
39. Zhang, T.; Zhang, P.; Hu, Z.; Qi, Q.; He, Y.; Zhang, J. New insight on Fe-bioavailability: Bio-uptake, utilization, and induce in optimizing methane production in anaerobic digestion. *Chem. Eng. J.* **2022**, *441*, 136099. [CrossRef]
40. Schindelin, H.; Kisker, C.; Schlessman, J.L.; Howard, J.B.; Rees, D.C. Structure of ADP·AIF4-stabilized nitrogenase complex and its implications for signal transduction. *Nature* **1997**, *387*, 370–376. [CrossRef]
41. Sing, K.S.W. Reporting physisorption data for gas/solid systems with special reference to the determination of surface area and porosity (Recommendations 1984). *Pure Appl. Chem.* **1985**, *57*, 603–619. [CrossRef]
42. Rouquerol, J.; Avnir, D.; Fairbridge, C.W.; Everett, D.H.; Haynes, J.H.; Pernicone, N.; Ramsay, J.D.F.; Sing, K.S.W.; Unger, K.K. Recommendations for the characterization of porous solids. *Pure Appl. Chem.* **1994**, *6*, 1739–1758. [CrossRef]
43. Lehmann, J.; Joseph, S. *Biochar for Environmental Management: Science, Technology and Implementation*, 2nd ed.; Routledge: London, UK, 2015. [CrossRef]
44. Qambrani, N.A.; Rahman, M.M.; Won, S.; Shim, S.; Ra, C. Biochar properties and eco-friendly applications for climate change mitigation, waste management, and wastewater treatment: A review. *Renew. Sustain. Energy Rev.* **2017**, *79*, 255–273. [CrossRef]
45. Choong, Y.Y.; Norli, I.; Abdullah, A.Z.; Yhaya, M.F. Impacts of trace element supplementation on the performance of anaerobic digestion process: A critical review. *Bioresour. Technol.* **2016**, *209*, 369–379. [CrossRef] [PubMed]
46. Zhang, L.; Jahng, D. Long-term anaerobic digestion of food waste stabilized by trace elements. *J. Waste Manag.* **2012**, *32*, 1151–1509. [CrossRef]
47. Chiappero, M.; Norouzi, O.; Hu, M.; Demichelis, F.; Berruti, F.; Di Maria, F.; Masek, O.; Fiore, S. Review of biochar role as additive in anaerobic digestion processes. *Renew. Sustain. Energy Rev.* **2020**, *131*, 110037. [CrossRef]
48. Montalvo, S.; Huiliñir, C.; Borja, R.; Sánchez, E.; Herrmann, C. Application of zeolites for biological treatment processes of solid wastes and wastewaters: A review. *Bioresour. Technol.* **2020**, *301*, 122808. [CrossRef]
49. Sun, M.T.; Yang, Z.M.; Lu, J.; Fan, X.L.; Guo, R.B.; Fu, S.F. Improvement of bacterial methane elimination using porous ceramsite as biocarrier. *J. Chem. Technol. Biotechnol.* **2018**, *93*, 2406–2414. [CrossRef]
50. Yu, Y.; Lu, X.; Wu, Y. Performance of an anaerobic baffled filter reactor in the treatment of algae-laden water and the contribution of granular sludge. *Water* **2014**, *6*, 122–138. [CrossRef]

fermentation

Article

Sequencing Batch Reactor Performance Evaluation on Orthophosphates and COD Removal from Brewery Wastewater

Siphesihle Mangena Khumalo [1,2,*], Babatunde Femi Bakare [2], Emmanuel Kweinor Tetteh [1] and Sudesh Rathilal [1]

1 Green Engineering Research Group, Department of Chemical Engineering, Faculty of Engineering and the Built Environment, Steve Campus, Durban University of Technology, S3 L3, P.O. Box 1334, Durban 4000, South Africa; emmanuelk@dut.ac.za (E.K.T.); rathilals@dut.ac.za (S.R.)
2 Environmental Pollution and Remediation Research Group, Department of Chemical Engineering, Faculty of Engineering, Mangosuthu University of Technology, P.O. Box 12363, Jacobs, Durban 4026, South Africa; bfemi@mut.ac.za
* Correspondence: khumalo.sm@outlook.com

Abstract: The discharge of industrial effluent constituting high orthophosphates and organic pollutants in water receiving bodies compromises freshwater quality and perpetuates eutrophication. In this study, an anaerobic–aerobic sequencing batch reactor (SBR) under activated sludge was investigated for orthophosphates and chemical oxygen demand (COD) removal from brewery wastewater. Raw brewery wastewater samples were collected on a daily basis for a period of 4 weeks. The findings of the study are reported based on overall removal efficiencies recording 69% for orthophosphates and 54% for total COD for a sludge retention time (SRT) of 7 days and hydraulic retention time of 18 h at mesophilic temperature conditions of ±25 °C. Moreover, the SBR system showed stability on orthophosphate removal at a SRT ranging from 3 to 7 days with a variation in organic volumetric loading rate ranging from 1.14 to 4.83 kg COD/m^3.day. The anaerobic reaction period was experimentally found to be 4 h with the aerobic phase lasting for 14 h. The SBR system demonstrated feasibility on orthophosphates and COD removal with variation in organic loading rate.

Keywords: orthophosphates; chemical oxygen demand; sequencing batch reactor; brewery wastewater; solid retention time

Citation: Khumalo, S.M.; Bakare, B.F.; Tetteh, E.K.; Rathilal, S. Sequencing Batch Reactor Performance Evaluation on Orthophosphates and COD Removal from Brewery Wastewater. *Fermentation* **2022**, *8*, 296. https://doi.org/10.3390/fermentation8070296

Academic Editor: Sanjay Nagarajan

Received: 30 April 2022
Accepted: 20 June 2022
Published: 23 June 2022

1. Introduction

The issue of freshwater scarcity perpetuated by environmental pollution among many other factors has become a global phenomenon, particularly in the sub-Saharan region [1–3]. The substantial increase in biological nutrients particularly phosphorus and nitrogenous compounds in water bodies results in eutrophic waters [4–6]. The environmentally detrimental eutrophic waters are characterized by high concentrations of aquatic weeds and algae, which eventually die, sink to the bottom, and decay, thus reducing the levels of dissolved oxygen in the water killing fish [4,6–8]. Moreover, eutrophic waters can cause adverse effects on human society, such as drinking water problems (i.e., taste and odor) and promotion of toxic phytoplankton species [8]. Phosphorus is one of the essential nutrients for plant growth, enriching water bodies with phosphorus results in the stimulation of toxic cyanobacterial (algal blooms) [5,8]. The occurrence of excess phosphorus and other biological nutrients in aquatic ecosystems perpetuated by environmental population necessitates the need to reduce biological nutrient loads entering the environment. One of the human activities perpetuating the formation of eutrophic waters is the discharge of voluminous untreated industrial wastewater into water-receiving bodies [6,9,10]. The brewing industry is not an exception, the beer-producing process is characterized by the use of large volumes of fresh water and generate voluminous amounts of wastewater [1,11,12], which require treatment prior to being discharged into water-receiving bodies.

Wastewater emanating from the brewery is characterized by high concentration levels of chemical oxygen demand (COD) (Table 1) which results from the high organic compounds found in brewery wastewater such as sugars, yeast, and soluble starch [11–14].

Furthermore, industrial wastewater generated from the brewery also contains phosphorus and nitrogenous pollutants; however, their concentrations depend greatly on the type of chemicals that are used during the cleaning process (i.e., caustic soda, phosphoric acid, and nitric acid) and the amount of yeast in wastewater [1,12,15]. From the data presented in Table 1, it is clear that the brewery's wastewater contains a significant percentage of COD in terms of pollutant composition, which can be harmful to the environment. It is worth mentioning that most nations, including South Africa (SA) and the European Union (EU), have dewatering regulations that the brewing business is obligated to uphold. Dewatering regulations are designed to manage and/or eradicate environmental issues associated with the discharge of untreated industrial effluent [1]. Hence, breweries should be able to manage their impacts on the environment, by developing wastewater treatment processes that can effectively treat their effluent to meet dewatering limits set by national and international environmental entities.

Table 1. Brewery wastewater composition and dewatering limits in SA and EU.

	Present Study		[1]	[16]	[17]	
Parameter	**Mean ± SD**	**Range**	**Range**	**Range**	**SA Discharge Limits**	**EU Discharge Limits**
Temperature, °C	31 ± 3.7	25.3–37	18–40	-	<44	-
pH	6.5 ± 2.4	4.4–6.17	3–12	3–12	5.0–9.5	-
Turbidity, NTU	570 ± 164	303–1039	-	-	-	-
Total COD, mg/L	7687 ± 2030	3447–11,813	2000–6000	1800–5000	75	125
BOD_5, mg/L	-	-	1200–3600	1005–3800	-	25
Phosphates, mg/L	343 ± 64	229–424	10–50	10–50	10	1–2
TS, mg/L	5951 ± 3387	2942–14,981	5100–8750	50–6000	-	-
VSS, mg/L	1799 ± 571	1043–2572	-	-	-	-

There are reported studies conducted on brewery wastewater treatment using a SBR with the common goal of minimizing wastewater-related environmental issues, namely anaerobic SBR [18], aerobic SBR [19], aerobic/anoxic SBR [20], and suspended and attached growth SBR [21]. Shao et al. [18] evaluated the performance of an anaerobic SBR in COD removal from raw brewery wastewater. Shao and co-workers reported 90% COD removal for a HRT and STR of 24 h and 60 days, respectively, for an OVLR range of 1.5 to 5.0 kg COD/m^3.day. Moreover, Wang et al. [19] reported 88% COD removal from brewery wastewater using a SBR for HRT and STR of 15 h and 90 days, respectively. It is worth noting that for the work reported by Wang and co-workers, the reactor effluent had a total COD concentration of 346 mg/L which is above the dewatering limits as indicated in Table 1.

The removal of phosphorus biologically from waste streams is achieved by introducing waste streams into an anaerobic environment in which phosphorus is released followed by an aerobic environment in which phosphorus is taken up by polyphosphate-accumulating organisms (PAOs) [22–24]. Ge et al. [25] investigated the performance of a SBR for phosphorus removal in abattoir wastewater. The findings of the investigation reported 90% phosphorus removal for a HRT and SRT of 0.5 to 1 day and 2 to 2.5 days, respectively, at an OVLR of between 2 to 3 g COD/L.

It should be noted that there are emerging advanced oxidation processes (AOPs) for wastewater treatment such as photocatalytic degradation known as photocatalysis [26]. The photocatalysis process involves the use of solids (photocatalysts) which can promote reactions in the presence of light without being consumed in the overall reaction [27], titanium oxide and zinc oxide are the most widely used photocatalysts [26]. Previous studies [28–30] have indicated that photocatalysis demonstrated good performance in degrading organic pollutants in wastewater streams. However, there are some drawbacks associated with the photocatalysis process such as high energy requirements for photocatalysts with a wide band gap energy [31], low light absorption abilities which hinder the overall photocatalytic quantum efficiency [26], and high costs of recycling and recuperating suspended photocatalysts [32]. On the other hand, biological methods have cemented their application in wastewater treatment because they are economically attractive and mostly used in industry [33].

Despite the advancement in wastewater treatment processes, biological methods are still widely used in wastewater treatment works as reported by Chen et al. [33]. It should be noted that the application of SBRs in brewery wastewater treatment has been investigated extensively [20,21,23,24]; however, to our knowledge, none of the reported studies have reported on the performance evaluation of a SBR for simultaneous COD and orthophosphates removal from brewery wastewater relative to the microbial population growth rate. The current study aims to evaluate the performance of a SBR for simultaneous COD and orthophosphates removal from brewery wastewater. The SBR system is selected on the basis that the settling and reaction phase takes place in the same vessel, which makes it easy to operate and economically attractive. Furthermore, the findings of the study will provide wastewater-producing industries with practical and technical reference information to assist in developing the most effective in-house wastewater treatment systems to reduce phosphorus and carbon pollutants, thus reducing the environmental pollution in water-receiving bodies. Moreover, the current study will give an insight into the substrate utilization rate relative to substrate concentration as well as microbial population growth rate relative to substrate utilization rate.

2. Materials and Methods

2.1. Sample Collection and Preparation

Brewery wastewater samples were collected at the effluent stream of the brewery on a daily basis for a period of 28 days using sterile glass sampling bottles. Samples were transported to the laboratory in a cooler box full of ice to maintain a temperature of 4 °C. Samples were collected mainly for the operation of a laboratory-scale SBR to investigate the performance of the SBR system on orthophosphates and COD removal from brewery wastewater. Upon arrival at the laboratory, samples were allowed to warm up to room temperature and sample composition analyses were conducted within 48 h from the time of sampling by standard methods [34]. Thereafter, charged into the reactor to commence treatment immediately.

2.2. Activated Sludge

Activated sludge was harvested from an anaerobic digester at a local brewery wastewater treatment plant. The microbial population was harvested using a 10 L bucket and then transported to the laboratory. In preparing the harvested microbial population for treatment, no chemicals were added to the sludge nor into the raw brewery wastewater to balance the N:C:P ratio. Only the condensed almost granular sludge was used for treatment since granular sludge is associated with good settleability, which is imperative for optimum treatment efficiencies.

2.3. Sequencing Batch Reactor Design

The laboratory-scale SBR, as shown in Figure 1, was made of transparent polyvinyl chloride, having a total volume of 22 L with a conical base having a slope of 60° for easy drainage of bio-solids. For experimental runs, the working volume was set at 13 L with the microbial population occupying 4 L and raw brewery wastewater occupying 9 L. This working volume was based on the selected HRT and SRT since they are both affected by the reactor working volume.

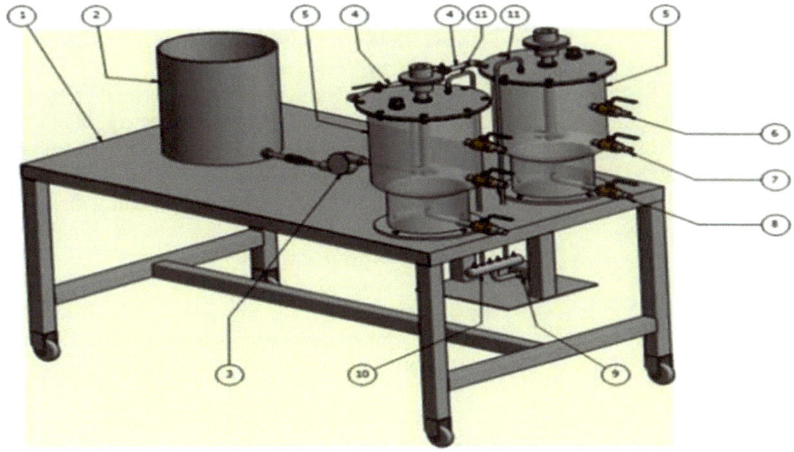

Figure 1. SBR isometric view: (**1**) table, (**2**) SBR holding tank, (**3**) centrifugal pump, (**4**) influent feed stream, (**5**) SBR vessel, (**6**) effluent sample point 1, (**7**) effluent sample point 2, (**8**) sludge discharge stream, (**9**) aerator pump, (**10**) manifold, and (**11**) oxygen aerator pipe [35].

Moreover, the reactor was not utilized into its maximum working volume to accommodate sludge bulking since the microbial growth rate is directly proportional to the substrate utilization rate [23]. The conical bottom of the reactor allowed a quiescent and easy gravitational settling mechanism. The reactor had a portable shaft mixer which was operated continuously to keep bio-solids suspended inside the reactor, thus allowing perfect mixing. Both the mixer shaft and impeller blades were made of stainless steel, with a drive motor mounted at the top of the reactor tank in a rubber gasket operating at 10 W.

2.4. Experimental Approach

The experimental approach which was adopted in this study is similar to the work reported by Shabangu and Bakare [36] which includes a sequence of operational steps which are defined as follows:

- The filling phase—This was considered the first operational phase of the SBR system. The reactor was first seeded with 4 L of activated sludge under anaerobic conditions. Raw brewery wastewater was fed into the holding tank where suspended solids were allowed to settle by gravitational force for a period of 2 h. After the settling phase, 9 L of raw brewery wastewater supernatant was pumped into the reactor. The filling phase took place under anaerobic conditions; however, the stirrer was switched on and set to operate at 350 rpm to allow mixing. According to Tchobanoglous [23], only mixing during the filling stage promotes filamentous growth control thus improving sludge settling and thickening. The agitation speed of the stirrer was set to be at 350 rpm because it was observed that higher agitation speed resulted in sludge bulking, thus compromising the solids' settleability. The filling phase on average for all experimental runs lasted for 5 min.
- Reaction phase—After the filling phase, the system was allowed to undergo an anaerobic phase which favored the polyphosphate-accumulating organisms, which lasted for a period of 4 h and thereafter the reaction phase was instigated. Oxygen was supplied using an aerator pump as depicted in Figure 1 at a flow rate of 7.5 L/min, maintaining a dissolved oxygen concentration of 3 mg/L. It is worth noting that for the current work, the effect of dissolved oxygen was not investigated. During the reaction phase, microorganisms consume substrate, i.e., orthophosphates under a controlled pH which was kept within the range of 4 to 9.5. According to Tchobanoglous [23], microbial activities are hindered at pH levels less than 4 and pH levels more than 9.5.

The aeration duration and anaerobic phase duration were predetermined experimentally which lasted for 14 h and 4 h, respectively. Moreover, the SBR was operated at mesophilic temperature of ±25 °C.

- Settling phase—During this phase, bio-solids were allowed to separate gravitationally from the treated liquid under quiescent conditions resulting in a clear clarified supernatant. During this phase, the stirrer was switched off as well as the aeration system, and no influent was charged into the reactor nor effluent drawn. The settling period lasted for 2 h to enhance optimum settling of bio-solids containing biodegradable organic and biological pollutants, thus resulting in a clear clarified supernatant with minimum suspended solids.
- Drawing phase—This phase was considered the final treatment operational stage for the SBR system. During this phase, the clarified supernatant was sampled as the treated reactor effluent by tapping the reactor effluent into a 250 mL sterile glass bottle for laboratory analysis.

2.5. Laboratory Analysis

Orthophosphates (PO_4^{3-}), total COD (TCOD), total solids (TS), volatile suspended solids (VSS), temperature, and pH were measured in accordance with the Standard Methods for the Examination of Water and Wastewater [34] standard method. Orthophosphate concentration was measured colorimetrically using a DR3900 spectrophotometer manufactured by Hach South Africa Pty Ltd, Johannesburg, supplied by Universal Water Supplies, from South Africa. The molybdovanadate method was implemented, in which orthophosphate reacts with molybdate in an acid medium to produce a mixed phosphate/molybdate complex. In the presence of vanadium, a yellow molybdovanadophosphiric acid is formed. The intensity of the yellow color is proportional to the phosphate concentration. Samples were measured at a wavelength of 430 nm. TCOD was measured as a quick indicator of organic pollutants in industrial wastewater emanated from the brewery. The TCOD was expressed in milligrams of oxygen per liter, which is the amount of oxygen consumed per liter of brewery wastewater. This parameter was measured spectrophotometrically (Hach DR3900) using the colorimetric method. According to the Standard methods for the Examination of Water and Wastewater [34], total solids are total dissolved solids plus suspended and settleable solids in water. In the case of brewery wastewater, dissolved solids consist of nitrate, phosphorus, and other particles. On the other hand, suspended solids include fine organic debris and other particulate matter.

Total solids were measured gravimetrically in mg TS/L, a well-mixed sample was dried at 105 °C for 24 h, the TS fraction was given by the weight of the residue after drying. Both temperature and pH were measured online, i.e., during the treatment process. Temperature and pH monitoring inside the reactor was carried out using a calibrated Thermo Scientific Orion Star A215 pH/conductivity meter manufactured by Thermo Fisher Scientific, Johannesburg, supplied by Universal Water Supplies, from South Africa.

2.6. Data Analysis

For data credibility, samples were measured in triplicates and statistically validated at a 95% confidence level. The removal efficiency for the SBR was calculated using Equation (1) below:

$$\text{Removal efficiency (\%)} = [(C_0 - C_f)/C_0] \times 100\% \quad (1)$$

where C_0 and C_f are the substrate concentrations (mg/L) in the SBR influent and effluent streams, respectively.

Substrate utilization/uptake rate and the microbial growth rate were monitored by using the Michaelis–Menten and the Monod empirical models as presented by Equations (2) and (3) [18]:

$$r_{su} = kXS/(K_s + S) \quad (2)$$

$$r_g = \mu_m XS/(K_s + S) \quad (3)$$

where r_{su} and r_g are the substrate utilization rate and bacteria growth rate from substrate utilization per unit of reactor volume, g/m^3.day, respectively, k is the maximum specific substrate utilization rate, g-substrate/g-microorganisms.day, X is the biomass concentration, g/m^3, S is the growth-limiting substrate concentration in a solution, g/m^3, K_s is the half-velocity constant, which is the substrate concentration at one-half the maximum specific substrate utilization rate, g/m^3, and μ_m is the maximum specific bacteria growth rate, g-biomass/g-biomass.day.

Moreover, a descriptive statistical analysis was conducted to calculate the mean and standard deviation (SD) using Equations (4) and (5), respectively, as well as the range.

$$\overline{x} = \frac{\Sigma X}{n} \quad (4)$$

$$SD = \sqrt{\frac{\Sigma(X - \overline{x})^2}{n - 1}} \quad (5)$$

where $\overline{x}$ is the mean, X is the numerical value of each sample, and n is the total number of samples analyzed. The *SD* was used to measure how far each of the measured physiochemical properties lies from the mean. It should be noted that high *SD* values mean that the value of the measured physiochemical property lies generally far from the mean, while low *SD* values mean the measured physiochemical property values are clustered close to the mean. Moreover, the *SD* was calculated by reducing the sample size from n to $n - 1$, this was implemented to avoid a biased estimate when using a sample size of n, thus underestimating the variability [37].

3. Results and Discussion

3.1. Effect of Hydraulic Retention Time (HRT) on Orthophosphate Removal

For the current study, the HRT was determined experimentally and the results obtained are presented in Figure 2. It can be seen that there was a significant increase in orthophosphate concentration during the first 4 h of the anaerobic phase. The increase in orthophosphate production was an indication that the PAOs were favored which are essential for orthophosphate removal in the aerobic phase. The PAOs are favored in the anaerobic environment because they do not require oxygen as an electron donor. However, they consume readily biodegradable substrates in wastewater using energy made available from stored phosphorus as polyphosphates, thus enabling PAOs to become dominant.

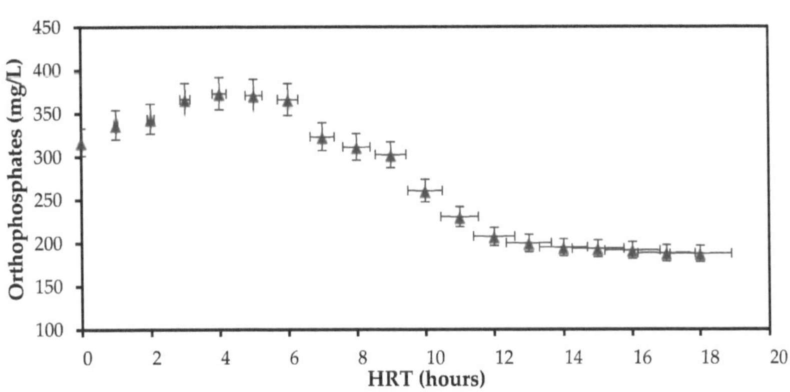

Figure 2. Orthophosphate concentration as a function of HRT profile.

Furthermore, it is evident from Figure 2 that at a HRT of 5 to 12 h there was a significant removal of orthophosphates and it reduced significantly at a HRT of between 13 and

18 h. Orthophosphate removal was achieved in the aerobic phase because in an aerobic environment, microorganisms grow new biomass and take up orthophosphates, typically more than the amount they released in the anaerobic environment [23]. Moreover, the orthophosphate removal mechanism is characterized by a faster orthophosphate release rate than the subsequent orthophosphate uptake rate in the aerobic phase. Thus, the aerobic phase last longer than the anaerobic phase for maximum orthophosphate removal, as presented by the orthophosphate concentration profile in Figure 2. Therefore, for the current study, a HRT of 18 h was considered with the anaerobic phase lasting 4 h and the aerobic phase duration being 14 h.

3.2. Effect of Solid Retention Time (STR) on Orthophosphate and COD Removal

Figures 3 and 4 present the findings of the study on orthophosphate and COD removal profile with variation in the SRT. It was observed that at a SRT of 3 days and above there was a significant removal in orthophosphates, recording a percentage removal of 70% and above. The significant removal at a SRT of 3 days and above was an indication that the sludge in the reactor was well acclimated to PAOs which are essential for orthophosphate removal. Furthermore, the findings of the study explicitly indicated that the system gained stability at a SRT of between 5 and 7 days, recording a maximum orthophosphate percentage removal of 80%. It was observed that operating at a SRT of 7 days and above promoted the growth of "glycogen-accumulating organisms", which cause a decrease in the growth rate of PAOs. Chan et al. [22] reported that longer SRTs of more than 10 days have the advantage of promoting the growth of "glycogen-accumulating organisms", causing a decrease in PAOs.

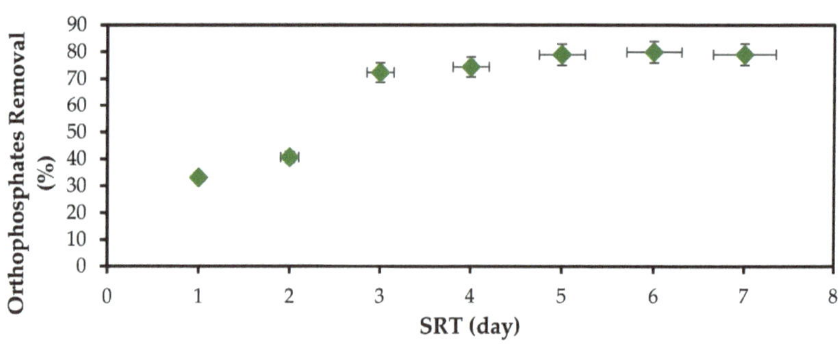

Figure 3. Orthophosphate removal with SRT variation.

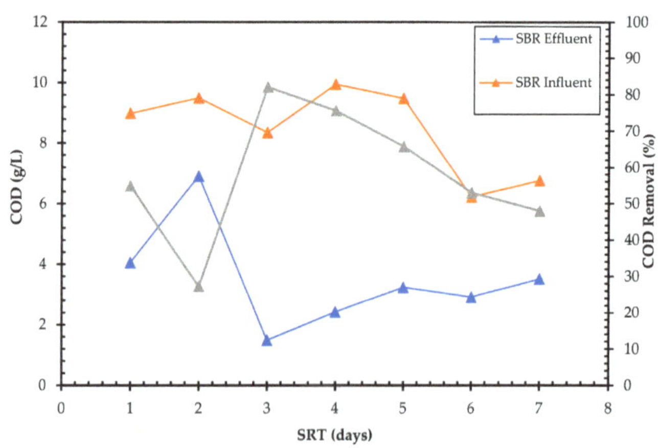

Figure 4. COD removal with SRT variation.

From Figure 4, it is apparent that biodegradation in terms of COD removal from brewery wastewater was taking place in the SBR, which is confirmed by the low COD concentration in the SBR effluent stream as compared to the SBR influent stream. However, the system under investigation did not yield conclusive findings on the relationship between SRT and COD removal. From the results presented in Figure 4, the lowest and highest COD removal efficiencies were recorded at a SRT of 2 and 3 days, respectively. The variation in COD removal is attributed to the variation in COD concentrations in the SBR influent stream. The lowest COD removal at a SRT of 2 days suggests that the SBR influent had a high fraction of slowly biodegradable COD, which constitutes particulate COD which is not explicitly accounted for in the current study. On the other hand, the highest COD removal at a SRT of 3 days suggests that the SBR influent stream has a high fraction of readily biodegradable COD, which constitutes soluble COD which is not explicitly accounted for in this study.

Moreover, the results presented in Figures 3 and 4 suggest that the SBR system under investigation needs to be combined with another wastewater treatment technology such as coagulation or advanced oxidation processes to comply with the dewatering limits presented in Table 1.

3.3. Orthophosphate and COD Removal with Variation in Organic Volumetric Loading Rate (OVLR)

According to Carucci et al. [38], fluctuations in organic loads in influent streams compromise the treatment efficacy of biological nutrient removal systems. In this research study, the orthophosphate and COD percentage removals were investigated with a variation in OVLR, and the findings of the study are presented in Figures 5 and 6. However, when analyzing Figure 5, it can be seen that the variation in OVLR had an insignificant effect on the orthophosphate percentage removal. This may be attributed to the basis that the microbial population used in this study was harvested from an anaerobic digester treating wastewater with high-strength organic loads; therefore, the microbial population adapted to variations in OVLR. Microbial populations in nature, when subjected to certain environmental conditions over a period of time, turn out to adapt to particular conditions, this period is referred to as the acclimation period. Orthophosphate removal of up to 80% was achieved in this study which was an indication that the microbial population in the reactor was well acclimated to a microbial population which was not affected by the variation in organic loads. Furthermore, the findings presented in Figure 5 suggests that the organic load in wastewater samples investigated in this study was within a range which did not have a negative effect on the system's microbial activities in orthophosphate removal.

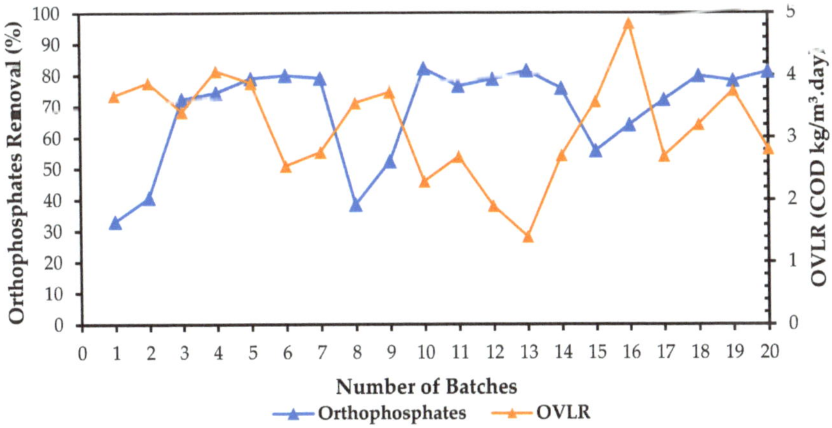

Figure 5. Orthophosphate removal with variation in OVLR.

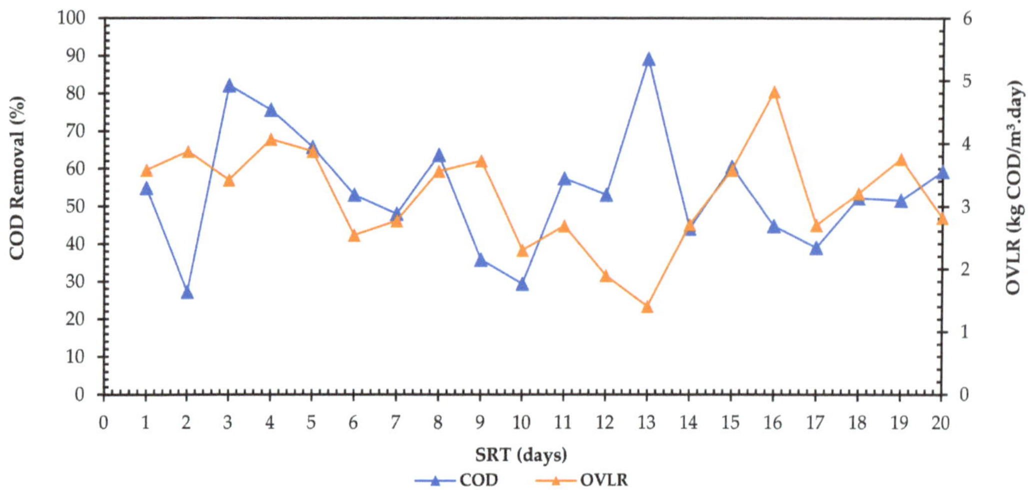

Figure 6. COD removal with variation in OVLR.

Figure 5 shows the findings of the current study, indicating that there is a variation in COD removal with OVLR variation as a function of SRT. As it was indicated in Section 3.2, the variation in COD removal is attributed to the variation in SBR influent stream composition. The results presented in Figure 6 suggest that the microbial community was able to biodegrade readily biodegradable COD which is not explicitly accounted for in this work. Moreover, the results presented in Figure 6 suggest that the brewery effluent under investigation constitutes a high composition of slowly biodegradable and particulate COD which can be removed by other advanced wastewater treatment processes.

3.4. Orthophosphate and Total Chemical Oxygen Demand (TCOD) Removal

The TCOD is a combination of the particulate COD and soluble COD. The findings of the study on orthophosphate and TCOD removal are presented in Figure 7. The current study on average achieved a TCOD removal efficiency of 54% which was lower than the orthophosphate percentage removal of 69%. Shabangu [35] and Bakare et al. [39] reported that higher COD efficiencies of up to 90% in SBR systems operated at mesophilic temperature conditions ranging between 20 and 25 °C can be achieved at longer HRTs ranging from 5 to 7 days. However, long HRT of up to 7 days may not be feasible for regions experiencing freshwater scarcity such as the southern part of Africa. Moreover, high HRTs of up to 7 days can result in high operation costs in terms of aeration.

Furthermore, the lower TCOD concentrations in the effluent stream when compared to the influent stream was an indication that indeed microbial activities were taking place inside the reactor during treatment. According to Tchobanoglous [23], during the anaerobic phase, POAs consume readily biodegradable organic substrates (e.g., biodegradable COD) with the aid of energy made available from stored phosphorus. Thus, enriching the sludge with PAOs. Based on organic substrate consumption mechanisms on orthophosphate removal systems reported by Tchobanoglous [23], it can be said that the 54% TCOD removal represents the fraction of readily biodegradable TCOD.

Moreover, for the current work, the pH range of the SBR influent stream was between 4.9–8.4 for different batches. The range in pH is attributed to varying brewery wastewater composition, depending on the brewery activities taking place inside the brewing house. The pH was left adjusted, as it was alluded in Section 2.4 that metabolic activities for microorganisms are inhibited at pH values of 9.5 and above or pH values below 4 [23]. The effect of pH on microbial activities is not explicitly accounted for in the current study;

however, it can be seen from Figures 2–7 that microbial activities were not inhibited despite the pH variation for different batches.

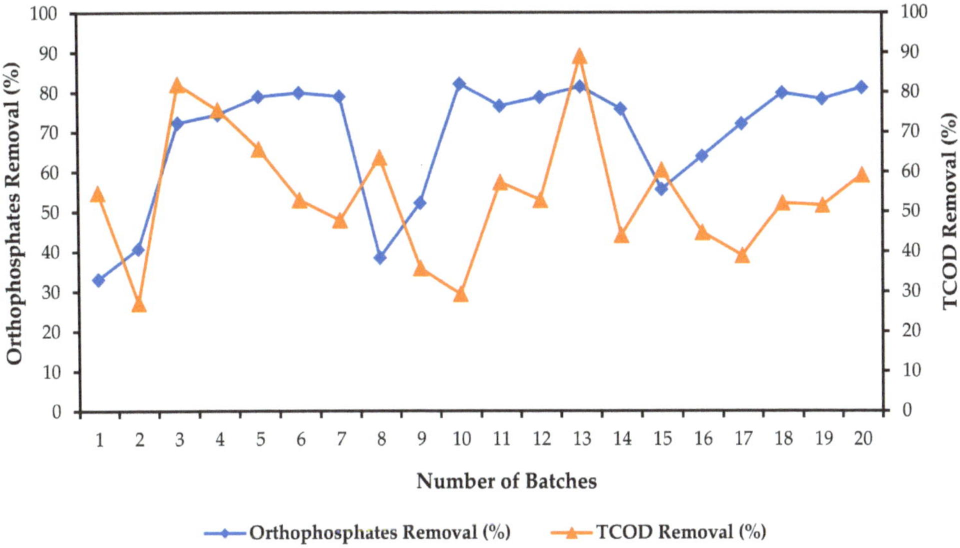

Figure 7. Orthophosphate and TCOD removal profile.

Table 2 presents a summary of similar work conducted on brewery wastewater treatment using a SBR system. The majority of reported studies do not focus on simultaneous COD and phosphorus removal. The high COD removal [18,19,39] compared to the current study, is attributed to high SRT and soluble COD [18,19]. Bakare et al. [39] reported high COD removal efficiency at a HRT of 120 days, which suggests that the brewery wastewater investigated had a high fraction of slowly biodegradable COD. Based on the results presented in Table 2, it is evident that a lot of work must be carried out aimed at investigating simultaneous COD and phosphorus removal from brewery wastewater.

Table 2. Summary of studies on brewery wastewater treatment using a SBR system.

Treatment Method	COD, %	TP, %	HRT, Hours	SRT, Days	OVLR, kg COD/m^3.day	Reference
Anaerobic SBR	>90		24	60	1.5–5	[13]
Aerobic SBR	88.7	-	15	90	3.5	[14]
Aerobic SBR	90		120	-	-	[26]
Aerobic–anaerobic SBR	54	69	18	7	1.4–4.1	Present study

3.5. Substrate Utilisation Rate and Microbial Population Growth Late

From Table 1, it is apparent that the brewery wastewater used in this study had a high COD composition compared to orthophosphates. Hence, COD was considered the microbial substrate. According to the Michaelis–Menten empirical model presented in Equation (2), the substrate utilization rate is directly proportional to the substrate concentration. The findings of the study presented in Figure 8 are congruent to the Michaelis–Menten's principle on substrate utilization rate. Note that the correlation between the utilization rate and the substrate (COD) concentration gave a coefficient of determination of less than 0.9. This is attributed to the fact that the first three system points which are beneath the trend line gave low substrate utilization rates relative to high COD concentrations. It is worth mentioning that the substrate utilization rate is a function of volatile suspended solids which are considered to be organic bio-solids. Therefore, low substrate utilization rates relative to high COD concentration could be attributed to the reactor influent stream having a high composition of inorganic bio-solids compared to organic bio-solids. Such bio-solid ratios result in low VSS fractions, consequently resulting in low substrate utilization rates despite high COD concentrations. This is accounted for in the findings presented in Table 1 showing a higher TS range compared to VSS.

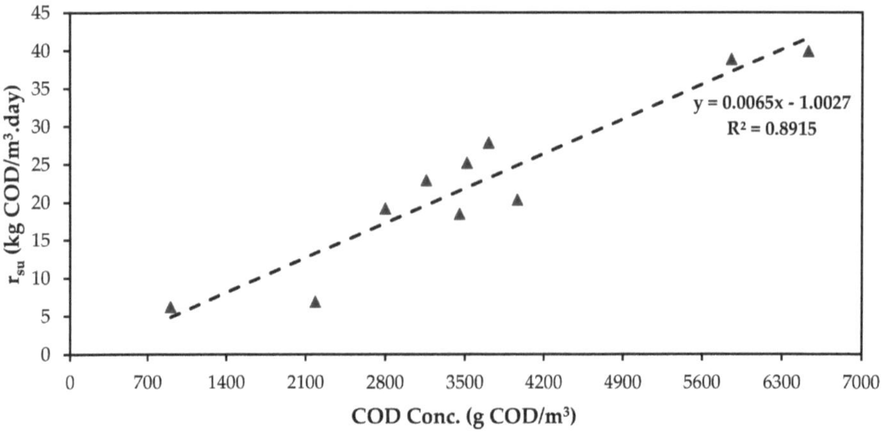

Figure 8. Substrate utilization rate as a function of COD concentration (g COD/m^3).

The relationship between the substrate utilization rate and microbial population growth rate was investigated using the Monod empirical model presented as Equation (3). Figure 9 presents the findings of the current study on the microbial growth rate, which indicates a strong correlation between microbial growth rate and substrate utilization rate. This is explicitly accounted for by the coefficient of determination $R^2 = 0.9582$. Moreover, the findings are congruent to the Monod's principle on microbial growth rate which clearly states that the substrate utilization rate is directly proportional to the microbial growth. The findings of the study recorded an average microbial growth rate of 16.86 kg/m^3.day.

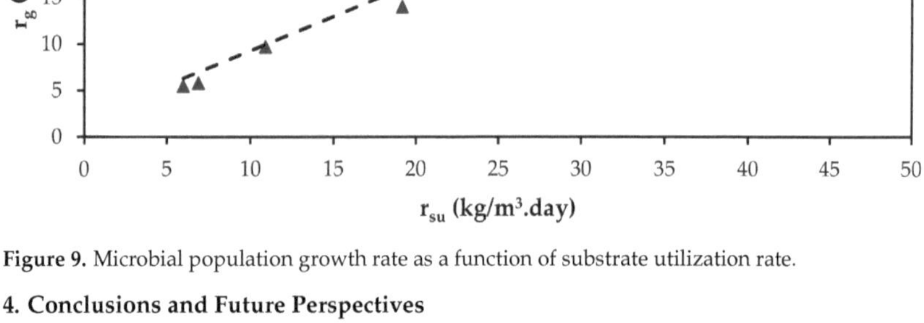

Figure 9. Microbial population growth rate as a function of substrate utilization rate.

4. Conclusions and Future Perspectives

In this study, the performance of an anaerobic–aerobic SBR for orthophosphate and COD removal from brewery wastewater was investigated. The findings of the study demonstrated good removal efficiencies on orthophosphates ranging from 33 to 81%, recording an overall treatment efficiency of 69%. Additionally, an average COD removal efficiency of 54% was recorded. Moreover, high removal efficiencies of orthophosphates and COD were obtained at a SRT of 3 to 7 days and a HRT of 18 h under mesophilic temperature of ±25 °C. Furthermore, the system did not show any negative effect on orthophosphate removal with the variation in organic volumetric loading rate which ranged from 1.14 to 4.83 kg COD/m^3.day. The low removal efficiency of COD maybe attributed to brewery effluent having particulate as well as slowly bio-degradable COD which can be removed by chemical methods. Based on the findings of the study, it can be concluded that the SBR demonstrated good treatment efficiency on orthophosphate removal from brewery wastewater with high-strength organic pollutants. However, the findings on COD suggests that the SBR performance can be improved by incorporating the SBR system with a chemical process aimed at eradicating pollutant fractions which cannot be removed by biological processes.

Despite the good performance of the SBR for the current study and from previous studies as reported in Table 2 on brewery wastewater treatment, none of the reported studies recorded a SBR effluent meeting the dewatering limits as presented in Table 1. This suggests that a lot of work still needs to be carried out on biological COD and orthophosphate removal from brewery wastewater. There are limited studies reporting on optimizing the OVLR to microbial population ratio for optimal biodegradation of organic pollutants. Moreover, more work needs to be carried out on techno-economic analysis for biological wastewater treatment processes integrated with AOPs since it is a promising emerging technology in wastewater treatment.

Author Contributions: Conceptualization, S.M.K., B.F.B. and S.R.; methodology, S.M.K. and E.K.T.; validation, S.M.K., B.F.B., S.R. and E.K.T.; formal analysis, S.M.K. and E.K.T.; investigation, S.M.K.; resources, B.F.B. and S.R.; data curation, S.M.K.; writing—original draft preparation, S.M.K.; writing—review and editing, B.F.B., S.R. and E.K.T.; visualization, S.M.K. and E.K.T.; supervision, B.F.B. and S.R.; project administration, E.K.T.; funding acquisition, B.F.B. and S.R. All authors have read and agreed to the published version of the manuscript.

Funding: This research was funded by the South African National Research Foundation under the Research Development Grants for Y-Rated Researchers, grant number 137765.

Institutional Review Board Statement: Not applicable.

Informed Consent Statement: Not applicable.

Data Availability Statement: Not applicable.

Acknowledgments: The authors wish to express their appreciation to the department of chemical engineering at the Durban University of Technology for proving necessary equipment to make this study a success. The authors also wish to express their appreciation to the department of chemical engineering at the Mangosuthu University of Technology for granting access to their well-equipped research laboratory.

Conflicts of Interest: The authors declare no conflict of interest.

References

1. Simate, G.S.; Cluett, J.; Iyuke, S.E.; Musapatika, E.T.; Ndlovu, S.; Walubita, L.F.; Alvarez, A.E. The treatment of brewery wastewater for reuse: State of the art. *Desalination* **2011**, *273*, 235–247. [CrossRef]
2. Dhakal, N.; Salinas-Rodriguez, S.G.; Hamdani, J.; Abushaban, A.; Sawalha, H.; Schippers, J.C.; Kennedy, M.D. Is Desalination a Solution to Freshwater Scarcity in Developing Countries? *Membranes* **2022**, *12*, 381. [CrossRef] [PubMed]
3. Abd-Elaty, I.; Kuriqi, A.; Shahawy, A.E. Environmental rethinking of wastewater drains to manage environmental pollution and alleviate water scarcity. *Nat. Hazards* **2022**, *110*, 2353–2380. [CrossRef] [PubMed]
4. Loza, V.; Perona, E.; Mateo, P. Specific responses to nitrogen and phosphorus enrichment in cyanobacteria: Factors influencing changes in species dominance along eutrophic gradients. *Water Res.* **2014**, *48*, 622–631. [CrossRef]
5. Qin, B.; Gao, G.; Zhu, G.; Zhang, Y.; Song, Y.; Tang, X.; Xu, H.; Deng, J. Lake eutrophication and its ecosystem response. *Chin. Sci. Bull.* **2013**, *58*, 961–970. [CrossRef]

6. Schindler, D.W.; Carpenter, S.R.; Chapra, S.C.; Hecky, R.E.; Orihel, D.M. Reducing phosphorus to curb lake eutrophication is a success. *Environ. Sci. Technol.* **2016**, *50*, 8923–8929. [CrossRef]
7. Liu, D.H.; Lipták, B.G. *Environmental Engineers' Handbook*, 2nd ed.; CRC Press Company: London, UK, 1997.
8. Homayoun Aria, S.; Asadollahfardi, G.; Heidarzadeh, N. Eutrophication modelling of Amirkabir Reservoir (Iran) using an artificial neural network approach. *Lakes Reserv. Res. Manag.* **2019**, *24*, 48–58. [CrossRef]
9. Atalay, S.; Ersöz, G. Advanced Oxidation Processes. In *Novel Catalysts in Advanced Oxidation of Organic Pollutants*; Springer: Cham, Swizerland, 2016; pp. 23–34.
10. Lu, X.; Zhen, G.; Estrada, A.L.; Chen, M.; Ni, J.; Hojo, T.; Kubota, K.; Li, Y.Y. Operation performance and granule characterization of upflow anaerobic sludge blanket (UASB) reactor treating wastewater with starch as the sole carbon source. *Bioresour. Technol.* **2015**, *180*, 264–273. [CrossRef]
11. Chen, H.; Chang, S.; Guo, Q.; Hong, Y.; Wu, P. Brewery wastewater treatment using an anaerobic membrane bioreactor. *Biochem. Eng. J.* **2016**, *105*, 321–331. [CrossRef]
12. Corsino, S.F.; di Biase, A.; Devlin, T.R.; Munz, G.; Torregrossa, M.; Oleszkiewicz, J.A. Effect of extended famine conditions on aerobic granular sludge stability in the treatment of brewery wastewater. *Bioresour. Technol.* **2017**, *226*, 150–157. [CrossRef]
13. Enitan, A.M.; Adeyemo, J.; Kumari, S.K.; Swalaha, F.M.; Bux, F. Characterization of brewery wastewater composition. *Int. J. Environ. Ecol. Eng.* **2015**, *9*, 1073–1076. [CrossRef]
14. Khumalo, S.M. Biological Nutrient Removal from Industrial Wastewater Using a Sequencing Batch Reactor. Master's Thesis, Durban University of Technology, Durban, South Africa, 2018.
15. Goldammer, T. *The Brewers' Handbook*; Keen Vision Publishing, LLC: Jordan, MN, USA, 1999.
16. Werkneh, A.A.; Beyene, H.D.; Osunkunle, A.A. Recent advances in brewery wastewater treatment; approaches for water reuse and energy recovery: A review. *Environ. Sustain.* **2019**, *2*, 199–209. [CrossRef]
17. Enitan, A.M.; Swalaha, F.M.; Adeyemo, J.; Bux, F. Assessment of brewery effluent composition from a beer producing industry in KwaZulu-Natal, South Africa. *Fresenius Environ. Bull.* **2014**, *23*, 693–701.
18. Shao, X.; Peng, D.; Teng, Z.; Ju, X. Treatment of brewery wastewater using anaerobic sequencing batch reactor (ASBR). *Bioresour. Technol.* **2008**, *99*, 3182–3186. [CrossRef]
19. Wang, S.G.; Liu, X.W.; Gong, W.X.; Gao, B.Y.; Zhang, D.H.; Yu, H.Q. Aerobic granulation with brewery wastewater in a sequencing batch reactor. *Bioresour. Technol.* **2007**, *98*, 2142–2147. [CrossRef]
20. Rodrigues, A.C.; Brito, A.G.; Melo, L.F. Posttreatment of a brewery wastewater using a sequencing batch reactor. *Water Environ. Res.* **2001**, *73*, 45–51. [CrossRef]
21. Ling, L.; Lo, K. Brewery wastewater treatment using suspended and attached growth sequencing batch reactors. *J. Environ. Sci. Health A* **1999**, *34*, 341–355. [CrossRef]
22. Chan, C.; Guisasola, A.; Baeza, J.A. Enhanced Biological Phosphorus Removal at low Sludge Retention Time in view of its integration in A-stage systems. *Water Res.* **2017**, *118*, 217–226. [CrossRef]
23. Tchobanoglous, G. *Wastewater Engineering: Treatment and Resource Recovery Volume 2*; McGraw-Hill: New York, NY, USA, 2014.
24. Wang, X.; Wang, S.; Xue, T.; Li, B.; Dai, X.; Peng, Y. Treating low carbon/nitrogen (C/N) wastewater in simultaneous nitrification-endogenous denitrification and phosphorous removal (SNDPR) systems by strengthening anaerobic intracellular carbon storage. *Water Res.* **2015**, *77*, 191–200. [CrossRef]
25. Ge, H.; Batstone, D.J.; Keller, J. Biological phosphorus removal from abattoir wastewater at very short sludge ages mediated by novel PAO clade Comamonadaceae. *Water Res.* **2015**, *69*, 173–182. [CrossRef]
26. Koe, W.S.; Lee, J.W.; Chong, W.C.; Pang, Y.L.; Sim, L.C. An overview of photocatalytic degradation: Photocatalysts, mechanisms, and development of photocatalytic membrane. *Environ. Sci. Pollut. Res.* **2020**, *27*, 2522–2565. [CrossRef] [PubMed]
27. Bhatkhande, D.S.; Pangarkar, V.G.; Beenackers, A.A.C.M. Photocatalytic degradation for environmental applications—A review. *J. Chem. Technol. Biotechnol.* **2002**, *77*, 102–116. [CrossRef]
28. Jia, C.; Wang, Y.; Zhang, C.; Qin, Q. UV-TiO_2 photocatalytic degradation of landfill leachate. *Water Air Soil Pollut.* **2011**, *217*, 375–385. [CrossRef]
29. Cailotto, S.; Massari, D.; Gigli, M.; Campalani, C.; Bonini, M.; You, S.; Vomiero, A.; Selva, M.; Perosa, A.; Crestini, C. N-Doped Carbon Dot Hydrogels from Brewing Waste for Photocatalytic Wastewater Treatment. *ACS Omega* **2022**, *7*, 4052–4061. [CrossRef]
30. Wani, A.; Bhasarkar, J.B.; Gaikwad, R. Photocatalytic Degradation of Sugar and Distillery Industry Effluent. *J. Inst. Eng. Ser. E* **2021**, *103*, 1–8. [CrossRef]
31. Micheal, K.; Ayeshamariam, A.; Boddula, R.; Arunachalam, P.; AlSalhi, M.S.; Theerthagiri, J.; Prasad, S.; Madhavan, J.; Al-Mayouf, A.M. Assembled composite of hematite iron oxide on sponge-like BiOCl with enhanced photocatalytic activity. *Mater. Sci. Energy Technol.* **2019**, *2*, 104–111. [CrossRef]
32. Ghernaout, D.; Elboughdiri, N. Advanced oxidation processes for wastewater treatment: Facts and future trends. *Open Access Libr. J.* **2020**, *7*, 1–15. [CrossRef]
33. Chen, D.; Cheng, Y.; Zhou, N.; Chen, P.; Wang, Y.; Li, K.; Huo, S.; Cheng, P.; Peng, P.; Zhang, R.; et al. Photocatalytic degradation of organic pollutants using TiO2-based photocatalysts: A review. *J. Clean. Prod.* **2020**, *268*, 121725. [CrossRef]
34. APHA. *Standard Methods for the Examination of Water and Wastewater*; American Public Health Association: Washington, DC, USA, 2005.
35. Shabangu, K.P. Aerobic Sequencing Batch Reactor for the Treatment of Industrial Wastewater from the Brewery. Master's Thesis, Durban University of Technology, Durban, South Africa, 2017.

36. Shabangu, K.P.; Bakare, B.F. Study of an SBR Treating Brewery Wastewater: Case of COD-HRT and BOD Removal. In Proceedings of the World Congress on Engineering and Computer Science, San Francisco, CA, USA, 22–24 October 2019.
37. Khumalo, S.M.; Bakare, B.F.; Rathilal, S.; Tetteh, E.K. Characterization of South African Brewery Wastewater: Oxidation-Reduction Potential Variation. *Water* **2022**, *14*, 1604. [CrossRef]
38. Carucci, A.; Lindrea, K.; Majone, M.; Ramadori, R. Different mechanisms for the anaerobic storage of organic substrates and their effect on enhanced biological phosphate removal (EBPR). *Water Sci. Technol.* **1999**, *39*, 21–28. [CrossRef]
39. Bakare, B.; Shabangu, K.; Chetty, M. Brewery wastewater treatment using laboratory scale aerobic sequencing batch reactor. *S. Afr. J. Chem. Eng.* **2017**, *24*, 128–134. [CrossRef]

fermentation

MDPI

Review

Intensification of Acidogenic Fermentation for the Production of Biohydrogen and Volatile Fatty Acids—A Perspective

Sanjay Nagarajan *, Rhys Jon Jones, Lucy Oram, Jaime Massanet-Nicolau and Alan Guwy

Sustainable Environment Research Centre, Faculty of Computing Engineering and Science, University of South Wales, Pontypridd CF37 1DL, UK; rhys.jones@southwales.ac.uk (R.J.J.); lucy.oram@southwales.ac.uk (L.O.); jaime.massanet@southwales.ac.uk (J.M.-N.); alan.guwy@southwales.ac.uk (A.G.)

* Correspondence: sanjay.nagarajan2@southwales.ac.uk

Abstract: Utilising 'wastes' as 'resources' is key to a circular economy. While there are multiple routes to waste valorisation, anaerobic digestion (AD)—a biochemical means to breakdown organic wastes in the absence of oxygen—is favoured due to its capacity to handle a variety of feedstocks. Traditional AD focuses on the production of biogas and fertiliser as products; however, such low-value products combined with longer residence times and slow kinetics have paved the way to explore alternative product platforms. The intermediate steps in conventional AD—acidogenesis and acetogenesis—have the capability to produce biohydrogen and volatile fatty acids (VFA) which are gaining increased attention due to the higher energy density (than biogas) and higher market value, respectively. This review hence focusses specifically on the production of biohydrogen and VFAs from organic wastes. With the revived interest in these products, a critical analysis of recent literature is needed to establish the current status. Therefore, intensification strategies in this area involving three main streams: substrate pre-treatment, digestion parameters and product recovery are discussed in detail based on literature reported in the last decade. The techno-economic aspects and future pointers are clearly highlighted to drive research forward in relevant areas.

Keywords: biohydrogen production; volatile fatty acids; intensification; pre-treatment; digester; product recovery; techno-economic aspects

Citation: Nagarajan, S.; Jones, R.J.; Oram, L.; Massanet-Nicolau, J.; Guwy, A. Intensification of Acidogenic Fermentation for the Production of Biohydrogen and Volatile Fatty Acids—A Perspective. *Fermentation* **2022**, *8*, 325. https://doi.org/10.3390/fermentation8070325

Academic Editor: Diomi Mamma

Received: 17 June 2022
Accepted: 6 July 2022
Published: 11 July 2022

1. Introduction

There is a need to address the ever-increasing energy and materials demand sustainably. Simultaneously, growing anthropogenic activities have led to an increase in global CO_2 levels, and there is, therefore, a pressing need to reduce emissions to control the global warming potential. Cumulative global CO_2 emissions have risen by ~64% over the past three decades [1]. The major contributors (>60%) to global emissions have been the electricity, heat and transportation sectors. While this has been the global trend, national emissions vary significantly between countries due to the difference in implementation of environmental policies, population density, per capita income and per capita emissions. For instance, in the UK, the cumulative CO_2 emissions have fallen by ~38% in the past three decades [1]. In particular, the electricity and heat sectors have recently managed to curb their CO_2 emissions significantly. With the implementation of the UK Net Zero strategy to achieve zero CO_2 emission targets by 2050, the cumulative emissions are expected to decrease more rapidly in the coming years. However, to achieve such stringent targets, it is important that emissions in all sectors are mitigated appropriately. For instance, the major contributor to emissions in the UK currently is the transportation sector (~35% of national emissions) (Figure 1). To address this issue directly, the use of sustainable and cleaner fuels is required in the transportation sector. This includes the use of both gaseous and liquid biofuels, such as biogas, biohydrogen and bioethanol. In addition to biofuels, electric vehicles also have a significant role to play in reducing emissions. The source of electricity will however be influential in determining the emission potential.

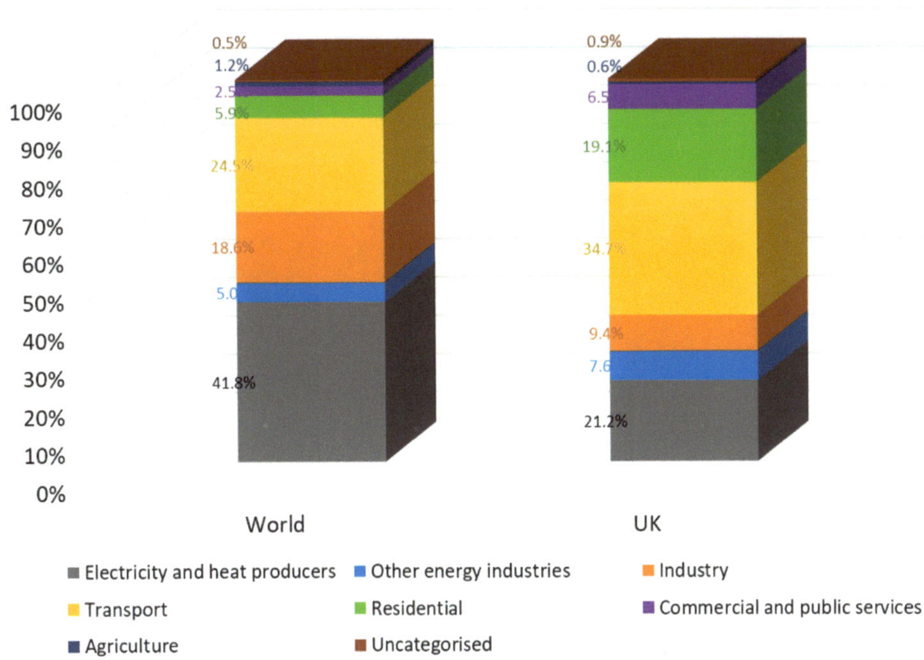

Figure 1. Global and UK CO_2 emissions by sector in 2019. Data obtained from International Energy Agency, Data and Statistics website [1].

Beyond the use of sustainable renewable energy as a means to mitigate CO_2 emissions, the sustainable production of chemicals and materials of high value is also necessary. Currently, the production of platform and commodity chemicals is highly reliant on fossil fuels, and whilst being an economically favourable route, it does not achieve the triple bottom line performance of being socio-economically and environmentally beneficial when produced from these materials. The production of these chemicals from biomass however offers extensive prospects where both renewable energy and high-value platform chemicals may be produced either simultaneously or sequentially in 'biorefineries'. The production of multiple products from a feedstock would also lead to approaching a circular bioeconomy which is critical to achieving the Net Zero targets.

1.1. 'Waste' to 'Value' for Approaching a Circular Economy

The backbone of a circular economy is "*to generate, utilise and recycle*", ensuring that wastes generated do not exit the loop. In this context, the utilisation of 'wastes' as 'resources' is critical to minimise reverting back to a linear economy framework. The utilisation of waste biomass is of particular interest to this perspective. All 'waste' biomasses are second-generation feedstocks, which neither interfere with the food chain nor compete for space with agricultural land. Examples of such 'waste' biomasses include agri and forest residues, food waste, paper and pulp industry wastes, distillery waste, wastewater and sludge. All these organic-matter-rich streams are originally 'waste streams' that have the potential to be valorised to biofuels and high-value chemicals. Utilising these 'wastes' as 'resources' would ensure that the feedstock-dependent end product pricing is reduced while ensuring sustainability and process circularity. It is however crucial to ensure that the yield of desired product per unit mass of the waste is sufficiently high to minimise net emissions.

Multiple routes to biomass valorisation are currently available (Figure 2). Typically, biomass streams can be valorised either via biochemical pathways or thermochemical pathways [2]. Biochemical pathways include anaerobic digestion (AD) to produce biogas, biohydrogen, volatile fatty acids (VFAs) and fertilisers, fermentation for the production of solvents and biofuels (e.g., bioethanol, acetone, butanol), and value-added chemicals (e.g., succinic acid, citric acid, lactic acid). Thermochemical valorisation routes mentioned in Figure 2 are typically used to produce bio-oil, bio-coal, biochar and syngas. Physico-chemical valorisation routes have also gained attention recently [3–7]. These routes often utilise biomass and its derivates as sacrificial electron donors for the production of renewable hydrogen or oxidised products such as sugars and short chain acids [4,8].

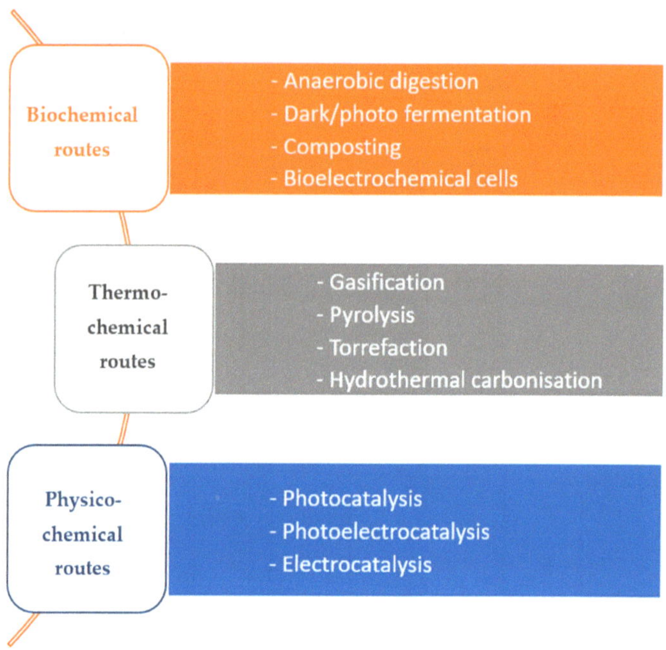

Figure 2. Biomass valorisation routes.

AD is a well-established technology widely used in the secondary stage of wastewater treatment. Its popularity in this process is due to its ability to remediate waste streams whilst generating energy vectors in the form of biogas [9]. While the technology is mature, limitations such as long residence times (>4 weeks), leading to large reactor volumes in the order of thousands of m^3, slower digestion kinetics, and sub-optimal carbon conversion leads to process inefficiencies and high capital expenditure [2]. In terms of revenue generation, biomethane (upgraded from biogas) is a low-value product (~EUR 0.5/kg [10]), and hence, allied products such as concentrated fertilisers from digestate are required to generate additional revenue [11]. Intensification strategies such as optimising operating parameters, and biomass pre-treatment to address feedstock complexity, have been proposed as effective routes to overcome these inefficiencies and maximise biomass conversion [2] and hence remain as the future perspectives for progressing the field. Thermochemical biomass conversion routes have similarly been extensively discussed in the literature [12–14] and the future direction for this route remains clear in maximising the techno-economics and understanding the life cycle impacts of the process. While AD and thermochemical routes are already commercially exploited, physico-chemical routes are on the lower end of the technology readiness level (TRL) spectrum (<TRL 3). This is primarily due to the heterogeneity of the biomass, with both the catalyst and biomass being solids suspended in a liquid phase,

leading to mass transfer limitations [15], non-specificity of biomass breakdown [4,8] and lower biomass conversion rates [6]. While current literature has extensive information on the conversion of biomass derivatives to value-added compounds [16–18], information on direct conversion is limited. Biomass derivatives and 'whole' biomass are different in nature due to their structure, physico-chemical bonding, solubility and reactivity; therefore, physico-chemical valorisation routes investigating the valorisation of biomass derivatives and whole biomass are not comparable. The research direction in this area is therefore clear in identifying novel routes for direct valorisation leading to higher TRL applications.

Most of the aforementioned biomass valorisation routes have clear pathways and future directions. There has been renewed focus on biohydrogen and VFA production from AD, as opposed to biomethane and fertiliser, and so, a critical analysis of this recent work is required. In addition, when biohydrogen or VFAs are produced via AD, the fertiliser potential of the digestate is not compromised and can still yield additional revenue. Increased recent interest in hydrogen is mainly due to its potential to decarbonise a variety of sectors that are generally hard to 'electrify' and its capability in the accelerated achievement of Net Zero goals. VFAs, on the other hand, are platform chemicals which find their use in a variety of industries, including food and beverages, cosmetics, chemicals and pharmaceutical industries [19]. The current fossil-based route for VFA synthesis is unsustainable, and therefore, a waste valorisation technology for the production of VFAs is ideal to decarbonise a number of these end-use sectors.

The UK recently devised a ten-point plan for a green industrial revolution to achieve net neutrality by 2050 [20] and to meet the Carbon Budget Six (CB6) targets. One of the key aspects of the plan is to enhance the production of low carbon hydrogen. It is expected that the low carbon hydrogen capacity of the UK will reach 1 GW in 2025 and 5 GW by 2030, leading to savings of ~41 $MtCO_2e$ (equivalent to ~9% of UK emissions in 2018). With the projected enhanced capacity, it is expected that 20–35% of the total energy consumption in the UK will be based on low-carbon hydrogen by 2050 [21].

To align with the Net Zero targets, the UK was the first country to develop an industrial decarbonisation strategy and aimed to reduce industrial carbon emissions by over 90% of 2018 levels [22]. Moreover, resource efficiency has also been a focus, as per the 25 year environment plan to mitigate the amount of waste generated [23]. Therefore, it is vital that a circular economy model incorporating intersectoral integration is established. Since most wastes are organic in nature (with cellulose in biomass being the world's most abundant organic material), biochemical valorisation routes are promising options to minimise waste and maximise value via such intersectoral integration approaches with a possibility of developing a multi product biorefinery platform. An example is the valorisation of waste biomass to biohydrogen and VFAs which are discussed in this review.

1.2. Green Hydrogen and VFAs—Need for Process Intensification

The majority of global hydrogen is produced from steam methane reforming (SMR), termed grey hydrogen, or from coal gasification, known as brown hydrogen, due to the low cost of production and high efficiency. SMR, however, has a high carbon footprint of 9–12 kg CO_2/kg H_2 [24–26] and requires a consistent supply of methane that is often derived from fossil fuels. Alternatively, SMR using biomethane (derived from conventional AD) is being pursued as a renewable and much cleaner option [27]. Biomethane-based SMR can fulfil the hydrogen generation needs intermittently until a cost-competitive complementary technology to electrolysis is established. The transition from grey and brown hydrogen is of utmost importance to achieve net neutrality and therefore, blue hydrogen (with carbon capture) and green hydrogen (renewables based) are gaining more interest. Grey or brown hydrogen coupled with carbon capture and storage (CCS) results in the production of blue hydrogen which has the potential to reduce carbon emissions by 35–85% [28]. The issues surrounding blue hydrogen however are the intensive capital requirement and potential CO_2 leakage [29]. Carbon capture storage and utilisation (CCSU) has therefore been proposed as an alternative strategy for blue hydrogen production [29].

In this case, the captured CO_2 may be used as a secondary feedstock for the production of high-value compounds (e.g., gas fermentation, microbial electrosynthesis) [30].

Green hydrogen, analogous to low carbon hydrogen, can be produced from water electrolysis or biomass electrolysis powered by renewable electricity. It has a near-zero carbon footprint (<0.6 kg CO_2/kg H_2) [31] and can greatly boost the acceleration towards achieving net neutrality. With water electrolysis largely favoured due to its higher TRL levels, the problems at scale depend on the consistent supply of renewable electricity as well as the use of critical raw materials as catalysts. Global green hydrogen trends (based on electrolysis) are currently shifting focus towards the use of non-critical, earth-abundant raw materials to de-stress the supply chain.

Other promising routes to produce green hydrogen include dark fermentation (via AD) and photo fermentation of biomass. Photo fermentation is an attractive pathway; however, it suffers from low hydrogen production rates and stringent reactor design to maximise light distribution within the bioreactors [32]. Photo fermentation is especially limited when wastewater or lignocellulosic biomass is used as a feedstock due to the light scattering, shielding and loss of photons caused by the selective absorption of light by coloured wastewater, thereby limiting the light harvesting efficiencies and metabolic rates. Dark fermentation can overcome these challenges. The biochemical pathway leading to the production of hydrogen in dark fermentation is an intermediate step (acidogenesis) in the conventional AD pathway. While the product of interest in conventional AD is biomethane, the methanogenic activity needs to be suppressed to ensure that biohydrogen is derived as the end product in acidogenic fermentation. Assuming $C_6H_{12}O_6$ (hexose) as the model molecular formula of the biomass and accounting for fractional biomass utilisation for its growth and energetic needs, ~0.1 kg H_2/kg biomass could be produced stoichiometrically. This corresponds to ~12 MJ energy recovered from 1 kg of biomass. While the green hydrogen productivity, especially via the AD route, is attractive, it is currently not cost competitive compared to grey hydrogen. One of the primary reasons is that hydrogen production in AD is affected by simultaneous VFA production. While both VFAs and hydrogen can be produced via methanogenesis suppressed AD, it is only possible to produce either VFAs or biohydrogen with higher yields at a given point of time in the digester. This is mainly because the yields of VFAs and hydrogen are interlinked, and often, their concentrations are inversely proportional to each other. The acidogenic and acetogenic stages of AD lead to the production of VFAs. In addition, homoacetogens present in the microbial consortia can further utilise hydrogen for acetic acid production. The VFA product mixture in AD typically includes acetic acid, propionic acid, butyric acid and valeric acid in varying proportions. Therefore, understanding the digestion kinetics and optimising the system for the production of desired products is important when acidogenic fermentation is the focus.

There are similarities between VFA and biohydrogen production in terms of biochemical pathways utilising waste organic matter as feedstock, mode of operation and scale-up. Furthermore, the prospect of retrofitting existing infrastructure to support hydrogen storage and transport as well as high VFA productivities are major advantages of the acidogenic fermentation route for biomass valorisation. These can directly address the challenges such as '*technological uncertainty*' and '*enabling infrastructure*' [21]; however, challenges around '*affordable costs*' still exist. This is predominantly due to the productivity that is linked to two aspects, namely feedstock heterogeneity (complexity due to recalcitrant inter and intramolecular bonding) and product selectivity (type of VFA or choice between biohydrogen, VFA and biomethane). These are the two main components that influence the carbon footprint as well as the techno-economics of the process.

This review therefore aims to bring together the intensification strategies that can predominantly enhance the productivities of desired products, namely biohydrogen and VFAs. While a number of challenges exist in improving the product yields, the scope of this review is restricted to three main areas that specifically influence the process scale up, namely:

1. Pre-digestion—strategies to address feedstock heterogeneity and improve the bioavailability of the biomass;
2. Anaerobic digestion—strategies to improve the bioconversion of biomass to desired products;
3. Product recovery—strategies to maximise recovery and purity of desired products.

The influence of these strategies on the techno-economics and the life cycle of the process are also pointed out. Finally, the future research focus in this particular area of AD is also discussed from our point of view.

2. Pre-Digestion

Pre-digestion in this context refers to the steps involved in preparing the feedstock for acidogenic fermentation. Predominantly, with the feedstock type and composition being the detrimental factors influencing the productivity of the desired product, it is important to ensure that the organic matter in the feedstock is highly bioavailable to the microbial consortia for digestion. In conventional AD, the rate limiting steps could either be the hydrolysis step or the methanogenic stage [33]. The former is linked to feedstock complexity, whereas the latter is linked to the growth rate of the methanogenic archaea. Therefore, once the substrate is readily bioavailable, the biochemical pathways will be initiated toward the desired product formation. It is therefore critical to ensure that the feedstock complexity is addressed to speed up the hydrolysis stage. Whether the desired product is biohydrogen or VFAs, the hydrolysis stage is the common precursory step. The branching out of biochemical pathways happens after the hydrolysis stage, so the discussion around addressing feedstock complexity in this section is common to both these fermentative pathways.

Overlapping with conventional AD, acidogenic fermentation can utilise any organic feedstock (e.g., wastewater, sewage sludge, agri-forest residue, food waste). The availability of the feedstock (quantity availability and frequency of feedstock production) and its composition are detrimental factors that can impact the digestion process. Irrespective of these factors, it is first important to assess the biochemical hydrogen potential (BHP) and the VFA potential (VFAP) of each feedstock separately. While BHP can be analogous to conventional biochemical methanation potential (BMP) tests, VFAP has never been performed before. Therefore, there is immense scope to develop standardised tests for BHP and VFAP via AD. This needs to be performed stoichiometrically first to determine the theoretical potential of the feedstock (similar to the Buswell–Muller methane yields for conventional AD). While the theoretical limits are indicators of the digestion ability of the feedstock, these can never be experimentally achieved due to the utilisation of a fraction of the feedstock for microbial growth and metabolism as well as the recalcitrance posed by a fraction of the feedstock (e.g., lignin). It is however paramount that the maximum achievable yield of the desired product is targeted by intensifying the digestion process. If the feedstock has high soft suspended solid content, such as sewage sludge or food waste, improving the degree of disintegration leading to an increased soluble COD concentration is required to have a positive impact on the digestion process. If the feedstock is lignocellulosic in nature such as agricultural or forest residues, improving the bioavailability of holocellulose (and/or delignification) to enhance microbial hydrolysis is important. As lignin provides structural integrity to the biomass, delignification or at least the exposure of the holocellulose to hydrolytic bacteria is critical. With either of the feedstock categories, inhibitor formation as a result of pre-treatment should be suppressed to prevent negatively influencing anaerobic digestion. In addition, the preferred pre-treatment should also be net positive (energetically and economically) and scalable if high TRL hydrogen and VFA production are targeted.

A number of researchers have investigated a wide range of pre-treatment methods for acidogenic fermentation, falling broadly under four categories: physical, chemical, biological and physico-chemical [34–36]. The mode of action of each pre-treatment on lignocellulosic biomass is shown in Figure 3. The most utilised methods in the past decade are however exclusively discussed in this review to restrict the scope purposefully to recent work (Tables 1 and 2).

Physical pre-treatment methods often comprise of techniques such as shredding, comminution or homogenisation, which target the reduction in particle size and increase in specific surface area. It has also been shown to affect the crystallinity index of the cellulose upon milling [37]. The reduced particle size of the feedstock favours faster hydrolysis rates and, in turn, higher desired product yields. While physical pre-treatment is known to generate no microbial inhibitors, it is often limited by its high specific energy consumption (thereby high OPEX). The specific energy consumption is directly related to the comminution ratio (i.e., the ratio of final particle size to the initial particle size) and the moisture content of the feedstock. The milling energy is reduced considerably when the moisture content in the biomass is reduced. To achieve this, an additional drying step (consuming more energy) is required. For instance, Miao et al. [38] compared the milling of a variety of lignocellulosic biomass by hammer mills and knife mills. To achieve the same final particle size of 1 mm with Miscanthus, the hammer mill required ~200 kJ/kg TS which is ~3.5 fold lower than a knife mill. Similarly, the dried biomass consumed nearly 50% less energy (~950 kJ/kg TS) as compared to biomass with 15% moisture. When valorisation via anaerobic digestion is desired, the moisture content in the feedstock is required, which helps with the mass transfer and hydrodynamics of the digester. Even with dry digestion, a significant amount of moisture is still retained in the feed slurry (~15–20% TS). Therefore, drying the feedstock prior to milling and then rehydrating the feedstock is neither resource nor energy efficient. Thus, when opting for physical treatment, it is of utmost importance to consider the type of biomass, moisture content and the comminution ratio required.

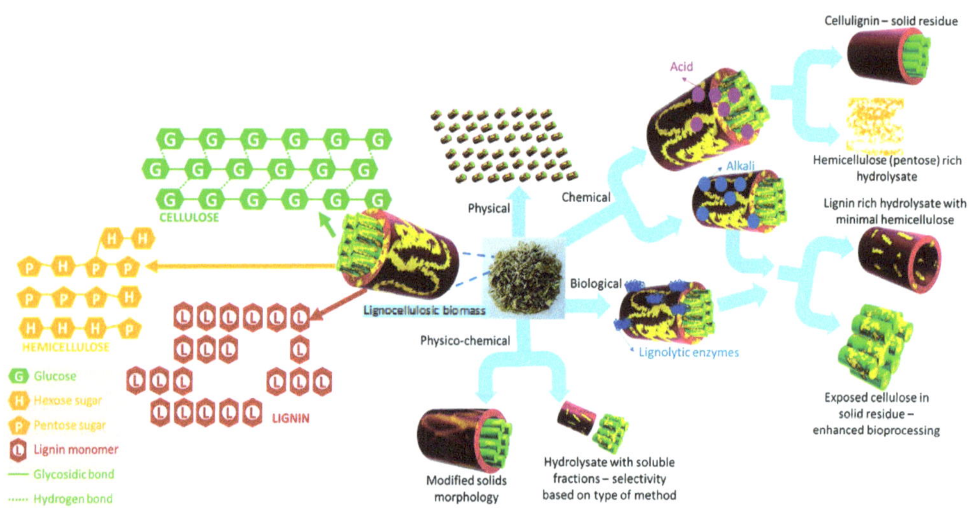

Figure 3. An overview of lignocellulosic biomass pre-treatment methods. Reprinted with permission from Konde et al. [39], Copyright Royal Society of Chemistry.

Chemical pre-treatment utilises acid, alkali or oxidising agents to depolymerise, hydrolyse or delignify the biomass [40]. Acid hydrolysis, when employed as a lignocellulosic biomass pre-treatment method often generates fermentation inhibitors such as furans and furfurals, while alkali hydrolysis results in partial delignification, leading to the increase in the concentration of soluble phenolics. Oxidative pre-treatment using O_3 or H_2O_2 are non-specific pre-treatment methods and tend to depolymerise any fraction of the biomass but have the tendency to delignify the biomass predominantly. These compounds may be inhibitory to the metabolism of the microbial consortium and can reduce the desired product yield [33,41]. It has been reported that furans are more inhibitory to acidogenic bacteria than soluble phenolics [33]. The inhibition of the metabolic activity of these classes of bacteria may result in the increased abundance of non-hydrogen/VFA producers and divert the digester towards the production of lactate or ethanol [33]. When the end product of digestion is conventional biogas, these alternative end products may be beneficial; however, when acidogenic fermentation is in focus, care has to be taken to ensure such inhibition is avoided. Additionally, from an engineering perspective, equipment corrosion can also occur and impede the process operation when acids are used in the system. This can indirectly lead to an increased concentration of heavy metals in the solution, thereby further reducing the desired product yields.

Biological pre-treatment often consists of the fungal or enzymatic pre-treatment of feedstock. White rot fungi or brown rot fungi are the commonly deployed species due to their ability to secrete extracellular ligninolytic and cellulolytic enzymes or perform hemicellulose hydrolysis, respectively [41]. Another filamentous fungus, *Trichoderma reesei*, has also been reported to pre-treat lignocellulosic biomass due to its ability to hydrolyse cellulose [42]. While biological substrate pre-treatment is effective in enhancing product yields, slow kinetics of pre-treatment (ranging from days to weeks) requiring large reactor volumes (incapable of efficient scale up) and its lack of capacity to continuously pre-treat the feedstock are seen as limitations. Alternatively, enzymatic hydrolysis of the feedstock has been proposed as a targeted biological pre-treatment strategy, but the cost of enzyme production/recovery still has to be considered prior to scale up.

The final category of pre-treatment consists of physico-chemical methods. In this kind of pre-treatment, the biomass is pre-treated to achieve the combined effect of both the physical as well as chemical methods, i.e., particle size reduction or increase in the surface area along with partial hydrolysis of polymers. Physico-chemical methods overcome the disadvantages posed by physical or chemical pre-treatment methods; they consume considerably less energy as compared to physical methods and generate insignificant quantities of inhibitors. For instance, steam explosion works on the basis of applying compressed steam to the biomass slurry followed by rapid depressurisation and consumes ~70% less energy than physical pre-treatment methods [43]. To favour enhanced product yield by avoiding inhibitors, liquid hot water pre-treatment has been suggested as an alternative to steam explosion [44]. Another physico-chemical pre-treatment method that is gaining attention in the area of anaerobic digestion is hydrodynamic cavitation [2,45,46]. Cavitation is the phenomenon of generation, growth and implosion of vaporous cavities. Acoustic cavitation, commonly known as ultrasonication, has been reported extensively in the literature for the pre-treatment of biomass; however, due to the handling volumes being limited to mL scale and high specific energy inputs (at times, higher than physical pre-treatment), they cannot be scaled up. Hydrodynamic cavitation, on the other hand, is less energy intensive and has been reported to be scaled up [45,47] with low specific energy inputs and high net energy gains. For instance, Nagarajan and Ranade [48] reported that the specific energy required to pre-treat sugarcane bagasse (at a low solid loading of 1%) was 0.5 MJ/kg TS; however, the net energy gain as a result of enhanced biomethane generation was reported to be ~1.4 MJ/kg TS. Overtreatment can however result in the generation of inhibitors, higher energy consumption and reduced desired product yield. Therefore, it is important to optimise the process to maximise the product yields and energy efficiency. The use of physico-chemical pre-treatment seems to be a promising method; however, with limited literature in the area of acidogenic fermentation, it needs more research. While a general overview of pre-treatment was presented so far, the following sections will discuss specific examples from literature in the past decade that has reported an impact on intensifying acidogenic fermentation.

2.1. Substrate Pre-Treatment to Enhance Biohydrogen Production

Table 1 shows an overview of reported data on various pre-treatment methods used to enhance biohydrogen yield. Physical pre-treatment is the most common conventional pre-treatment to enhance digestion yields. In the last decade, however, the use of milling-based methods to intensify biohydrogen production has dwindled. It is understandable that this decrease could have been due to the inability of these methods to achieve a net positive energy gain. One of the few papers that reported the use of physical pre-treatment was by Yukesh Kannah et al. [49], who used chopped rice straw as the feedstock and high-speed dispersion as the pre-treatment methodology. A 2 L batch pre-treatment was performed with a 2% straw concentration. The gap between the rotor and stator was 0.3 mm, which enabled the effective disintegration of straw at an optimum speed of 12,000 rpm. At a specific energy consumption of ~1.47 MJ/kg TS, a degree of disintegration of ~9.5% was observed. Upon mesophilic batch digestion, this corresponded to a 7.3-fold increase in hydrogen yield as compared to the untreated straw that generated 8 mL H_2/g COD. While an increase in hydrogen yield was observed, a net positive energy gain could not be achieved with this kind of pre-treatment.

Deng et al. [50] reported the use of 2% sulphuric acid pre-treatment on grass silage (2%) at an elevated optimum temperature of 135 °C for 15 min. At these conditions, a hydrolysis efficiency of ~50% was observed and resulted in a three-fold increase in hydrogen yield. Amongst the reducing sugars formed due to acid hydrolysis, xylose dominated the hydrolysate with ~70% concentration, suggesting that hemicellulose hydrolysis was predominantly achieved with acid pre-treatment leaving behind a cellulolignin solid residue. Reilly et al. [51], on the other hand, utilised alkali pre-treatment of wheat straw as a strategy to enhance hydrogen production. The substrate was soaked in 80 mM lime for 2 days, and it was determined that ~36% of the hemicellulose and minimal lignin were solubilised, whilst cellulose remained unaffected. The pre-treated solid residue upon neutralisation was subjected to digestion with the bioreactors supplemented with an Accelerase enzyme cocktail. The optimal conditions resulted in a biohydrogen yield, which was ~29-fold higher (59 mL H_2/g VS) than the untreated straw. The hydrolysate, when added to the solid residue for digestion, resulted in the inhibition of hydrogen production due to the presence of enhanced concentrations of $CaCO_3$. Instead of lime, a stronger alkali, NaOH was used to pre-treat milled corn cobs by Kucharska et al. [52]. They also supplemented the pre-treated slurry with an enzymatic cocktail to intensify the hydrolysis process and enhance the biohydrogen yields by >5 fold. Unlike acid and alkali pre-treatment, oxidative degradation pre-treatment has also been reported to enhance biohydrogen production. For example, Wu et al. [53] reported the use of ozone pre-treatment to intensify biohydrogen production from milled wheat straw. Since oxidative pre-treatments are non-specific in nature, they tend to degrade polymers in their vicinity. With lignin present in the cell wall of the biomass structure, it tends to be degraded first via oxidative pre-treatments. Accordingly, at optimal conditions, nearly 40% delignification was reported, corresponding to a 2.5-fold increase in biohydrogen yield.

Biological pre-treatment in its current state, while being efficient, is not scalable. With conventional enzymatic hydrolysis systems, this is related to the enzyme recovery costs. With the cost of enzyme production becoming relatively cheaper, there is still potential to explore this area. For instance, Leaño and Babel [54] reported the enzymatic hydrolysis of cassava wastewater using various commercially available enzymes. They used OPTIMASH BG©, which is commonly used in the bioethanol industry and α-amylase in separate experiments to determine the effect on biohydrogen production. In addition to an increased yield of biohydrogen (~50%), a reduced lag time in hydrogen production was also observed, suggesting that the complexity of the wastewater was reduced during enzymatic hydrolysis. To improve the effectiveness of the biological pre-treatment systems, novel and innovative strategies have also been reported. For example, Chandrasekhar and Venkata Mohan [55] reported the use of bioelectrochemical hydrolysis as an unconventional means of biological pre-treatment to intensify biohydrogen production. With a 10 h HRT and graphite electrodes (without an external voltage supply), they pre-treated blended food waste. The overflow from the bioelectrochemical cell was used as the feedstock for the fed-batch mesophilic digester that operated with an HRT of 72 h. At optimum conditions, ~35% increase in biohydrogen yields was observed. The fed-batch operation of pre-treatment is a promising strategy and can pave the way for a step-wise, modular scale up; however, intensive research is required to achieve this.

Physico-chemical pre-treatment is a promising strategy to enhance biohydrogen yields due to its ability to process wet feedstock and scalability. For instance, hot compressed water was used to pre-treat sake brewery waste (sake lees). Compressed hot water at a high temperature of 130 °C and pressure of 3 bars was used to treat 10% biomass at a holding time of 1 h [56]. A reduction in lag time for biohydrogen generation was reported as a result of pre-treatment. Asadi and Zilouei [57] reported the use of an organosolv pre-treatment of rice straw to enhance biohydrogen production. In their case, the biomass was blended with an ethanol–water mix (45% *v*/*v*). One percent sulphuric acid was used to catalyse the hydrolysis process at an optimal temperature of 180 °C at a 30 min holding time. Upon treatment, a sequential enzymatic hydrolysis step was also carried out using 5% Cellic CTec2 to further increase the reducing sugar yield. At these conditions, the glucose concentration was enhanced by >4-fold and positively influenced the biohydrogen yield. Other researchers have also reported such complex and sequential pre-treatment processes [58,59] to improve hydrogen yield. While such processes may be beneficial in enhancing biohydrogen yields, the need to use a complex process requires justification both economically and environmentally. Alternatively, other researchers have reported simpler processes involving heat/irradiation and mild acid to achieve similar, if not higher, biohydrogen yields [60,61]. Cavitation, mainly sonication, is another physico-chemical method that has been reported to enhance biohydrogen yields. For instance, Hu et al. [62] reported the use of sonication followed by alkali treatment of antibiotic fermentation residue to enhance biohydrogen yields by 79%. Enhanced soluble carbohydrate release, resulting in reduced lag time, was attributed to the increase in biohydrogen yields. A similar increase in biohydrogen was also reported by Gadhe et al. [63], who sonicated food waste at an optimum, but high specific energy input of 13.5 MJ/kg TS.

Table 1. Pre-treatment methods reported to enhance biohydrogen production.

Feedstock	Pre-Treatment Conditions	Digestion Conditions	Influence on H_2 Yield	Reference
		Physical Pre-treatment		
Chopped dried rice straw	20 g/L straw, 2 L, *high-speed disperser* at 12,000 rpm, 30 min, 0.3 mm gap between rotor and stator	1 L batch, 25% inoculum, 70% straw slurry, 37 °C, 100 rpm, 10 days	7.3-fold increase in specific H_2 yield	[49]
		Chemical Pre-treatment		
Air-dried and milled corn cobs	5 g biomass, 0.1 L pH 11.5 *NaOH*, 25 °C, 6 h, followed by enzymatic hydrolysis with Viscozyme L and glucosidase (0.001 L/g biomass), 42 °C, 24 h	1 L batch, 10% *v*/*v* inoculum, pH 7, 37 °C, 320 rpm, 116 h	>5-fold increase in H_2 yield	[52]
Grass silage	2% silage, 0.1 L, 2% *H_2SO_4*, 135 °C, 15 min	1% silage, 0.2 L batch, 0.02 L inoculum, pH 7, 4 days (1st stage of a 2-stage system)	3-fold increase in H_2 yield	[50]
Milled wheat straw	5 g straw, 40% water, 0.75 bars, *0.63 LPM O_3*, 45 min	0.08 L, 2 g TS, pH 6, 1.9% inoculum (*v*/*v*), 1 mL hydrolytic enzyme mix, 35 °C, 60 rpm, 8 days	~2.5-fold increase in cumulative H_2 yield	[53]
Milled wheat straw	5 g VS, 62.5 mL 80 mM *$Ca(OH)_2$*, 20 °C, 2 days	0.5 L, 8% w/v inoculum, 1 mL Accelerase-1500, pH 6.25, 35 °C, 16 days	~29-fold increase in specific H_2 yield	[51]

Table 1. *Cont.*

Feedstock	Pre-Treatment Conditions	Digestion Conditions	Influence on H_2 Yield	Reference
Biological Pre-treatment				
Blended food waste	0.5 L, *bioelectrochemical hydrolysis*, open to air graphite cathode, graphite anode, 0.075 L inoculum, 20 g COD/L, pH 7, 10 h HRT, 29 °C	0.25 L fed-batch, 0.075 L inoculum, 10 g/L, pH 6, 72 h HRT, 29 °C	~35% increase in cumulative H_2 yield	[55]
Cassava wastewater	0.2% *OPTIMASH BG®* enzyme, 0.22 L wastewater, pH 4, 60 °C, 45 rpm	0.06 L, substrate to inoculum ratio 5 (*v*/*v* basis), pH 7, 37 °C, 90 rpm, 10 days	Reduced lag time, 51% increase in specific H_2 yield	[54]
	0.2% *α-amylase* enzyme, 0.22 L wastewater, 37 °C, 45 rpm		Reduced lag time, 49% increase in specific H_2 yield	
Physico-chemical Pre-treatment				
Commercial Sake Lees	10% biomass, 0.1 L, *130 °C, 3 bars*, 1 h	0.11 L batch, 9% biomass, substrate to inoculum ratio of 1:1 *v*/*v*, pH 6, 75 rpm, 37 °C, 5 days	Reduction in lag time observed after pre-treatment	[56]
Marine macroalgae *Ulva reticulate*	*Acidic H_2O_2 induced microwave*, 0.5 L, 2% biomass, 0.024 g H_2O_2/g TS, 0.1 N H_2SO_4, pH 5, 40% microwave power, 10 min, 10.8 MJ/kg TS	0.15 L batch, 70% substrate, 25% inoculum, pH 5.5, 130 rpm, 37 °C,	7.7-fold increase in specific H_2 yield	[59]
Waste-activated sludge	0.15 L sludge, 0.3 g *sodium citrate*/g sludge, 1 h, 150 rpm, followed by *121 °C*, 30 min	0.2 L batch, substrate to inoculum ratio 3 (*v*/*v* basis), pH 7, 100 rpm, 37 °C	4.4-fold increase in specific H_2 yield	[58]
Antibiotic fermentation residue	0.2 L, 6 mm *sonication* probe, 30 min, 4 s ON 6 s OFF, followed by 5 M *NaOH* addition to reach pH 10, mixed for 24 h	0.2 L batch, substrate to inoculum ratio 3 (*v*/*v* basis), pH 7, 37 °C	79% increase in specific H_2 yield	[62]

2.2. Substrate Pre-Treatment to Enhance VFA Production

VFAs are platform chemicals and value-added products that are of growing interest due to their applicability in a variety of chemical and process industries. Intensification of VFA production via effective pre-treatment is therefore also gaining significant attention, especially if the feedstock of interest is a 'waste'. Similar to biohydrogen production, studies focusing on physical pre-treatment have moved away from energy-intensive milling methods. More recently, freezing and thawing as a pre-treatment was reported by She et al. [64] to intensify VFA production from waste-activated sludge. A higher degree of disintegration can be possible with such a pre-treatment strategy. Furthermore, it has been claimed that the formation of intracellular crystals during the freezing stage can lead to the breakage of cell membranes, leading to an enhanced soluble COD content upon pre-treatment. She et al. [64] performed five cycles of freezing and thawing (one cycle = −24 °C freezing for 8 h, 35 °C thawing for 2 h) with a 0.45 L batch of sludge and followed it up with fed-batch mesophilic digestion to achieve a 35% increase in VFA concentration compared to the controls. Zeng et al. [65], on the other hand, utilised waste-activated sludge in a bioelectrochemical cell with graphite electrodes at an applied potential of 12 V for 30 min, which suppressed biomethane production and improved the VFA yield by ~100-fold. The gradual shift of microbial communities upon pre-treatment showed that the digestion favoured VFA accumulation rather than methanogenesis.

Conventional alkali pre-treatment has been reported to enhance VFA yields. For instance, Pham et al. [66] pre-treated seaweed (40% TS) with 0.5 N NaOH to enhance the VFA yield by two-fold. Unconventional treatment possibilities have however also been explored, such as using alkaline ferrate [67], carbide slag [68] or tetrakis hydroxymethyl phosphonium sulphate [69]. At pH 10 (2 M NaOH) and 0.5 g/g VSS K_2FeO_4, increased solubilisation of waste-activated sludge coupled with extracellular polymeric substance release resulted in a 2.4-fold increase in VFA concentration [67]. Acetic acid was found to be the predominant product in the VFA mixture. In a first-of-its-kind work, Tao et al. [68] reported the use of carbide slag to pre-treat grass and intensify VFA yields. Carbide slag is an alkaline waste that is generated as a by-product of calcium carbide hydrolysis [70]. It may be used to produce cement; however, it has a high potential to pollute the atmosphere (dust) and water bodies (leaching). Due to its chemical composition and alkalinity, it may be used to pre-treat biomass [68]. In this study, 5% grass was pre-treated with 1.75% slag at 120 °C for 40 min. Upon treatment, the solid residue was separated, washed until a neutral pH was reached and subjected to enzymatic hydrolysis prior to mesophilic acidogenic digestion. Similar to most alkali treatment methods, the hemicellulose and lignin were solubilised to an extent, leaving behind a cellulose-rich solid residue. Enzymatic hydrolysis of the pre-treated residue resulted in a >6-fold increase in reducing sugars, thereby leading to an enhanced VFA production of up to 2.4-fold. Acetate dominated the VFA mixture, followed by butyrate and propionate. Another unconventional chemical pre-treatment that was reported was the use of a biocide tetrakis hydroxymethyl phosphonium sulphate on sludge [69]. In total, 20 mg/g of biocide was found to be optimum at room temperature; however, a 2-day treatment time was required. A 49% increase in soluble COD content was observed, leading to a four-fold increase in VFA concentration. Higher molecular weight fatty acids dominated the VFA mixture obtained from the pre-treated feedstock.

Fang et al. [71] reported the use of white rot fungi to pre-treat autoclaved solid digestate (obtained from a biogas plant digesting agricultural, fruit and vegetable residues). A 6-week pre-treatment period was required to increase the VFA concentration by 1.2-folds. This is a typical example of the long pre-treatment times taken by biological methods in breaking down lignocellulosic materials. Furthermore, they use dried, chopped and autoclaved substrates, all of which might have an impact on the biomass structure and composition. Therefore, this could be classified under combined pre-treatment methods rather than just 'biological' pre-treatment. They performed a similar exercise with mushroom residue and achieved a >70% increase in VFA yield [72]. Unlike traditional biological methods, Pham et al. [66] reported the use of *Vibrio* spp. to pre-treat seaweed samples. Seaweed lacks (or contain in negligible quantities) lignin in its cell wall; however, the complexity in digestion arises due to the presence of other polymers, such as alginate. The alginate lyase activity of the bacteria was effective in pre-treating the seaweed prior to VFA production, as reported by the authors. Bacterial treatment was found to be more effective than alkali pre-treatment in this case, and a 2.5-fold increase in VFA concentration was observed with the pre-treated seaweed. Acetate (53%) followed by propionate (27%) and butyrate (15%) dominated the mixture. Despite being effective, the bacterial- or fungal-based methods require a long time to hydrolyse the substrate. Enzymatic treatment can be an alternative if costs are not inhibitory. Bahreini et al. [73] reported the use of Novozym 50199 to pre-treat (10 min) primary sludge and enhance the maximum VFA concentration by 56% in a fed-batch digester. Similar results with a VFA increase of up to 39% were reported by Owusu-Agyeman et al. [74], who used an enzyme cocktail of α-amylase, lipase, cellulase, dextranase and protease to pre-treat primary sludge.

The use of physico-chemical pre-treatment methods for intensifying VFA production has been growing in the recent decade. Conventional hydrothermal treatment of thickened activated sludge at 190 °C, 12.5 bars and 10 min was reported to increase the maximum VFA concentration by three-fold [75]. Hydrothermal pre-treatment was effective in increasing the soluble COD content by almost 10-fold compared to the untreated sludge with a soluble COD of ~2 g/L. This corresponded to a decreased total suspended solid concentration of the sludge with a reduced particle size distribution. The specific energy consumption for this pre-treatment was reported to be 481 kJ/kg sludge. Another conventional method is a thermo-chemical pre-treatment method, namely autoclaving in the presence of alkali to enhance digestion efficiency. Suresh et al. [76] autoclaved lipid extracted 5% microalgal slurry in the presence of 1% NaOH and subjected the samples to mesophilic digestion and observed a 20% increase in VFA concentration with the pre-treated sample. They also pre-treated the lipid-extracted microalgae using a microwave-based method in the presence of 1% NaOH [76]. They achieved >50% solubilisation of the substrate; however, the increase in the maximum VFA concentration was only 10% and significantly less than the NaOH-autoclave pre-treatment. Microwave-assisted ionic-liquid-based pre-treatment of straw was found to produce VFA with a five-fold increase as compared to the untreated straw [77]. While the combined effect helped in enhancing the VFA yield, the microwave assistance helped to lower the required ionic liquid loading needed for pre-treatment. Suresh et al. [76] also investigated the use of sonication as a pre-treatment in the presence of alkali and reported that although the degree of solubilisation was similar to microwave-alkali pre-treatment but less than autoclave-alkali pre-treatment (~80%), the enhancement in the maximum VFA concentration was 30% when compared to the untreated substrate as well as higher than the other two reported methods. Sonication has also been used to pre-treat crushed food waste by Guo et al. [78], who, at an optimal specific energy input of 1.2 kJ/mL (37.7 kJ/g TS), achieved >55% degree of disintegration corresponding to a 4.3-fold increase in maximum VFA concentration. Liu et al. [79] however observed a 63% increase in VFA concentration from sonicated food waste at an optimal energy input of 1.8 kJ/mL (18 kJ/g TS). Beyond food waste, sonication has also been investigated for lignocellulosic biomass such as grass. Wang et al. [80] sonicated 2% dried and milled grass slurry in 0.75% lime solution with a specific energy input of 1.5 kJ/mL (7.8 kJ/g TS) in pulsed mode (5 s ON 5 s OFF). The solids and liquids were separated, neutralised and subjected to mesophilic digestion. Compared to the digestion of the slurry, the cumulative VFAs produced from the solid and liquid fractions were significantly higher and were found to be >2 fold higher than the untreated feedstock.

While sonication has been reported extensively to pre-treat biomass to enhance digestion efficiency by improving COD solubilisation and increasing the degree of disintegration, it is limited by its volume of operation and high specific energy requirements. To overcome these limitations, hydrodynamic cavitation is a suitable alternative. For the first time, Lanfranchi et al. [46] reported the use of hydrodynamic cavitation for pre-treating mixed organic waste (waste-activated sludge, vegetable and fruit waste) to improve VFA yields. They used a rotor-stator device at an inlet pressure of 2 bar, an inflow rate in the range of 80–100 LPM and a rotor speed in the range of 1450–1550 rpm. An optimum pre-treatment time of 50 min was found to be beneficial in increasing the soluble COD by 83%, corresponding to a nine-fold increase in maximum VFA concentration. The specific energy consumption was reported to be 3.7 MJ/kg TS, which is significantly lower than the acoustic cavitation-based pre-treatment methods. Nevertheless, due to the presence of moving parts, there are issues pertaining to clogging and maintenance at an industrial scale. Furthermore, compared to other hydrodynamic cavitation devices, rotor-stator devices are known to be relatively energy intensive [2]. Nonetheless, it is promising to see that a scalable biomass pre-treatment is being exploited for intensifying digestion yields.

A promising aspect of the recently reported pre-treatment methods is that despite the kind of pre-treatment used, desired product type or enhancement achieved, researchers are moving in the right direction of not only understanding the fundamentals of pre-treatment but also the interaction of the pre-treated substrate and the microbial consortia. Most of the discussed papers in this section have also reported omics studies, looking into the abundance and diversity of specific genus and their shifts as a result of pre-treatment. Such investigation will help to better understand the pre-treatment and digestion processes and lead to devising effective monitoring tools and scale-up strategies.

Table 2. Pre-treatment methods reported to enhance VFA production.

Feedstock	Pre-Treatment Conditions	Digestion Conditions	Influence on VFA Yield	Reference
Physical Pre-treatment				
Waste-activated sludge	0.45 L, 5 cycles of freezing and thawing, −24 °C freezing for 8 h, 35 °C thawing for 2 h	1 L fed-batch, sludge-to-inoculum ratio of 2 (w/w), 80 rpm, 25 days retention time, 35 °C	35% increase in maximum VFA concentration	[64]
Waste-activated sludge	0.5 L, graphite electrodes, 15 V, pH 6.7, 30 min, 25 °C	0.06 L sludge, 0.02 L inoculum, 35 °C, 60 rpm, 35 days	Suppressed CH4 production, ~100-fold increase in specific VFA yield	[65]
Chemical Pre-treatment				
Waste-activated sludge	0.8 L feedstock, pH 10 (2 M NaOH), 0.5 g/g VSS K_2FeO_4, 120 rpm, 60 min	0.4 L batch, 10% v/v inoculum, 160 rpm, 35 °C, 12 days	~2.4-fold increase in maximum VFA concentration	[67]
Air-dried and chopped macroalgae	40% TS, 0.5 N NaOH, 18 h	0.1 L batch, 4% TS feedstock, 10% inoculum, 35 °C, 150 rpm, 4 days	2-fold increase in maximum VFA concentration	[66]
Grass waste	0.2 L, 5% grass, 1.75% carbide slag, 120 °C, 40 min	0.25 L batch, substrate-to-inoculum ratio 2 (VS basis), pH 7, 100 rpm, 35 °C, 14 days	0.6–2.4-fold increase in maximum VFA concentration	[68]
Sludge	0.5 L sludge, 20 mg/g tetrakis hydroxymethyl phosphonium sulphate, 2 days, 150 rpm, 30 °C	0.35 L sludge batch, 0.03 L inoculum, pH 6, 2 days, 150 rpm, 30 °C	4-fold increase in maximum VFA concentration	[69]
Biological Pre-treatment				
Autoclaved solid digestate	100 g TS, 10 g white rot fungi Pleurotus Sajor-Caju, 25 °C, 70% relative humidity, 6 weeks	0.4 L batch, 15% TS, inoculum-to-substrate ratio 2 (TS basis), 30 °C, 18 days	1.2-fold increase in maximum VFA concentration	[71]
Air-dried and chopped macroalgae	4% TS, 0.09 L, *Vibrio* spp., 26–30 °C, 2 days	0.1 L batch, 4% TS feedstock, 10% inoculum, 35 °C, 150 rpm, 4 days	2.5-fold increase in maximum VFA concentration	[66]
Primary sludge	1% *Novozym 50199* to biomass, 300 rpm, 10 min	0.5 L fed-batch, 2-day retention time, 25 °C	56% increase in maximum VFA concentration	[73]
Physico-chemical Pre-treatment				
Crushed food waste	0.3 L feedstock, 8 mm 20 kHz sonication probe, 1 W/mL, 20 min	0.18 L batch, substrate to inoculum ratio 6 (VS basis), 180 rpm, 35 °C, 5 days	~4.3-fold increase in maximum VFA concentration	[78]
Lipid-extracted microalgae *Ettlia* sp.	10 mL of 5% microalgal slurry, 1% NaOH, 25% amplitude sonication	0.1 L batch, 3% TS, 20% v/v inoculum, pH 7.2, 150 rpm, 35 °C, 7 days	30% increase in maximum VFA concentration	[76]
	10 mL of 5% microalgal slurry, 1% NaOH, microwave		10% increase in maximum VFA concentration	
	10 mL of 5% microalgal slurry, 1% NaOH, autoclave 121 °C, 1 h, 1 bar		20% increase in maximum VFA concentration	
Thickened waste-activated sludge	1 L sludge, 190 °C, 10 min, 12.5 bars	0.3 L batch, 1 gTCOD/gVSS substrate to inoculum ratio, pH 5.5, 120 rpm, 37 °C, 3 days	3-fold increase in maximum VFA concentration	[75]

Table 2. *Cont.*

Feedstock	Pre-Treatment Conditions	Digestion Conditions	Influence on VFA Yield	Reference
Waste-activated sludge	0.2 L sludge, *0.01 g sodium dodecylbenzene sulfonate/g TS, 70 °C,* 1 h, 400 rpm	0.2 L batch, 150 rpm, 37 °C, 7 days	4-fold increase in maximum VFA concentration	[81]
Grass clippings	0.1 L, 2% grass, *0.75% $Ca(OH)_2$, sonication at 2.5 W/mL* for 10 min (5 s ON 5 s OFF pulse)	Solids and liquids were separated and fermented, 0.2 L batch, pH 7, 120 rpm, 35 °C, 12 days	~2.1-fold increase in maximum VFA concentration	[80]
Waste-activated sludge and vegetable/fruit waste	Rotor-stator hydrodynamic cavitation, 2 bars inlet pressure, 80–100 L/min inflow rate, 1450–1550 rpm rotor speed, 50 min	4 L batch, substrate-to-inoculum ratio 6–7 (VS basis), 37 °C, 14 rpm	~9-fold increase in maximum VFA concentration	[46]

3. Anaerobic Digestion for the Production of Biohydrogen or VFAs

AD shows promise as an industrially viable method for the production of biohydrogen and VFAs, due to its feasibility of utilising various organic wastes as potential feedstocks. Carbohydrates are the preferred carbon source for fermentation; however, the use of carbohydrate-rich substrates such as glucose, sucrose and starch are associated with high commercial costs and competition with human-food requirements [82]. There are however numerous waste streams that contain a wide spectrum of carbohydrates that can be obtained from industrial, agricultural and municipal sources that are available at little to no cost [83].

Conventional AD broadly proceeds through four main stages: hydrolysis, acidogenesis, acetogenesis and methanogenesis (Figure 4). During hydrolysis, complex organic macromolecules, such as carbohydrates, proteins and lipids, are broken down into their respective monomers (monosaccharides, amino acids and fatty acids) by hydrolytic bacteria.

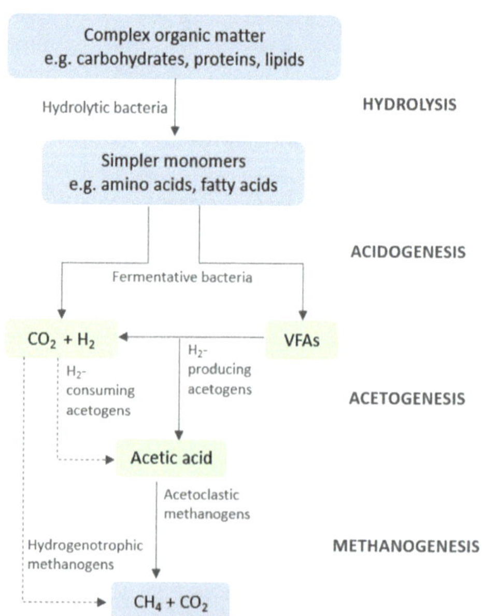

Figure 4. The four main stages of AD.

The products of hydrolysis are then fermented by acidogenic bacteria, which facilitate the formation of VFAs and hydrogen (Figure 5). Acidogenesis largely proceeds through the acetic (1) and butyric acid (3) pathways [84], resulting in hydrogen as a by-product. However, propionic acid is another common VFA produced in the AD of organic wastes, and this process is hydrogen-consuming (2).

$$C_6H_{12}O_6 + 2H_2O \rightarrow 2CH_3COOH + 4H_2 + 2CO_2 \quad (1)$$

$$C_6H_{12}O_6 + 2H_2 \rightarrow 2CH_3CH_2COOH + 2H_2O \quad (2)$$

$$C_6H_{12}O_6 + 2H_2O \rightarrow C_3H_7COOH + 2H_2 + 2CO_2 \quad (3)$$

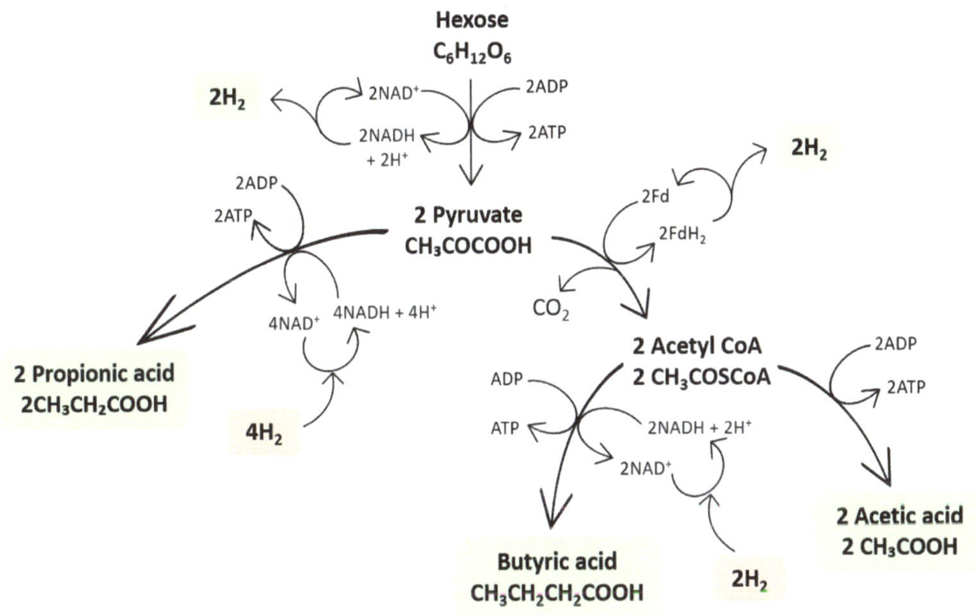

Figure 5. The metabolic pathways during acidogenesis that lead to hydrogen and VFAs from hexose. (NAD: Nicotinamide adenine dinucleotide; Fd: Ferredoxin; ATP: Adenosine triphosphate; ADP: Adenosine diphosphate).

The preferred pathway for acidogenic bacteria is the production of acetic acid since it provides the biggest energy yield for growth. However, as the partial pressure of hydrogen increases, the process that allows for the conversion of pyruvate to acetate becomes energetically unfavourable. The metabolic pathways therefore shift to produce VFAs that are more reduced than acetic acid, such as propionic and butyric acid [85]. Pyruvate is often the pivotal intermediate which can be converted into a spectrum of products, such as acetate, propionate, butyrate, lactate, propanol, butanol, hydrogen and carbon dioxide. The proportions of pyruvate directed to each pathway depend on several factors, including substrates, environmental conditions and microbial populations [86].

Hydrogen and acetate are produced during the acetogenic phase through the oxidation of the longer chain fatty acids. However, these catabolic reactions are endergonic and depend on low concentrations of acetate and hydrogen to drive the oxidation pathway forward. Acetogenesis also includes a hydrogen consumption process, known as homoacetogenesis, which is utilised to fix carbon dioxide into more acetic acid. Both hydrogen and VFAs are consumed in the final methanogenic stage, whereby acetoclastic and hydrogenotrophic methanogens either convert acetic acid into carbon dioxide and methane or oxidise hydrogen to reduce carbon dioxide to methane, respectively [87].

The yields of VFAs and hydrogen produced through AD are largely dependent on bioreactor conditions. Bioprocesses that favour the formation of either hydrogen or a VFA mixture will often produce the other as a by-product. In fact, single-stage fermentation for the co-production of hydrogen and VFAs have achieved bioconversion efficiencies of up to 64% [88]. Increased VFA concentrations however can negatively interfere with hydrogen production, either as a result of hydrogen consumption by homoacetogens or inhibition by undissociated acid molecules [89]. Higher partial pressures of hydrogen within bioreactors can also alter the proportions of VFAs obtained [90].

While the metabolic pathways leading to either H_2 or VFAs are similar, enhanced production of either of the products requires process optimisation. This is largely achieved by optimising operational conditions and bioreactor design to ensure the inhibition of methanogenic activity and that conditions are favourable for either hydrogen- or VFA-producing microorganisms. Process parameters that influence the acidogenic fermentation yield are discussed in detail in this section. Each of the subsections will discuss the influence of the parameter on both biohydrogen and VFA production.

3.1. Hydraulic Retention Time (HRT)

HRT is an important engineering parameter and is influenced by the reactor volume, V and the mass inflow rate, Q of the feedstock (4).

$$\text{HRT} = \frac{\text{V}}{\text{Q}} \tag{4}$$

HRT has been reported to play an important role in maximising hydrogen yield when fermenting organic wastes. Shorter HRTs can suppress homoacetogenesis and methanogenesis, which are both hydrogen-consuming pathways [91]. If HRTs are too short, however, biomass washout can occur [92]. HRTs as low as 6 h have proven successful in enhancing hydrogen yields of a galactose reactor [93]. However, this study only investigated three short HRTs (2 h, 3 h and 6 h), so it is unclear whether longer HRTs would have elicited higher yields. For instance, another study using a waste sugar feedstock discovered that an HRT between 14–15 h was optimal for biohydrogen production [84]. In a study investigating hydrogen production from three feedstocks at three different HRTs, Salem et al. [60] discovered that the optimum HRT for hydrogen yield using bean wastewater was 24 h (80 mL H_2/g VS), and for potato wastewater, it was 18 h (150 mL H_2/g VS). These results are consistent with a study by Massanet-Nicolau et al. [94] that utilised sewage biosolids as the feedstock and found that a 24 h HRT resulted in the most stable hydrogen producing period during which a hydrogen yield of 21.9 mL H_2/g VS was achieved. A 24 h HRT was also found to be optimum for a food waste reactor, producing a yield of 255.4 $\pm$ 33.6 mL H_2/g VS/d [95].

Several studies have also analysed the effect that HRT has on VFA production, with results varying depending on the feedstock. For instance, studies using synthetic and low-strength wastewater have found that HRTs as low as 6–8 h produce the maximum total VFA concentration [96,97]. Although with more recalcitrant feedstocks, particularly those utilised at an industrial scale, longer HRTs are generally more beneficial for VFA production. A 6-day HRT increased total VFA concentration from 2.4 g COD/L to over 50 g COD/L in a reactor fed with urban biowastes [98]. Using other organic waste streams, Jankowska et al. [99] found that for cheese whey, the optimum HRT was 20 days, producing a total VFA concentration of 16.3 g/L, and for mixed sludge fermentation, it was 12 days, producing a total VFA concentration of 12.5 g/L. Overall, the literature suggests that HRTs in the order of hours are beneficial for H_2 production, whereas in the order of days is required to obtain VFAs.

3.2. Organic Loading Rate (OLR)

The OLR within a bioreactor can aid in maximising production performance in a continuous system and is often altered by adjusting the HRT. Kumar et al. [93] reported a peak hydrogen production rate at a 3 h HRT and an OLR of 120 g galactose/L/d. However, when the HRT was adjusted to 6 h, the OLR was halved to 60 g galactose/L/d, and this resulted in the maximum hydrogen yield of 2.21 mol H_2/mol galactose. This study determined that at higher OLRs under lower HRTs, more H_2-producing bacteria, such as *Clostridia*, become dominant. Operating at a higher HRT, however, resulted in a stronger ability to retain active biomass in the system, leading to greater overall hydrogen yields. A similar hydrogen yield (2.1 mol H_2/mol glucose) was achieved at an OLR of 6.5 g COD/L/d in a glucose reactor at an HRT of 8 h [100]. In a different study, using industrial wastewater feedstock, Ferraz Júnior et al. [100] reported a maximum hydrogen yield (1.4 mol H_2/mol total carbohydrates) at an OLR of 72.4 g COD/L/d and HRT of 12 h. This study also reported that applying OLRs in excess of 100 g COD/L/d can result in significant reductions in biohydrogen yield and production rate, and such organic overloads can cause biomass washout in suspended-growth systems, e.g., continuously stirred tank reactors (CSTRs). Therefore, higher OLRs may only be suitable in immobilised-cell reactors, such as packed bed reactors, where the washout of active biomass is hindered [101].

Varying the OLR to enhance VFA production has also recently been investigated. Tang et al. [102] studied the impact of three separate OLRs on the VFA concentration in a food waste reactor (at a 5-day fixed HRT) and found higher VFA yields when the OLR was increased from 14 to 22 g TS/L/d. Iglesias-Iglesias et al. [103] also reported increases in VFA production at higher OLRs using sewage sludge as the substrate. Similarly, a stepwise increase in OLR from 3 to 12 g COD/L/d at an 8-day HRT enhanced VFA production in a microalgae biomass fermentation experiment [104]. These results are consistent with another study that found an optimum OLR value of 12.9 g COD/L/d at a 12-day HRT using olive mill solid residue as the feedstock [105,106]. Some recent studies have also reported the effect that OLR has on VFA composition, with results indicating that higher OLRs produce greater yields of longer chain VFAs, such as butyric, valeric and caproic acids [96,103,107,108].

3.3. pH

The pH level within a reactor is a key parameter that can influence the metabolic pathways of AD. Methanogens are most active between the pH range of 6.5–8.2 [109], and although the optimum pH for hydrolytic and acidogenic bacteria has been suggested as 5.4–6.5, pH levels as low as 4.0 and as high as 11.0 have been employed with various substrates [110].

For hydrogen production, it has been reported that a pH of 5.5 can be most beneficial due to the inhibition of both methanogenesis and homoacetogenesis [111]. Indeed, many studies have reported an optimum pH for hydrogen production from food waste within the range of 5.0–6.0 [112–114], whereas a pH value within the range of 6.0–7.0 has proven most successful with crop residues and agricultural wastes [115–117]. It is clear from the literature that the optimum pH for hydrogen production is largely dependent on the substrate used. For instance, a recent study by Tsigkou et al. [118] found that the optimum pH for hydrogen production when using a fruit/vegetable mixture was 6.5, but when using a mixed waste substrate, it was 7.5. Generally, the optimum pH for biohydrogen production from organic wastes is within the range of 5.0–7.0 since this favours the activity of hydrogenases and is also suitable for microbial growth and metabolism [119].

For VFA production, recent studies suggest a higher pH level is more beneficial. A more alkaline pH has been shown to not only improve hydrolysis efficiency but also enhance VFA yield when using complex feedstocks [120–122]. Cabrera et al. [120] reported an increased acetic acid concentration from 1.08 g/L to 3.14 g/L when the pH was increased from 5.0 to 9.0. A maintained pH level of 10.0 was also optimum for VFA production in a waste-activated sludge reactor [123], which is consistent with [124], which also reported a peak VFA production efficiency at pH 10.0. The pH level also plays a critical role in determining the VFA composition. Acidic conditions often result in a higher acetic acid concentration, and alkaline conditions result in a butyrate-dominant product mixture [124,125].

3.4. Temperature

Fermentation temperature is an important factor that can impact microbial metabolisms and the efficiency of substrate conversion to desired products [126]. Mesophilic fermentation often takes place within the range of 30–40 °C, whilst the thermophilic range is typically 50–60 °C. Recent research also suggests the use of psychrophilic temperatures (<20 °C) within anaerobic digesters, which could be particularly useful in countries with colder climates. Studies have shown that process parameters, including COD removal and biogas production, are comparable in a psychrophilic and mesophilic reactor [127]. The biotechnological potential of psychrophilic reactors is still under-utilised since certain disadvantages limit their use on a larger scale. These include the alteration of physical and chemical properties within biomass, thereby reducing substrate availability, inhibition of important cellular processes and the requirement to modify existing digester designs and, in some cases, use acclimated microbial biomass [128]. As a result, studies using mesophilic and thermophilic bioreactors still dominate the literature. A temperature of 55 °C has been reported as optimum for biohydrogen production from the fermentation of rice straw [129], food waste and manure [130] and sewage sludge [131]. Conversely, other studies have observed higher hydrogen yields at mesophilic temperatures. Ziara et al. [132] tested four different temperatures (35, 45, 50 and 55 °C) on a digester fed with lactate wastewater and found that biohydrogen production only occurred at 35 and 45 °C. These results are consistent with [126], who reported a maximum hydrogen yield (492.3 ± 5.1 mL/g TS) at a temperature of 36.6 °C.

Similar findings have been reported for VFA yields at various temperatures. Huang et al. [133] investigated the effect of eight temperatures between the range 25–65 °C on the AD of waste-activated sludge. They reported that the average acetate concentration increased with temperature until a peak at 40 °C, producing a yield of 0.29 g/L, beyond which the accumulation decreased. A study by Moretto et al. [98] also reported an optimum temperature of 37 °C, which resulted in a maximum total VFA concentration of 65 g/L. Mesophilic temperatures have also proven preferential for VFA production from sewage sludge [134] and the organic fraction of municipal solid waste (OFMSW) [135].

Although thermophilic conditions present advantages, including an increased rate of hydrolysis, enhanced pathogen destruction and a higher rate of organic matter destruction, mesophilic conditions are most promising for larger-scale digesters due to the lower energy requirements and more stable operation [129,136].

3.5. Operational Mode and Reactor Configuration

Batch bioreactors have been extensively used for both biohydrogen and VFA production to evaluate the viability of feedstocks and the effect of various process conditions. Batch experiments are widely used to optimise process parameters at the lab scale, and the results often indicate an initial increase in desired products, followed by a rapid decline as the feedstock is used up [137]. While the initial insight provided by batch systems is useful for understanding the digestion process, it is important to understand digester behaviour in (semi)continuous systems to aid scale up.

With respect to acidogenic digestion, various bioreactor configurations have been investigated at a laboratory scale, both under batch and (semi)continuous modes. The most extensively studied is the continuously stirred tank reactor (CSTR); however, other methods include anaerobic fluidised bed reactors (AFBRs), anaerobic sequencing batch reactors (ASBRs), anaerobic packed bed reactors (APBRs) and up flow anaerobic sludge blanket reactors (UASBRs). There is also a strong correlation between production efficiencies and the size of the microbial population present within the reactor. Therefore, cell retention strategies such as granulation and immobilisation systems have also been investigated to enhance overall product yields [138]. An overview of the typical reactor configurations used for biohydrogen and VFA production is shown in Table 3 and discussed in detail in this section.

Table 3. Advantages and disadvantages of typical reactor configurations used for biohydrogen and VFA production.

Reactor Configuration	Advantages	Disadvantages
CSTR	• Constant suspension and homogeneous mixing lead to efficient contact between substrates and microbes • Simple design and easy maintenance • Can be operated continuously • Low operating cost	• Susceptible to shear strain at high mixing speeds • Low HRTs can cause biomass washout • May require combination with an immobilisation system
AFBR	• Enhanced mass transfer • Likelihood of biomass washout is low • Operate at shorter HRTs (favours biohydrogen production) • Can be operated continuously	• High energy requirement to supply constant fluidisation • Support material required for adhesion of biomass • Less applicable for high solid biomass and longer HRTs • Difficulties in scale-up
ASBR	• Not reliant on HRT for active biomass retention • Ability to adjust SRT allows for an additional mechanism for microbial manipulation	• Unable to handle high solid biomass since they are susceptible to organic overload which can reduce performance • Less applicable to high-strength wastewaters
APBR	• Able to tolerate high OLRs • Low construction costs • Good feedstock retention	• Susceptible to excess biomass accumulation • Often require a recirculation loop to improve mass transfer and a support material for immobilisation
UASBR	• Good microbial retention without the use of biofilms or support materials • Simple method which has proven successful at large scale • Low energy requirement	• Extended start-up time • High pathogen, nutrient, COD and BOD levels remain in the effluent • Difficult to maintain proper HRT

3.5.1. CSTR

In a CSTR, the microbes and substrates are constantly suspended and mixed, which facilitates effective contact and higher mass transfer [139]. Their benefits include simple design, easy maintenance, homogenous mixing, and a well-maintained HRT. CSTRs are also the preferred reactor configuration when no differentiation between solids and liquid retention times are required. CSTRs are widely considered an effective and economical approach for the production of both biohydrogen [140–143] and VFAs [98,144,145] from organic waste streams. Although this kind of reactor is sensitive to operational parameters, including pH, temperature and HRT, limitations in mass transfer have proven a critical parameter for optimum performance. Research indicates that the concentration of desired products increases when the mixing speed increases until an optimum; exceeding this can result in shear strain that can damage floc particles and relevant microbial populations [146]. Their main drawback however is biomass washout at lower HRTs [92]. As a result of this, some studies have combined CSTRs with immobilised systems to retain more active biomass in the reactor. Keskin et al. [147] reported higher hydrogen yields and greater resistance to biomass washout in an immobilised bioreactor configuration compared to a conventional CSTR, particularly at higher OLRs.

3.5.2. AFBR

In AFBRs, a fluidisation medium (liquid or gas) is passed through the digester containing the feedstock, usually of high solid content, to ensure suspension. This enhances microbial activity via enhanced mass transfer and can cause greater degradation of wastewaters [148]. In comparison with CSTRs, the likelihood of biomass washout is lower, but more energy is required for constant fluidisation. Often, a support material is also required for biomass adhesion; examples from the literature include shredded tires [149], activated carbon [150], polystyrene and expanded clay [151]. The literature indicates the broader use of AFBRs for hydrogen production in comparison to studies focused on VFA production. This could be due to the fact that they can operate at shorter HRTs and higher OLRs which favours biohydrogen production (Section 3.2). For instance, Amorim et al. [152] reported an increase in hydrogen yield from 0.13 to 1.91 mol H_2/mol glucose when the HRT decreased from 8 to 2 h in an AFBR utilising cassava wastewater.

3.5.3. ASBR

This reactor process involves cycling through the stages of feeding, reaction, settling and decanting. Within this semi-batch process, the use of a settling stage allows for greater solid retention within the reactor, meaning that the solid retention time (SRT) becomes independent of HRT. Recent studies have obtained high hydrogen yields using ASBR systems. Maaroff et al. [153] tested a two-stage ASBR system for biohydrogen production from palm oil mill effluent and achieved yields as high as 2.52 mol H_2/mol sugar at an optimum HRT of 12 h. A similar HRT was utilised in a study by Santiago et al. [154], who reported a 16 h HRT and 55 h SRT as optimum for biohydrogen yields from organic waste. This study also examined the effect of SRT and HRT on VFA production and found that a similar SRT (60 h) but a longer HRT (48 h) was optimum for VFA yields. The ability to adjust the SRT in an ASBR provides an additional mechanism to manipulate microbial communities, which in turn can be used to enhance desired metabolic pathways. Through analysis of population dynamics within ASBRs, it is reported that hydrolytic bacteria are dominant at shorter SRTs, whilst acidogenic and acetogenic bacteria become more dominant at longer SRTs [155].

3.5.4. APBR

These reactors are often used with feedstocks of high organic content, and the beds can contain granules or biofilms to improve function at lower HRTs and increase tolerance to high OLRs [156]. APBRs are easy to operate and require lower construction costs when compared with other reactors [139]. Additionally, due to their ability to retain high feedstock concentrations within the reactor, high conversion rates can be achieved. Despite this, some studies have reported unstable operation, mainly attributed to excess biomass accumulation in the bed, which can lead to the proliferation of H_2- and VFA-consuming microbes [157]. Since there is no continuous mixing, such as in a CSTR, a recirculation loop is often implemented to improve mass transfer and enhance product yields [158]. Various support materials have been tested for their ability to immobilise relevant microbes within APBRs and therefore impact production performance. Muri et al. [159] analysed the impact of three different support materials (Mutag BioChip™, expanded clay and activated carbon) on hydrogen yields within an APBR fed with synthetic wastewater and reported the highest yield (1.80 mol H_2/mol glucose) when the reactor was packed with Mutag BioChip™.

3.5.5. UASBR

UASBRs are another reactor configuration that aims to retain microbes within the reactor. They do not use biofilms or support materials and instead rely on the formation of biological granules. UASBRs are best utilised with medium–high strength wastewaters since the feedstocks need to have good settling characteristics. They are a simple and reliable method for wastewater treatment, and many large-scale plants have been successfully operated [160]. They have proven successful for biohydrogen and VFA production from various organic waste streams [161–165]; however, some disadvantages include an extended start-up time and excess pathogen, nutrient and overall biomass content in the effluent.

3.6. Additives

The impact that additional chemicals and nutrients have on the production yields of hydrogen and VFAs has been extensively investigated. Adjusting the carbon to nitrogen (C/N) ratio of feedstocks is a common parameter used to enhance digester performance. Ratios that are too high (>30) can lead to insufficient nitrogen available to maintain microbial biomass, whilst ratios that are too low can increase ammonia production, which can inhibit microbial activity (Section 3.6) [166]. Studies suggest that increasing C/N ratios can enhance biohydrogen yield with ratios as high as 137 [157] and 173 [167], resulting in maximum hydrogen yields from reactors fed with sucrose and wheat powder solution, respectively. Argun et al. also reported a maximum VFA yield at the same C/N ratio, producing 11 g/L of total VFAs. A C/N ratio of 47 was reported as optimum for biohydrogen production in a sewage sludge reactor, whilst the same study determined a C/N ratio of 130 resulted in the maximum VFA yield [168].

Metals have been among the most widely employed additives in AD systems. The addition of iron and nickel (in ion or nanoparticle form) have shown significant enhancements in biohydrogen yields due to their ability to facilitate the acceleration of the electron transfer between ferredoxin and hydrogenase, which, in turn, drives hydrogen generation [169]. The addition of nickel ion and Ni^0 nanoparticles has been shown to effectively enhance biohydrogen yields [170], and the use of biologically synthesised iron nanoparticles improved biohydrogen yields by up to 44% when compared with no addition [171]. In contrast, the use of metal additives in studies focused on VFA production have produced varying results. Zhang et al. [172] reported higher VFA yields when adding iron to a cadmium-containing system compared to adding nickel. The addition of Co^{2+} and Zn^{2+} [173] lowered total VFA yields in comparison to the control, but yields of propionic acid were significantly increased (from 28.71 to 317 mg of propionic acid/g of COD).

Biochar addition has also recently been investigated in various anaerobic digesters. A 15 g/L biochar addition produced a maximum hydrogen yield of 3990 mL/L in Zhao et al. [174]. Sugiarto et al. [175] found that biochar addition not only increased hydrogen yields by 107% but also the primary elements present in biochar (Fe, K and Ca) were responsible for increasing both acetic acid and butyric acid concentrations. Similarly, acetic acid concentration increased from 0.18 to 0.36 g/L in response to a 0.6 g/L biochar addition in a study by Lu et al. [176].

The influence of salinity on biohydrogen and VFA yields has produced contrasting results in the recent literature. Taheri et al. [177] found that increasing NaCl concentration from 0.5 g/L to 30 g/L had a negative effect on hydrogen yield, decreasing it from 1.1 mol H_2/mol glucose to 0.3 mol H_2/mol glucose. However, Sarkar et al. [178] reported that a NaCl concentration as high as 40 g/L resulted in maximum hydrogen yields from a food waste reactor. This study also found that a 40 g/L NaCl addition improved total VFA yields by 1.35 times, producing 6.58 g/L compared with 4.84 g/L from the control experiment. He et al. [179] examined the impact of 4 NaCl concentrations (10, 30, 50 and 70 g/L) on VFA production, and although the highest yield was achieved at 10 g/L (0.542 g/g dry weight of food waste), yields remained high at 70 g/L (0.441 g/g dry weight).

The addition of antibiotics is associated with the suppressions of methanogenesis and therefore has been studied as a method to increase biohydrogen and VFA yields. Recent research indicates that certain antibiotics can have an inhibitory effect on each stage of AD; however, the more severe inhibition impacts acetogenesis and methanogenesis [180,181]. The addition of roxithromycin to a waste-activated sludge fermenter had the most severe inhibitory effect on methanogenesis, so the VFA yield more than doubled when the antibiotic concentration increased from 0 to 100 mg/kg TSS [182]. Huang et al. [181] reported maximum VFA yields when concentrations of clarithromycin reached 1000 mg/kg TSS.

Contrastingly, Tao et al. [183] found a negative correlation between the concentration of thiosulfate and yields of both VFAs and hydrogen. This study inferred that thiosulfate inhibited several metabolic pathways of acidogenesis and in particular restricted the activity of key enzymes, including butyryl CoA and NADH.

Although the addition of some chemicals and nutrients may have a positive impact on hydrogen and VFA yields, most of the studies reported yields from small-scale digesters (<1 L). It is therefore unclear how beneficial the addition of chemicals and antibiotics would be on a larger scale, particularly from an economic point of view.

3.7. *Undesired By-Products and Inhibitors*

The accumulation of hydrogen and VFAs within reactor systems can make it thermodynamically unfavourable for their continued production and result in inhibited digestion with reduced yields. Studies have shown that the continuous recovery of hydrogen and VFAs from fermentation broths can enhance the overall yields of both products. Jones et al. [184] utilised proton exchange membranes to remove hydrogen and electrodialysis to remove VFAs from a sucrose reactor to effectively enhance hydrogen yields by a factor of 3.75. In Hassan et al.'s work [185], the in situ recovery of VFAs from a food waste reactor almost doubled hydrogen yields and increased VFA concentration from 1.9 to 4.7 g/L. Further studies operating electrodialysis on reactors fed with grass waste [87] and food waste [145] were also effective in enhancing VFAs yield. Methods for hydrogen and VFA extraction are discussed further in Section 4.

While end-product inhibition is commonly observed with acidogenic digestion, inhibition due to the formation of undesired by-products should not be ignored. The formation of undesired by-products is influenced by the type of substrate, operating conditions (temperature and pH) and the diversity and abundance of the microbial community present in the digester. The most common inhibitor encountered in digesters is ammonia. The presence of excess ammonia, both in free (NH_3) and ionic (NH_4^+) form, can induce inhibition. Ammonia is able to pass through cell walls and react with protons to produce NH_4^+, which can alter intracellular pH and inhibit microbial activity [186]. Whilst methanogenesis is considered the most severely inhibited AD stage, excess ammonia can also have an inhibitory effect on fermentative bacteria [187]. The severity of ammonia inhibition is dependent on operational conditions, and studies reveal that thermophilic temperatures and a higher pH (>7) can increase the percentage of total ammonia present in digesters [188]. Methods to limit ammonia inhibition in reactors include adsorption using zeolites [189] or biochars [190], pH reduction [191] and adjustments in C/N ratios [192]. Hydrogen yields were increased by 10–26% in a study by [193] utilising zeolites for ammonia stripping. The use of nitrogen sparging and sulfuric acid absorption was used by Ye et al. [123] to remove ammonia from a waste-activated sludge digester, which enhanced total VFA yields by 21.7%.

Many other inhibitory substances and contaminants, such as sulphides, phenols and furans, may already be present in organic waste streams and methods to limit their inhibition include dilution, adsorption, acclimatisation and precipitation. Sulphate derivates, such as sulphides, can be toxic to fermentative microbes, and sulphate-reducing bacteria have been reported to outcompete hydrogen-producing acetogens through the consumption of substrates, such as butyrate and propionate [194,195]. It is therefore critical to control the levels of sulphates (and sulphides) within bioreactors. Desulphurisation techniques, including the use of bio scrubbers [196] and biofilms [197], have shown promise in large-scale systems. Additionally, Dhar et al. [198] reported that the Fe^{2+} addition was successful in limiting biohydrogen inhibition caused by sulphide in excess of 0.025 g/L. Metal ion addition has also been used to limit the inhibition effect of humic acid in a sludge anaerobic digester [199].

The presence of phenols and furans within bioreactors can negatively impact the growth and cell membrane function of fermentative bacteria [200], and hence, methods to extract them from bioreactors, as well as limit their production, have been investigated.

An adsorption resin was used by Trujillo-Reyes et al. [201] to successfully extract phenolic compounds from a raspberry waste stream, accumulating up to 2402 mg gallic acid equivalents/kg of feedstock which significantly enhanced biogas yields. Adsorption using activated carbon supported with a nano zero-valent iron material was also effective in removing phenolic compounds from organic wastewater, which improved the efficiency of hydrogen-producing bacteria and increased the conversion rate of organic matter [202]. As mentioned in Section 2, the choice of pre-treatment method is often one of the most important parameters that impact the presence of both phenolic and furan compounds. For instance, Kim and Karthikeyan [203] found that carbohydrate degradation to furan-derived compounds, in particular furfural, was much greater when applying an acid pre-treatment (873 ppm) compared with an alkali pre-treatment (375 ppm).

While the most common parameters influencing acidogenic fermentation were discussed in this section, it is worth noting that pre-digestion parameters such as inoculum treatment (to limit methanogenesis) and substrate pre-treatment (to enhance its bioavailability) as well as downstream processing (product recovery) are also key to enhance product yields. It is therefore important to approach acidogenic fermentation with a holistic approach with a specific focus on each of these aspects.

4. Product Recovery

Global hydrogen demand increased from 19.2 Mt in 1975 to 73.9 Mt in 2018 [204]. Increasing the production of hydrogen sustainably via low carbon technologies is therefore critical to meet the growing demand. While much focus has been given to biohydrogen production, equal attention has also been given to its recovery. It is important to advance on both these fronts to maximise the desired product yields. A range of recovery options for biohydrogen and VFAs are shown in Table 4.

Table 4. Examples of biohydrogen and VFA recovery strategies employed in the literature.

Recovery and Production Methods	Fermentation Conditions	Recovery and Production Data	Author
	Hydrogen Recovery		
Electrochemical proton exchange membrane and CO_2 scrubbing of bioreactor gas phase; Homoacetogenesis and end-product inhibition arrested	Sucrose inoculated with heat-treated AD digestate. 3.34 L, continuously fed CSTR; pH 5.5; 35 °C; 24 h HRT	1.79 mol H_2/mol hexose 7 cm_{H2}^3/min >99% purity of recovered H_2	[205]
As above with electrodialytic recovery of VFAs from the liquid phase to arrest end-product inhibition further	As above with 18 h HRT	0.90 mol H_2/mol hexose 3.47 cm_{H2}^3/min >99% purity of recovered H_2	[184]
	VFA Recovery		
Filtration and electrodialysis for in situ VFA recovery; Methanogenesis and end product inhibition arrested	1% TS food waste inoculated with heat-treated AD digestate. 100 L continuously fed CSTR; pH 5.5; 35 °C, 10 d HRT	17 g VFA/day recovered from bioreactor	[145]
Inline ultrasonic sieving, centrifugation, microfiltration, and electrodialysis for in situ VFA recovery; Methanogenesis and end product inhibition arrested	5% TS grass waste inoculated with heat-treated AD digestate. 100 L continuously fed CSTR; pH 5.5; 35 °C, 8.25 d HRT	VFAs continually recovered into an external 30 L solution of up to 4500 mg VFA/L VFA yields of 404 mg VFA/g VS achieved	[87]
As above with an additional pervaporation stage before electrodialysis to aid VFA selectivity	As above with 7 d HRT	VFAs continually recovered into an external 30 L solution of up to 4000 mg VFA/L VFA yields of 875 mg VFA/g VS achieved	[206]

Recent work has reported prolonging thermodynamically favourable fermentation conditions for continued and enhanced hydrogen production without subsequent methanogenesis [205]. In this study, a proton exchange membrane was used to recover hydrogen from the gas phase of a sucrose bioreactor in conjunction with carbon dioxide scrubbing. The in situ recovery of these gasses had two effects. The first was that homoacetogenesis was arrested by making the substrates (hydrogen and carbon dioxide) unavailable for homoacetogens to consume. The second effect was to arrest end-product inhibition by preventing hydrogen accumulation in the reactor headspace. This maintained thermodynamically favourable conditions for continued hydrogen production whilst continually recovering purified hydrogen as it was being produced. Hydrogen yields in a 4 L sucrose CSTR were increased from 0.07 to 1.79 mol H_2/mol hexose using this methodology over three 24 h HRTs. A similar methodology was employed by Jones et al. [184]; however, in addition to electrochemical hydrogen recovery and carbon dioxide scrubbing, electrodialysis was also used to alleviate end-product inhibition, increasing hydrogen yields almost four-fold from 0.24 to 0.90 mol H_2/mol hexose. Whilst no techno-economical work was carried out in this study, the hydraulic retention times were increased from a matter of hours to 2 days, resulting in greater substrate utilisation rates. This is of particular importance to waste remediation industries since, currently, the fermentation of organic wastes to hydrogen is not viable due to low hydraulic retention times resulting in either poor substrate utilisation rates or the requirement to add sparging gases such as nitrogen at economic and environmental costs. Shortening hydraulic retention times may also become problematic for waste treatment facilities, and the contaminated nature of waste streams means that it would be difficult to maintain conditions in which a targeted microbial consortium could thrive. The use of electrochemical means to recover hydrogen from the gas phase of bioreactors would be relatively straightforward to retrofit to existing infrastructure could be powered using excess renewables and increase hydraulic retention times.

In addition to the biological production of hydrogen and methane, the targeted biological production and recovery of VFAs in preference to these biogases have been investigated. Whilst the majority of research into the recovery of VFAs from fermentation broths is at the millilitre scale and could not yet be considered to be industrially applicable, there has been some research, as reported in more detail in Jones et al. [207] in which fermentations at larger scales using real waste, and waste analogues, have been carried out with subsequent VFA recovery. For example, acetic acid, propionic acid, butyric acid, and valeric acid were recovered from a 4 L fermentation of olive mill waste via a combination of filtration and electrodialysis into a solution with total VFA concentrations over 14 g/L [208]. Jones et al. [87,145,206] showed that the targeted recovery of VFAs from the liquid phase of 100 L bioreactors arrests methanogenesis and increases the yields of VFAs. In one of these studies, a combination of filtration and electrodialysis was used to recover a VFA solution of up to 4 g/L from a 100 L continuously fed food waste digester [145]. This VFA recovery also had the effect of increasing substrate utilisation rates and increased the rate of production of VFAs from 4 to 35 mg VFA/g VS d^{-1} by alleviating end-product inhibition. In the other two studies, similar methodologies were used to recover VFAs from continuously-fed 100 L grass fermentations [87,206]. In one, a combination of mechanical sieving, filtration and electrodialysis recovered a VFA solution of up to 4.8 g/L and increased yields from 287 to 404 mg VFA/g VS. In the other, the addition of a pervaporation stage to exclude non-volatile compounds from recovery, a VFA solution of 4.5 g/L was produced, and yields as high as 875 mg VFA/g VS were achieved. These sorts of results are promising when considering VFAs as a valorisation route for wastes since the methodologies employed would be easily retrofitted to existing infrastructure and are scalable. No techno-economic considerations were made in these bodies of work, and so more research to clarify this aspect of the process would be useful.

5. Techno-Economics and Process Life Cycle

Broadly, biohydrogen production can be split into photo and dark fermentation. Producing hydrogen in these ways has CO_2 emissions of 0.1–4.0 kg CO_2/kg H_2, compared with up to 38 kg CO_2/kg H_2 for fossil fuel derived hydrogen [204]. For comparison, power to hydrogen via electrolysis is carbon neutral at the point of use; however, the whole life carbon emissions of hydrogen as an energy vector are contingent upon its production pathway [209] (Figure 6).

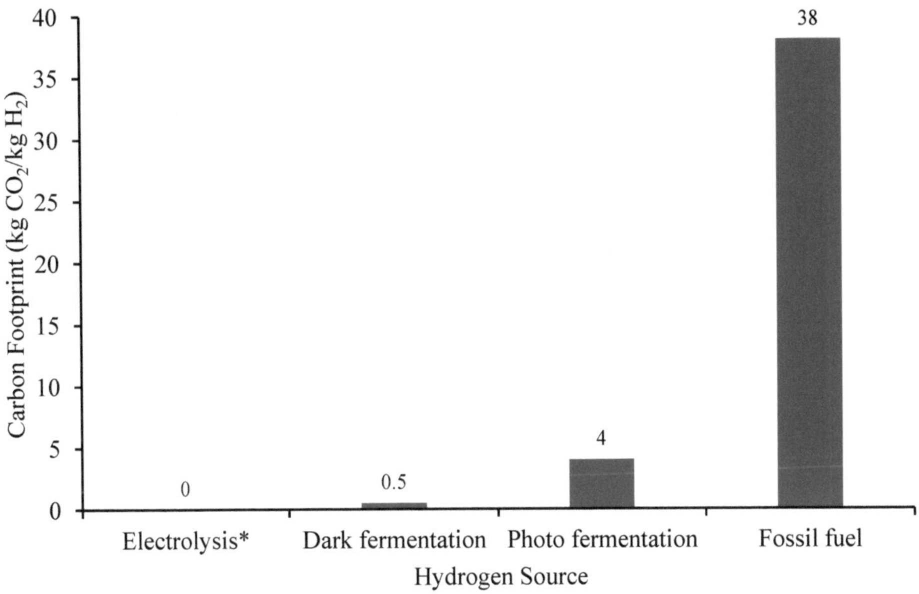

Figure 6. Carbon footprint of different pathways to hydrogen. * At point of use when powered by excess renewable energy.

According to IEA [210], coal, natural gas, LPG, gasoline, and diesel have kg CO_2 equivalents per kWh of 1.05, 0.55, 0.62, 0.69 and 0.73, respectively. These are still, globally, the most commonly used energy vectors [211], and renewable energy sources contribute to just 10% of global energy consumption, excluding transport, so it would be necessary to increase the global installed capacity of renewables from 2500 GW in 2020 [212] by a factor of six to meet Paris Agreement 2050 targets [213]. The intermittent nature of installed renewable energy sources, 90% of which is solar and wind, requires reliable and sustainable energy storage [211], and hydrogen is an obvious candidate for such storage. In Europe, it is reported that the production costs of hydrogen could fall to less than EUR 2.00/kg by 2050 [214]; however, this relies upon the continued decline of the cost of renewable energy.

Techno-economic analyses for photo fermentative biohydrogen production have briefly been investigated in recent articles. In one [215], a combination of biohydrogen production from 140 ha open pond and 14 ha photobioreactor sources yielded 1.2 million GJ/year of hydrogen, with an initial investment of USD 55 m. Whilst they posit that biohydrogen production units greater than 30 m^3 could be profitable, approximately 14% of that profit is derived from the co-production of CO_2. Fu et al. [216], however, note that photo fermentation is not without its own issues, including the costly nature of constant illumination and the opacity of the fermentation broth.

When considering dark-fermentation, modelling of biohydrogen production from the co-digestion of food and beverage wastewater found that a plant with a biohydrogen production rate of 63,000 m^3/year could have a production cost of USD 1/m^3 [217]. In the same article, it was reported that when agricultural residue was used as the substrate, hydrogen could be produced at USD 2.57 and 2.83/kg with dark and photo fermentation, respectively. Encouragingly, payback periods of between 5 and 10 years have been reported for hydrogen production from the fermentation, depending on scale, comparing favourably with gasification and steam methane reforming, which can have payback periods as high as 20 years [218].

Despite these promising production data, other researchers [219] conclude that the lack of synergy between academic research and industrial implementation makes a current jump to market unachievable for biohydrogen, especially where biohydrogen at the lab scale has been produced under very strict and steady fermentation conditions. Further, an economic feasibility threshold of USD 1.50/kg has been reported in other research [220], suggesting that the technology is not yet mature enough for commercial success. Other studies [221] also report costs as high as USD 6.98/kg for hydrogen production from dark fermentation, so there is considerable disparity amongst the literature.

When considering hydrogen production from wastewaters only, dark fermentation is shown to be the best economical path to hydrogen when compared with photo fermentation, electrolysis, electrodialysis, photocatalysis, photoelectrochemical methods and super water gasification [220]. There is however room for considerable improvements in biohydrogen production. A number of bottlenecks to biological hydrogen production have been established, and there is a growing body of lab-scale work to overcome these problems. Fu et al. [216] review anaerobic biohydrogen production from waste-activated sludge and note that whilst the literature widely reports high theoretical maxima for hydrogen yields from waste-activated sludge, real-world yields are always significantly lower. This is mostly due to hydrogen-producing microorganisms mostly metabolising soluble carbohydrates in favour of proteins, humic matter, VFAs and alcohols, all of which are typically abundant in industrial wastewaters. Further, the hydrogen produced is typically consumed rapidly by methanogens, acetogens and sulphate reducers to produce H_2S.

As mentioned in Sections 2 and 3, strategies to improve hydrogen production have long been employed. The techno-economic applicability of these strategies to industrial-scale hydrogen production however has yet to be investigated, and more work is called for. A recent study [222] notes that because these limitations of hydrogen production have yet to be addressed and tested on large scales, the techno-economics of biohydrogen production are still less attractive than producing hydrogen from even non-renewable sources such as fossil fuels. This is exacerbated by higher costs per unit of hydrogen at smaller scales, which may be reduced by up to eight times upon scale-up [204]. Similarly, Lepage et al. [223] state that whilst biohydrogen is a promising future possibility, it is currently not mature enough to compete with established thermochemical pathways despite the unsustainable nature of the latter. This highlights the need for greater work to demonstrate larger-scale hydrogen production utilising industrially applicable engineering solutions.

Another issue for the economic feasibility of biohydrogen is competition from the other sustainable pathways to hydrogen, notably electrolysis driven by excess wind, solar and hydropower. The decreasing costs of renewable energy as a means through which to power electrolysis, especially considering solar PV costs less than USD 1/W, thanks to falling silicon prices [224]. Competitive levelized costs of energy from hydrogen have been demonstrated in several studies. Off-grid hybrid systems of solar PV, wind and hydrogen fuel cells were simulated in various configurations for regularly and seasonally occupied households and found energy costs as low as USD 0.309/kWh [225], taking into account the capital expenditure of the technologies required to achieve such an energy scenario. On larger scales, similar and self-sufficient energy scenarios have been reported with energy costs of USD 0.394/kWh [226] and USD 0.334/kWh. It is also reported however that energy management systems and methods to integrate and coordinate sustainable hydrogen production systems need to be optimised before they can no longer be considered economic drawbacks to larger-scale projects [211].

A common theme in the literature is that the cost of electrolysis is the main economic limiting factor when considering hydrogen as a sustainably derived energy vector [214,227]. Storage has been posited as a means through which to make better use of existing electrolysis capacity without the installation of further electrolysis apparatus, especially when that hydrogen can be stored in major geological features such as salt caves [227]. In some European countries, the hydrogen capacity in such formations exceeds 10 EJ of hydrogen [214].

As alluded to above, storing hydrogen is becoming less of an unknown, especially with the emergence of solid-state storage in stand-alone use cases. It is reported however that more work is required to establish the availability, mass and degeneration of storage facilities [224]. It is also reported that storing hydrogen in its compressed form is already commercially viable [228], provided it is transported by road, and that liquefied hydrogen with a gravimetric energy density of 2.0 kWh/kg can be achieved at a cost of USD 6.00/kWh, lower than the USA Department of Energy's goal of USD 8.00/kWh.

In shifting from a carbon-based energy economy to a hydrogen-based one, changes to infrastructure need to be considered from a techno-economic point of view. The majority of the reported literature excludes calculations surrounding vehicular hydrogen use, which is of concern because transportation is currently the least diverse energy use sector, with 90% of its energy demand met by fossil fuels [229]. By 2050 however, the transportation sector is predicted to be the single greatest demand case for hydrogen as an energy vector [229]. To make this affordable for consumers however high transportation and refuelling station costs will need to be addressed [230]. The purity required for vehicular hydrogen will pose a problem for biological hydrogen, which will either need to be purified downstream or extracted from biogas mixtures in such a way that makes it inherently high-purity as per the electrochemical approach employed by Massanet-Nicolau et al. [205].

Another bottleneck for large-scale and industrial hydrogen production is likely to be international cooperation. Nazir et al. [228] summarise that a critical and often overlooked issue for any future zero-carbon economy is international cooperation and the requirement for legislation and regulations to be harmonious and transnational. Modelling has been undertaken [231] investigating the hypothetical outcomes of the implementation of a renewable hydrogen quota imposed upon European gas and electricity markets. This would increase electricity prices by 12%, with renewable energy producers being the biggest beneficiaries. Gas prices however would fall by 3% for the consumer; however, there would be a net benefit to energy producers. Policy recommendations for European hydrogen implementation have been suggested [232], including coordinated incentivisation of hydrogen vehicle purchases and refuelling infrastructure; harmonising the blending concentrations for hydrogen and natural gas; implementing certificates and obligations for products produced and obtained using low-carbon hydrogen; and promoting the production of low-carbon hydrogen by penalising polluting activities further.

As mentioned in the previous section, biological hydrogen production is similar in methodology to biological VFA production, and both can be produced from common substrates. As such, there will be direct competition for these substrates, and the likelihood is that the most economically viable product will be successful, especially considering there are competing, alternative, renewable pathways to hydrogen. According to the IEA [210], the demand for hydrogen is approximately 120 megatons per year, with a global market value of USD 130 bn. The demand for VFAs is 16,820–18,500 kilotons per year, with an approximate market size of USD 17.1 bn per year [87,233]; smaller in terms of both mass and market size. There is also a growing case for the economically viable production of VFAs from waste. The wholesale price for VFAs such as propionic acid is currently USD 6.00/kg, and assuming sympathetic government incentives and favourable conditions, it can be derived from industrial wastewater at USD 3.80/kg [233]. Bahreini et al. [234] also found that although VFA extraction inherently reduces biogas yields, the cost of this reduction is offset and exceeded by the income generated by the recovered VFAs, resulting in short payback periods ranging from 1.6 to 6.3 years. There is even now an emerging case for the utilisation of cash crops from which to derive VFAs from anaerobic digestion. Moscariello et al. [235] carried out millilitre scale batch digestions of pre-treated hemp biomass residue and calculated that with alkaline pre-treatment, revenues of EUR 9364/ha/year could be achieved. It is not clear however how appropriate this sort of pre-treatment would be at industrial scales, especially from an environmental standpoint. The versatility of VFAs as a platform chemical with which to valorise biowastes is highlighted in Huq et al. [236], who make the techno-economic case for the use of food-waste-derived VFAs in the synthesis of aviation fuel. In this case, production costs are as low as USD 0.30/kg, with a wholesale price of USD 2.50/gallon. This compares favourably with the cost of producing hydrogen from biomass, being several orders of magnitude less costly than the above-reported costs of biohydrogen production.

As it stands, the least costly means through which to produce hydrogen sustainably remains to use excess renewable energy, potentially leaving room for VFA production from bioremediation of wastes in preference to methane. This is especially true when considering the diverse, growing markets for VFAs, including disinfection, herbicides, bleaching, the food industry, the textile industry, pharmaceuticals, fungicides, polymer production, cosmetics and lubricants [207]. Given the increasing demand for bioplastic production and the finite nature of fossil fuels, PHA production is of particular interest here with recent research [237], demonstrating through life cycle analyses that the cumulative energy demand of PHAs produced using VFAs from an early stage, non-optimised grass-fed bioreactors are in the same order of magnitude as deriving polymers from fossil fuels.

Further work is called for to optimise the VFA production and recovery processes to fulfil the promise shown by this sustainable pathway to PHAs, framed in the context of decreasing demand for grass as a grazing crop but with little scope for viable alternative uses for that land.

6. Future Perspectives

There is an immense potential to utilise 'wastes' as 'resources' to generate value and contribute to a circular economy framework. Biorefineries are seen as one of the promising routes to contribute to sustainability. A multiple-product platform is attractive techno-economically; however, the challenges in achieving desired product yields to break even are not yet established at an industrial scale. With biogas from AD being a low-value product, research attention has been diverted towards the production of biohydrogen and VFAs as promising energy carriers and platform chemicals. While the biochemical pathways to produce biohydrogen and VFAs via AD overlap, there is a possibility of establishing a biorefinery where biohydrogen can be the primary product. The VFA mixture in the liquid phase could be recovered (as a mixture as opposed to a specific VFA with high purity) and valorised further to produce polyhydroxyalkanoates (PHAs) that can be used to make bioplastics [238]. Moreover, coupling allied processes such as photo fermentation and dark fermentation [239] or biochemical and bioelectrochemical systems [240] can pave the way for maximising product recovery via biorefineries.

A number of strategies relating to the intensification of AD for biohydrogen or VFA production were discussed throughout this review; however, there are gaps in several areas that still need more attention. For instance, with the feedstock composition and heterogeneity largely influencing the end product productivity and yields, focused research on scalable pre-treatment methods is needed. The general consensus is that physico-chemical methods are energetically and environmentally favourable when compared to their counterparts, and while this is a generality, evidence to support this is limited. Hydrodynamic cavitation has been utilised effectively as a pre-treatment method for enhancing biogas yields [2], and the preliminary evidence on positively influencing acidogenic fermentation yields is promising [46]. Similarly, thermal hydrolysis has shown immense potential for enhancing biogas yields (e.g., the CAMBI thermal hydrolysis process) from sewage sludge at a commercial level; however, its influence on acidogenic fermentation is unclear.

The choice of bioreactor and its configuration is also important in maximising product yields. While CSTRs have always been the first-choice reactor, alternative configurations with better offerings must be explored to improve the fluid dynamics and mass transfer within the system (Table 3). Another aspect that requires deep focus is the selection of appropriate product recovery methodologies. While there are established product recovery strategies for recovering biohydrogen and VFAs from bioreactors, the combination of technology, scale and mode of operation and productivity must not lead to detrimental techno-economics. Furthermore, to achieve a triple bottom line performance of social-economic-environmental benefits, in addition to techno-economic feasibility studies, an LCA investigation is also required. In particular, with the market demand for sustainable VFAs being high, a limited investigation at scale and LCA is needed.

Besides these aspects that require extensive focus, some of the emerging areas in AD include two-stage digestion for biohythane production (maximising energy recovery), on-demand production of VFAs or biohydrogen (feeding the supply chain depending on market needs) and intersectoral coupling to enable efficient waste management (unconventional wastes as resources). Collective efforts in multiple innovative areas as required in addition to intensifying the existing promising options to meet the goals of net-zero emissions by 2050.

7. Conclusions

Shifting focus from conventional biogas production towards biohydrogen or VFAs has been receiving renewed attention recently. Advancements in technology and policy support have paved the way for initiating this revived interest. This presented a need to analyse the literature critically and establish the current status of acidogenic fermentation. This review has specifically addressed this in three main areas; pre-digestion focusing on substrate pre-treatment as a measure to address feedstock heterogeneity, parameter optimisation for improving product yields and product recovery. The techno-economic aspects of biohydrogen and VFA production were also discussed as appropriate. Finally, gaps in these areas were highlighted and the future research direction required was identified. Amongst the areas discussed, three main conclusions were drawn: a scalable pre-treatment that is both effective and energetically favourable is needed, reactor designs specific to handling the feedstock of interest must be exploited and energy-efficient and scalable product recovery options need to be integrated downstream of the fermenter to maximise product yields and economics.

Author Contributions: Conceptualization, S.N.; writing—original draft preparation, S.N., R.J.J. and L.O.; writing—review and editing, S.N., J.M.-N. and A.G.; funding acquisition, J.M.-N. and A.G. All authors have read and agreed to the published version of the manuscript.

Funding: This research was funded by Sêr Cymru II—WEFO ERDF Programme 80761 for the VFA Factory project.

Institutional Review Board Statement: Not applicable.

Informed Consent Statement: Not applicable.

Data Availability Statement: Not applicable.

Acknowledgments: The authors would like to acknowledge the support of Maria Ramos-Suarez for providing her comments on the manuscript.

Conflicts of Interest: The authors declare no conflict of interest.

References

1. International Energy Agency. Data and Statistics. Available online: https://www.iea.org/data-and-statistics/data-browser?country=UK&fuel=CO2%20emissions&indicator=CO2BySector (accessed on 3 July 2022).
2. Nagarajan, S.; Ranade, V.V. Valorizing Waste Biomass via Hydrodynamic Cavitation and Anaerobic Digestion. *Ind. Eng. Chem. Res.* **2021**, *60*, 16577–16598. [CrossRef]
3. Demir, M.E.; Chehade, G.; Dincer, I.; Yuzer, B.; Selcuk, H. Design and analysis of a new system for photoelectrochemical hydrogen production from wastewater. *Energy Convers. Manag.* **2019**, *199*, 111903. [CrossRef]
4. Nagarajan, S.; Skillen, N.; Robertson, P.; Lawton, L. Cellulose Photocatalysis for Renewable Energy Production. In *Metal, Metal-Oxides and Metal Sulfides for Batteries, Fuel Cells, Solar Cells, Photocatalysis and Health Sensors*; Rajendran, S., Karimi-Maleh, H., Qin, J., Lichtfouse, E., Eds.; Springer International Publishing: Cham, Switzerland, 2021; pp. 1–34.
5. Nagarajan, S.; Skillen, N.C.; Irvine, J.T.S.; Lawton, L.A.; Robertson, P.K.J. Cellulose II as bioethanol feedstock and its advantages over native cellulose. *Renew. Sustain. Energy Rev.* **2017**, *77*, 182–192. [CrossRef]

6. Chang, C.; Skillen, N.; Nagarajan, S.; Ralphs, K.; Irvine, J.T.S.; Lawton, L.; Robertson, P.K.J. Using cellulose polymorphs for enhanced hydrogen production from photocatalytic reforming. *Sustain. Energy Fuels* **2019**, *3*, 1971–1975. [CrossRef]
7. Wakerley, D.W.; Kuehnel, M.F.; Orchard, K.L.; Ly, K.H.; Rosser, T.E.; Reisner, E. Solar-driven reforming of lignocellulose to H_2 with a CdS/CdOx photocatalyst. *Nat. Energy* **2017**, *2*, 17021. [CrossRef]
8. Nagarajan, S. Development of Photocatalytic Reactor Technology for the Production of Fermentable Sugars. Ph.D. Thesis, Queen's University, Belfast, Ireland, 2017.
9. Meynell, P.J. *Methane: Planning a Digester*; Schocken Books: New York, NY, USA, 1976.
10. NNFCC. Biomethane RTFC Calculator. Available online: https://www.nnfcc.co.uk/publications/tool-biomethane-rtfc-calculator (accessed on 6 April 2022).
11. Vaneeckhaute, C.; Lebuf, V.; Michels, E.; Belia, E.; Vanrolleghem, P.A.; Tack, F.M.G.; Meers, E. Nutrient Recovery from Digestate: Systematic Technology Review and Product Classification. *Waste Biomass Valorization* **2017**, *8*, 21–40. [CrossRef]
12. Pang, S. Advances in thermochemical conversion of woody biomass to energy, fuels and chemicals. *Biotechnol. Adv.* **2019**, *37*, 589–597. [CrossRef]
13. Kang, K.; Klinghoffer, N.B.; ElGhamrawy, I.; Berruti, F. Thermochemical conversion of agroforestry biomass and solid waste using decentralized and mobile systems for renewable energy and products. *Renew. Sustain. Energy Rev.* **2021**, *149*, 111372. [CrossRef]
14. Ong, H.C.; Chen, W.-H.; Farooq, A.; Gan, Y.Y.; Lee, K.T.; Ashokkumar, V. Catalytic thermochemical conversion of biomass for biofuel production: A comprehensive review. *Renew. Sustain. Energy Rev.* **2019**, *113*, 109266. [CrossRef]
15. Nagarajan, S.; Stella, L.; Lawton, L.A.; Irvine, J.T.S.; Robertson, P.K.J. Mixing regime simulation and cellulose particle tracing in a stacked frame photocatalytic reactor. *Chem. Eng. J.* **2017**, *313*, 301–308. [CrossRef]
16. Huang, C.-W.; Nguyen, B.-S.; Wu, J.C.S.; Nguyen, V.-H. A current perspective for photocatalysis towards the hydrogen production from biomass-derived organic substances and water. *Int. J. Hydrogen Energy* **2020**, *45*, 18144–18159. [CrossRef]
17. Lu, X.; Xie, S.; Yang, H.; Tong, Y.; Ji, H. Photoelectrochemical hydrogen production from biomass derivatives and water. *Chem. Soc. Rev.* **2014**, *43*, 7581–7593. [CrossRef] [PubMed]
18. Ibrahim, N.; Kamarudin, S.K.; Minggu, L.J. Biofuel from biomass via photo-electrochemical reactions: An overview. *J. Power Sources* **2014**, *259*, 33–42. [CrossRef]
19. Ramos-Suarez, M.; Zhang, Y.; Outram, V. Current perspectives on acidogenic fermentation to produce volatile fatty acids from waste. *Rev. Environ. Sci. Bio/Technol.* **2021**, *20*, 439–478. [CrossRef]
20. UK BEIS. The Ten Point Plan for a Green Industrial Revolution. 2020; Volume 38. Available online: https://assets.publishing.service.gov.uk/government/uploads/system/uploads/attachment_data/file/936567/10_POINT_PLAN_BOOKLET.pdf (accessed on 7 April 2022).
21. UK BEIS. UK Hydrogen Strategy. 2021. Available online: https://assets.publishing.service.gov.uk/government/uploads/system/uploads/attachment_data/file/1011283/UK-Hydrogen-Strategy_web.pdf (accessed on 7 April 2022).
22. UK BEIS. Industrial Decarbonisation Strategy. 2021. Available online: https://www.gov.uk/government/publications/industrial-decarbonisation-strategy (accessed on 7 April 2022).
23. UK DEFRA. A Green Future: Our 25 Year Plan to Improve the Environment. 2018. Available online: https://www.gov.uk/government/publications/25-year-environment-plan (accessed on 7 April 2022).
24. European Commission. A Hydrogen Strategy for a Climate-Neutral Europe. 2020. Available online: https://ec.europa.eu/energy/sites/ener/files/hydrogen_strategy.pdf (accessed on 8 April 2022).
25. Woodward, E.; Han, O.; Forbes, C. *Enabling the European Hydrogen Economy*; Aurora Energy Research: Oxford, UK, 2021.
26. International Energy Agency. *Global Hydrogen Review*; IEA: Paris, France, 2021; p. 224.
27. Braga, L.B.; Silveira, J.L.; da Silva, M.E.; Tuna, C.E.; Machin, E.B.; Pedroso, D.T. Hydrogen production by biogas steam reforming: A technical, economic and ecological analysis. *Renew. Sustain. Energy Rev.* **2013**, *28*, 166–173. [CrossRef]
28. Gorski, J.; Jutt, T.; Tam Wu, K. Carbon intensity of blue hydrogen production. *Pembin. Inst.* **2021**. Available online: https://www.pembina.org/reports/carbon-intensity-of-blue-hydrogen-revised.pdf (accessed on 16 June 2022).
29. Ali Khan, M.H.; Daiyan, R.; Neal, P.; Haque, N.; MacGill, I.; Amal, R. A framework for assessing economics of blue hydrogen production from steam methane reforming using carbon capture storage & utilisation. *Int. J. Hydrogen Energy* **2021**, *46*, 22685–22706.
30. Sun, X.; Atiyeh, H.K.; Huhnke, R.L.; Tanner, R.S. Syngas fermentation process development for production of biofuels and chemicals: A review. *Bioresour. Technol. Rep.* **2019**, *7*, 100279.
31. Ewing, M.; Israel, B.; Jutt, T.; Talebian, H.; Stepanik, L. *Hydrogen on the Path to Net-Zero Emissions–Costs and Climate Benefits*; Pembina Institute: Toronto, ON, Canada, 2020; p. 6.
32. Hitam, C.N.C.; Jalil, A.A. A review on biohydrogen production through photo-fermentation of lignocellulosic biomass. *Biomass Convers. Biorefin.* **2020**, 1–19. [CrossRef]
33. Monlau, F.; Sambusiti, C.; Barakat, A.; Quéméneur, M.; Trably, E.; Steyer, J.P.; Carrère, H. Do furanic and phenolic compounds of lignocellulosic and algae biomass hydrolyzate inhibit anaerobic mixed cultures? A comprehensive review. *Biotechnol. Adv.* **2014**, *32*, 934–951.
34. Bundhoo, Z.M.A. Potential of bio-hydrogen production from dark fermentation of crop residues: A review. *Int. J. Hydrogen Energy* **2019**, *44*, 17346–17362.

35. Rajesh Banu, J.; Merrylin, J.; Mohamed Usman, T.M.; Yukesh Kannah, R.; Gunasekaran, M.; Kim, S.-H.; Kumar, G. Impact of pretreatment on food waste for biohydrogen production: A review. *Int. J. Hydrogen Energy* **2020**, *45*, 18211–18225. [CrossRef]
36. Sun, J.; Zhang, L.; Loh, K.-C. Review and perspectives of enhanced volatile fatty acids production from acidogenic fermentation of lignocellulosic biomass wastes. *Bioresour. Bioprocess.* **2021**, *8*, 68. [CrossRef]
37. Zheng, Y.; Zhang, Q.; Zhang, Z.; Jing, Y.; Hu, J.; He, C.; Lu, C. A review on biological recycling in agricultural waste-based biohydrogen production: Recent developments. *Bioresour. Technol.* **2022**, *347*, 126595. [PubMed]
38. Miao, Z.; Grift, T.E.; Hansen, A.C.; Ting, K.C. Energy requirement for comminution of biomass in relation to particle physical properties. *Ind. Crops Prod.* **2011**, *33*, 504–513. [CrossRef]
39. Konde, K.S.; Nagarajan, S.; Kumar, V.; Patil, S.V.; Ranade, V.V. Sugarcane bagasse based biorefineries in India: Potential and challenges. *Sustain. Energy Fuels* **2021**, *5*, 52–78. [CrossRef]
40. Hendriks, A.T.W.M.; Zeeman, G. Pretreatments to enhance the digestibility of lignocellulosic biomass. *Bioresour. Technol.* **2009**, *100*, 10–18. [CrossRef]
41. Bundhoo, M.A.Z.; Mohee, R.; Hassan, M.A. Effects of pre-treatment technologies on dark fermentative biohydrogen production: A review. *J. Environ. Manag.* **2015**, *157*, 20–48. [CrossRef]
42. Cheng, X.-Y.; Liu, C.-Z. Fungal pretreatment enhances hydrogen production via thermophilic fermentation of cornstalk. *Appl. Energy* **2012**, *91*, 1–6. [CrossRef]
43. Sun, Y.; Cheng, J. Hydrolysis of lignocellulosic materials for ethanol production: A review. *Bioresour. Technol.* **2002**, *83*, 1–11. [CrossRef]
44. Agbor, V.B.; Cicek, N.; Sparling, R.; Berlin, A.; Levin, D.B. Biomass pretreatment: Fundamentals toward application. *Biotechnol. Adv.* **2011**, *29*, 675–685. [CrossRef]
45. Garuti, M.; Langone, M.; Fabbri, C.; Piccinini, S. Monitoring of full-scale hydrodynamic cavitation pretreatment in agricultural biogas plant. *Bioresour. Technol.* **2018**, *247*, 599–609. [CrossRef] [PubMed]
46. Lanfranchi, A.; Tassinato, G.; Valentino, F.; Martinez, G.A.; Jones, E.; Gioia, C.; Bertin, L.; Cavinato, C. Hydrodynamic cavitation pre-treatment of urban waste: Integration with acidogenic fermentation, PHAs synthesis and anaerobic digestion processes. *Chemosphere* **2022**, *301*, 134624. [CrossRef] [PubMed]
47. Nagarajan, S.; Ranade, V.V. Pre-treatment of distillery spent wash (vinasse) with vortex based cavitation and its influence on biogas generation. *Bioresour. Technol. Rep.* **2020**, *11*, 100480. [CrossRef]
48. Nagarajan, S.; Ranade, V.V. Pretreatment of Lignocellulosic Biomass Using Vortex-Based Devices for Cavitation: Influence on Biomethane Potential. *Ind. Eng. Chem. Res.* **2019**, *58*, 15975–15988. [CrossRef]
49. Yukesh Kannah, R.; Kavitha, S.; Sivashanmugham, P.; Kumar, G.; Nguyen, D.D.; Chang, S.W.; Rajesh Banu, J. Biohydrogen production from rice straw: Effect of combinative pretreatment, modelling assessment and energy balance consideration. *Int. J. Hydrogen Energy* **2019**, *44*, 2203–2215. [CrossRef]
50. Deng, C.; Lin, R.; Cheng, J.; Murphy, J.D. Can acid pre-treatment enhance biohydrogen and biomethane production from grass silage in single-stage and two-stage fermentation processes? *Energy Convers. Manag.* **2019**, *195*, 738–747. [CrossRef]
51. Reilly, M.; Dinsdale, R.; Guwy, A. Mesophilic biohydrogen production from calcium hydroxide treated wheat straw. *Int. J. Hydrogen Energy* **2014**, *39*, 16891–16901. [CrossRef]
52. Kucharska, K.; Rybarczyk, P.; Hołowacz, I.; Konopacka-Łyskawa, D.; Słupek, E.; Makoś, P.; Cieśliński, H.; Kamiński, M. Influence of alkaline and oxidative pre-treatment of waste corn cobs on biohydrogen generation efficiency via dark fermentation. *Biomass Bioenergy* **2020**, *141*, 105691. [CrossRef]
53. Wu, J.; Upreti, S.; Ein-Mozaffari, F. Ozone pretreatment of wheat straw for enhanced biohydrogen production. *Int. J. Hydrogen Energy* **2013**, *38*, 10270–10276. [CrossRef]
54. Leaño, E.P.; Babel, S. Effects of pretreatment methods on cassava wastewater for biohydrogen production optimization. *Renew. Energy* **2012**, *39*, 339–346. [CrossRef]
55. Chandrasekhar, K.; Venkata Mohan, S. Bio-electrohydrolysis as a pretreatment strategy to catabolize complex food waste in closed circuitry: Function of electron flux to enhance acidogenic biohydrogen production. *Int. J. Hydrogen Energy* **2014**, *39*, 11411–11422. [CrossRef]
56. Choiron, M.; Tojo, S.; Chosa, T. Biohydrogen production improvement using hot compressed water pretreatment on sake brewery waste. *Int. J. Hydrogen Energy* **2020**, *45*, 17220–17232. [CrossRef]
57. Asadi, N.; Zilouei, H. Optimization of organosolv pretreatment of rice straw for enhanced biohydrogen production using Enterobacter aerogenes. *Bioresour. Technol.* **2017**, *227*, 335–344. [CrossRef] [PubMed]
58. Yang, G.; Wang, J. Enhancing biohydrogen production from disintegrated sewage sludge by combined sodium citrate-thermal pretreatment. *J. Cleaner Prod.* **2021**, *312*, 127756. [CrossRef]
59. Dinesh Kumar, M.; Kaliappan, S.; Gopikumar, S.; Zhen, G.; Rajesh Banu, J. Synergetic pretreatment of algal biomass through H_2O_2 induced microwave in acidic condition for biohydrogen production. *Fuel* **2019**, *253*, 833–839. [CrossRef]
60. Salem, A.H.; Brunstermann, R.; Mietzel, T.; Widmann, R. Effect of pre-treatment and hydraulic retention time on biohydrogen production from organic wastes. *Int. J. Hydrogen Energy* **2018**, *43*, 4856–4865. [CrossRef]
61. Yin, Y.; Wang, J. Pretreatment of macroalgal Laminaria japonica by combined microwave-acid method for biohydrogen production. *Bioresour. Technol.* **2018**, *268*, 52–59. [CrossRef]

62. Hu, Y.; Shen, Y.; Wang, J. Pretreatment of antibiotic fermentation residues by combined ultrasound and alkali for enhancing biohydrogen production. *J. Clean. Prod.* **2020**, *268*, 122190. [CrossRef]
63. Gadhe, A.; Sonawane, S.S.; Varma, M.N. Ultrasonic pretreatment for an enhancement of biohydrogen production from complex food waste. *Int. J. Hydrogen Energy* **2014**, *39*, 7721–7729. [CrossRef]
64. She, Y.; Hong, J.; Zhang, Q.; Chen, B.-Y.; Wei, W.; Xin, X. Revealing microbial mechanism associated with volatile fatty acids production in anaerobic acidogenesis of waste activated sludge enhanced by freezing/thawing pretreatment. *Bioresour. Technol.* **2020**, *302*, 122869. [CrossRef]
65. Zeng, Q.; Zan, F.; Hao, T.; Khanal, S.K.; Chen, G. Sewage sludge digestion beyond biogas: Electrochemical pretreatment for biochemicals. *Water Res.* **2022**, *208*, 117839. [CrossRef] [PubMed]
66. Pham, T.N.; Um, Y.; Yoon, H.H. Pretreatment of macroalgae for volatile fatty acid production. *Bioresour. Technol.* **2013**, *146*, 754–757. [CrossRef] [PubMed]
67. Li, L.; He, J.; Wang, M.; Xin, X.; Xu, J.; Zhang, J. Efficient Volatile Fatty Acids Production from Waste Activated Sludge after Ferrate Pretreatment with Alkaline Environment and the Responding Microbial Community Shift. *ACS Sustain. Chem. Eng.* **2018**, *6*, 16819–16827. [CrossRef]
68. Tao, X.; Zhang, P.; Zhang, G.; Nabi, M.; Wang, S.; Ye, J.; Bao, S.; Zhang, Q.; Chen, N. Carbide slag pretreatment enhances volatile fatty acid production in anaerobic fermentation of four grass biomasses. *Energy Convers. Manag.* **2019**, *199*, 112009. [CrossRef]
69. Wu, Q.-L.; Guo, W.-Q.; Bao, X.; Yin, R.-L.; Feng, X.-C.; Zheng, H.-S.; Luo, H.-C.; Ren, N.-Q. Enhancing sludge biodegradability and volatile fatty acid production by tetrakis hydroxymethyl phosphonium sulfate pretreatment. *Bioresour. Technol.* **2017**, *239*, 518–522. [CrossRef]
70. Han, F.; Wu, L. Comprehensive Utilization of Carbide Slag. In *Industrial Solid Waste Recycling in Western China*; Han, F., Wu, L.E., Eds.; Springer: Singapore, 2019; pp. 357–391.
71. Fang, W.; Zhang, P.; Zhang, X.; Zhu, X.; van Lier, J.B.; Spanjers, H. White rot fungi pretreatment to advance volatile fatty acid production from solid-state fermentation of solid digestate: Efficiency and mechanisms. *Energy* **2018**, *162*, 534–541. [CrossRef]
72. Fang, W.; Zhang, X.; Zhang, P.; Carol Morera, X.; van Lier, J.B.; Spanjers, H. Evaluation of white rot fungi pretreatment of mushroom residues for volatile fatty acid production by anaerobic fermentation: Feedstock applicability and fungal function. *Bioresour. Technol.* **2020**, *297*, 122447. [CrossRef]
73. Bahreini, G.; Nazari, L.; Ho, D.; Flannery, C.C.; Elbeshbishy, E.; Santoro, D.; Nakhla, G. Enzymatic pre-treatment for enhancement of primary sludge fermentation. *Bioresour. Technol.* **2020**, *305*, 123071. [CrossRef]
74. Owusu-Agyeman, I.; Balachandran, S.; Plaza, E.; Cetecioglu, Z. Co-fermentation of municipal waste streams: Effects of pretreatment methods on volatile fatty acids production. *Biomass Bioenergy* **2021**, *145*, 105950. [CrossRef]
75. Kakar, F.l.; Koupaie, E.H.; Razavi, A.S.; Hafez, H.; Elbeshbishy, E. Effect of Hydrothermal Pretreatment on Volatile Fatty Acids Production from Thickened Waste Activated Sludge. *BioEnergy Res.* **2020**, *13*, 591–604. [CrossRef]
76. Suresh, A.; Seo, C.; Chang, H.N.; Kim, Y.-C. Improved volatile fatty acid and biomethane production from lipid removed microalgal residue (LRμAR) through pretreatment. *Bioresour. Technol.* **2013**, *149*, 590–594. [CrossRef] [PubMed]
77. Yu, Z.; Ma, H.; Boer, E.d.; Wu, W.; Wang, Q.; Gao, M.; Vo, D.-V.N.; Guo, M.; Xia, C. Effect of microwave/hydrothermal combined ionic liquid pretreatment on straw: Rumen anaerobic fermentation and enzyme hydrolysis. *Environ. Res.* **2022**, *205*, 112453. [CrossRef] [PubMed]
78. Guo, W.; Wu, Q.; Yang, S.; Luo, H.; Peng, S.; Ren, N. Optimization of ultrasonic pretreatment and substrate/inoculum ratio to enhance hydrolysis and volatile fatty acid production from food waste. *RSC Adv.* **2014**, *4*, 53321–53326. [CrossRef]
79. Liu, N.; Jiang, J.; Yan, F.; Gao, Y.; Meng, Y.; Aihemaiti, A.; Ju, T. Enhancement of volatile fatty acid production and biogas yield from food waste following sonication pretreatment. *J. Environ. Manag.* **2018**, *217*, 797–804. [CrossRef]
80. Wang, S.; Tao, X.; Zhang, G.; Zhang, P.; Wang, H.; Ye, J.; Li, F.; Zhang, Q.; Nabi, M. Benefit of solid-liquid separation on volatile fatty acid production from grass clipping with ultrasound-calcium hydroxide pretreatment. *Bioresour. Technol.* **2019**, *274*, 97–104. [CrossRef]
81. Wan, J.; Fang, W.; Zhang, T.; Wen, G. Enhancement of fermentative volatile fatty acids production from waste activated sludge by combining sodium dodecylbenzene sulfonate and low-thermal pretreatment. *Bioresour. Technol.* **2020**, *308*, 123291. [CrossRef]
82. Baeyens, J.; Zhang, H.; Nie, J.; Appels, L.; Dewil, R.; Ansart, R.; Deng, Y. Reviewing the potential of bio-hydrogen production by fermentation. *Renew. Sustain. Energy Rev.* **2020**, *131*, 110023. [CrossRef]
83. Elsharnouby, O.; Hafez, H.; Nakhla, G.; El Naggar, M.H. A critical literature review on biohydrogen production by pure cultures. *Int. J. Hydrogen Energy* **2013**, *38*, 4945–4966. [CrossRef]
84. Krupp, M.; Widmann, R. Biohydrogen production by dark fermentation: Experiences of continuous operation in large lab scale. *Int. J. Hydrogen Energy* **2009**, *34*, 4509–4516. [CrossRef]
85. Mosey, F.E. *Mathematical Modelling of the Anaerobic Digestion Process: Regulatory Mechanisms for the Forma Tion of Short-Chain Volatile Acids From Glucose*; IWA: Nanjing, China, 1983; pp. 209–232.
86. Zhou, M.; Yan, B.; Wong, J.W.C.; Zhang, Y. Enhanced volatile fatty acids production from anaerobic fermentation of food waste: A mini-review focusing on acidogenic metabolic pathways. *Bioresour. Technol.* **2018**, *248*, 68–78. [CrossRef]
87. Jones, R.J.; Massanet-Nicolau, J.; Fernandez-Feito, R.; Dinsdale, R.M.; Guwy, A.J. Recovery and enhanced yields of volatile fatty acids from a grass fermentation via in-situ solids separation and electrodialysis. *J. Clean. Prod.* **2021**, *296*, 126430. [CrossRef]

88. Greses, S.; Tomás-Pejó, E.; González-Fernández, C. Short-chain fatty acids and hydrogen production in one single anaerobic fermentation stage using carbohydrate-rich food waste. *J. Clean. Prod.* **2021**, *284*, 124727. [CrossRef]
89. Van Ginkel, S.; Logan, B.E. Inhibition of biohydrogen production by undissociated acetic and butyric acids. *Environ. Sci. Technol.* **2005**, *39*, 9351–9356. [CrossRef] [PubMed]
90. Zhang, F.; Chen, Y.; Dai, K.; Shen, N.; Zeng, R.J. The glucose metabolic distribution in thermophilic (55 °C) mixed culture fermentation: A chemostat study. *Int. J. Hydrogen Energy* **2015**, *40*, 919–926. [CrossRef]
91. Kirli, B.; Karapinar, I. The effect of HRT on biohydrogen production from acid hydrolyzed waste wheat in a continuously operated packed bed reactor. *Int. J. Hydrogen Energy* **2018**, *43*, 10678–10685. [CrossRef]
92. Nagarajan, S.; Prasad Sarvothaman, V.; Knörich, M.; Ranade, V.V. A simplified model for simulating anaerobic digesters: Application to valorisation of bagasse and distillery spent wash. *Bioresour. Technol.* **2021**, *337*, 125395. [CrossRef]
93. Kumar, G.; Sivagurunathan, P.; Park, J.H.; Park, J.H.; Park, H.D.; Yoon, J.J.; Kim, S.H. HRT dependent performance and bacterial community population of granular hydrogen-producing mixed cultures fed with galactose. *Bioresour. Technol.* **2016**, *206*, 188–194. [CrossRef]
94. Massanet-Nicolau, J.; Dinsdale, R.; Guwy, A. Hydrogen production from sewage sludge using mixed microflora inoculum: Effect of pH and enzymatic pretreatment. *Bioresour. Technol.* **2008**, *99*, 6325–6331. [CrossRef]
95. Moreno-Andrade, I.; Carrillo-Reyes, J.; Santiago, S.G.; Bujanos-Adame, M.C. Biohydrogen from food waste in a discontinuous process: Effect of HRT and microbial community analysis. *J. Hydrogen Energy* **2015**, *30*, 17246–17252. [CrossRef]
96. Khan, M.A.; Ngo, H.H.; Guo, W.; Liu, Y.; Nghiem, L.D.; Chang, S.W.; Nguyen, D.D.; Zhang, S.; Luo, G.; Jia, H. Optimization of hydraulic retention time and organic loading rate for volatile fatty acid production from low strength wastewater in an anaerobic membrane bioreactor. *Bioresour. Technol.* **2019**, *271*, 100–108. [CrossRef]
97. Lv, N.; Cai, G.; Pan, X.; Li, Y.; Wang, R.; Li, J.; Li, C.; Zhu, G. pH and hydraulic retention time regulation for anaerobic fermentation: Focus on volatile fatty acids production/distribution, microbial community succession and interactive correlation. *Bioresour. Technol.* **2022**, *347*, 126310. [CrossRef] [PubMed]
98. Moretto, G.; Valentino, F.; Pavan, P.; Majone, M.; Bolzonella, D. Optimization of urban waste fermentation for volatile fatty acids production. *Waste Manag.* **2019**, *92*, 21–29. [CrossRef] [PubMed]
99. Jankowska, E.; Duber, A.; Chwialkowska, J.; Stodolny, M.; Oleskowicz-Popiel, P. Conversion of organic waste into volatile fatty acids–The influence of process operating parameters. *Chem. Eng. J.* **2018**, *345*, 395–403. [CrossRef]
100. Ferraz Júnior, A.Ô.D.N.; Zaiat, M.; Gupta, M.; Elbeshbishy, E.; Hafez, H.; Nakhla, G. Impact of organic loading rate on biohydrogen production in an up-flow anaerobic packed bed reactor (UAnPBR). *Bioresour. Technol.* **2014**, *164*, 371–379. [CrossRef]
101. Fuess, L.T.; Mazine Kiyuna, L.S.; Garcia, M.L.; Zaiat, M. Operational strategies for long-term biohydrogen production from sugarcane stillage in a continuous acidogenic packed-bed reactor. *Int. J. Hydrogen Energy* **2016**, *41*, 8132–8145. [CrossRef]
102. Tang, J.; Wang, X.; Hu, Y.; Zhang, Y.; Li, Y. Lactic acid fermentation from food waste with indigenous microbiota: Effects of pH, temperature and high OLR. *Waste Manag.* **2016**, *52*, 278–285. [CrossRef]
103. Iglesias-Iglesias, R.; Campanaro, S.; Treu, L.; Kennes, C.; Veiga, M.C. Valorization of sewage sludge for volatile fatty acids production and role of microbiome on acidogenic fermentation. *Bioresour. Technol.* **2019**, *291*, 121817. [CrossRef]
104. Magdalena, J.A.; Greses, S.; González-Fernández, C. Impact of Organic Loading Rate in Volatile Fatty Acids Production and Population Dynamics Using Microalgae Biomass as Substrate. *Sci. Rep.* **2019**, *9*, 18374. [CrossRef]
105. Zulkhairi, M.; Yusoff, M.; Rahman, N.a.A.; Abd-Aziz, S.; Ling, C.M.; Hassan, M.A.; Shirai, Y. The Effect of Hydraulic Retention Time and Volatile Fatty Acids on Biohydrogen Production from Palm Oil Mill Effluent under Non Sterile Condition. *Aust. J. Basic Appl. Sci.* **2010**, *4*, 577–587.
106. Rincón, B.; Sánchez, E.; Raposo, F.; Borja, R.; Travieso, L.; Martín, M.A.; Martín, A. Effect of the organic loading rate on the performance of anaerobic acidogenic fermentation of two-phase olive mill solid residue. *Waste Manag.* **2008**, *28*, 870–877. [CrossRef]
107. Malinowsky, C.; Nadaleti, W.; Debiasi, L.R.; Gonçalves Moreira, A.J.; Bayard, R.; Borges de Castilhos Junior, A. Start-up phase optimization of two-phase anaerobic digestion of food waste: Effects of organic loading rate and hydraulic retention time. *J. Environ. Manag.* **2021**, *296*, 113064. [CrossRef] [PubMed]
108. De Groof, V.; Coma, M.; Arnot, T.C.; Leak, D.J.; Lanham, A.B. Adjusting Organic Load as a Strategy to Direct Single-Stage Food Waste Fermentation from Anaerobic Digestion to Chain Elongation. *Processes* **2020**, *8*, 1487. [CrossRef]
109. Mao, C.; Feng, Y.; Wang, X.; Ren, G. Review on research achievements of biogas from anaerobic digestion. *Renew. Sustain. Energy Rev.* **2015**, *45*, 540–555. [CrossRef]
110. Wainaina, S.; Lukitawesa; Kumar Awasthi, M.; Taherzadeh, M.J. Bioengineering of anaerobic digestion for volatile fatty acids, hydrogen or methane production: A critical review. *Bioengineered* **2019**, *10*, 437–458. [CrossRef]
111. Luo, G.; Karakashev, D.; Xie, L.; Zhou, Q.; Angelidaki, I. Long-term effect of inoculum pretreatment on fermentative hydrogen production by repeated batch cultivations: Homoacetogenesis and methanogenesis as competitors to hydrogen production. *Biotechnol. Bioeng.* **2011**, *108*, 1816–1827. [CrossRef]
112. Chinellato, G.; Cavinato, C.; Bolzonella, D.; Heaven, S.; Banks, C.J. Biohydrogen production from food waste in batch and semi-continuous conditions: Evaluation of a two-phase approach with digestate recirculation for pH control. *Int. J. Hydrogen Energy* **2013**, *38*, 4351–4360. [CrossRef]

113. Doi, T.; Matsumoto, H.; Abe, J.; Morita, S. Feasibility study on the application of rhizosphere microflora of rice for the biohydrogen production from wasted bread. *Int. J. Hydrogen Energy* **2009**, *34*, 1735–1743. [CrossRef]
114. Kim, S.H.; Han, S.K.; Shin, H.S. Feasibility of biohydrogen production by anaerobic co-digestion of food waste and sewage sludge. *Int. J. Hydrogen Energy* **2004**, *29*, 1607–1616. [CrossRef]
115. Dareioti, M.A.; Vavouraki, A.I.; Kornaros, M. Effect of pH on the anaerobic acidogenesis of agroindustrial wastewaters for maximization of bio-hydrogen production: A lab-scale evaluation using batch tests. *Bioresour. Technol.* **2014**, *162*, 218–227. [CrossRef]
116. Yokoyama, H.; Waki, M.; Moriya, N.; Yasuda, T.; Tanaka, Y.; Haga, K. Effect of fermentation temperature on hydrogen production from cow waste slurry by using anaerobic microflora within the slurry. *Appl. Microbiol. Biotechnol.* **2007**, *74*, 474–483. [CrossRef]
117. Zhang, M.L.; Fan, Y.T.; Xing, Y.; Pan, C.M.; Zhang, G.S.; Lay, J.J. Enhanced biohydrogen production from cornstalk wastes with acidification pretreatment by mixed anaerobic cultures. *Biomass Bioenergy* **2007**, *31*, 250–254. [CrossRef]
118. Tsigkou, K.; Tsafrakidou, P.; Athanasopoulou, S.; Zafiri, C.; Kornaros, M. Effect of pH on the Anaerobic Fermentation of Fruit/Vegetables and Disposable Nappies Hydrolysate for Bio-hydrogen Production. *Waste Biomass Valorization* **2020**, *11*, 539–551. [CrossRef]
119. Guo, X.M.; Trably, E.; Latrille, E.; Carrre, H.; Steyer, J.P. Hydrogen production from agricultural waste by dark fermentation: A review. *Int. J. Hydrogen Energy* **2010**, *35*, 10660–10673. [CrossRef]
120. Cabrera, F.; Serrano, A.; Torres, Á.; Rodriguez-Gutierrez, G.; Jeison, D.; Fermoso, F.G. The accumulation of volatile fatty acids and phenols through a pH-controlled fermentation of olive mill solid waste. *Sci. Total Environ.* **2019**, *657*, 1501–1507. [CrossRef]
121. Jankowska, E.; Chwialkowska, J.; Stodolny, M.; Oleskowicz-Popiel, P. Volatile fatty acids production during mixed culture fermentation–The impact of substrate complexity and pH. *Chem. Eng. J.* **2017**, *326*, 901–910. [CrossRef]
122. Lu, Y.; Zhang, Q.; Wang, X.; Zhou, X.; Zhu, J. Effect of pH on volatile fatty acid production from anaerobic digestion of potato peel waste. *Bioresour. Technol.* **2020**, *316*, 123851. [CrossRef]
123. Ye, M.; Luo, J.; Zhang, S.; Yang, H.; Li, Y.-Y.; Liu, J. In-situ ammonia stripping with alkaline fermentation of waste activated sludge to improve short-chain fatty acids production and carbon source availability. *Bioresour. Technol.* **2020**, *301*, 122782. [CrossRef]
124. Atasoy, M.; Eyice, O.; Schnürer, A.; Cetecioglu, Z. Volatile fatty acids production via mixed culture fermentation: Revealing the link between pH, inoculum type and bacterial composition. *Bioresour. Technol.* **2019**, *292*, 121889. [CrossRef]
125. Begum, S.; Anupoju, G.R.; Sridhar, S.; Bhargava, S.K.; Jegatheesan, V.; Eshtiaghi, N. Evaluation of single and two stage anaerobic digestion of landfill leachate: Effect of pH and initial organic loading rate on volatile fatty acid (VFA) and biogas production. *Bioresour. Technol.* **2018**, *251*, 364–373. [CrossRef]
126. Nguyen, P.K.T.; Kim, J.; Das, G.; Yoon, H.H.; Lee, D.H. Optimization of simultaneous dark fermentation and microbial electrolysis cell for hydrogen production from macroalgae using response surface methodology. *Biochem. Eng. J.* **2021**, *171*, 108029. [CrossRef]
127. Connaughton, S.; Collins, G.; O'Flaherty, V. Psychrophilic and mesophilic anaerobic digestion of brewery effluent: A comparative study. *Water Res.* **2006**, *40*, 2503–2510. [CrossRef] [PubMed]
128. Tiwari, B.R.; Rouissi, T.; Brar, S.K.; Surampalli, R.Y. Critical insights into psychrophilic anaerobic digestion: Novel strategies for improving biogas production. *Waste Manag.* **2021**, *131*, 513–526. [CrossRef] [PubMed]
129. Chen, H.; Wu, J.; Huang, R.; Zhang, W.; He, W.; Deng, Z.; Han, Y.; Xiao, B.; Luo, H.; Qu, W. Effects of temperature and total solid content on biohydrogen production from dark fermentation of rice straw: Performance and microbial community characteristics. *Chemosphere* **2022**, *286*, 131655. [CrossRef]
130. Karlsson, A.; Vallin, L.; Ejlertsson, J. Effects of temperature, hydraulic retention time and hydrogen extraction rate on hydrogen production from the fermentation of food industry residues and manure. *Int. J. Hydrogen Energy* **2008**, *33*, 953–962. [CrossRef]
131. Zheng, H.-S.; Guo, W.-Q.; Yang, S.-S.; Feng, X.-C.; Du, J.-S.; Zhou, X.-J.; Chang, J.-S.; Ren, N.-Q. Thermophilic hydrogen production from sludge pretreated by thermophilic bacteria: Analysis of the advantages of microbial community and metabolism. *Bioresour. Technol.* **2014**, *172*, 433–437. [CrossRef] [PubMed]
132. Ziara, R.M.M.; Miller, D.N.; Subbiah, J.; Dvorak, B.I. Lactate wastewater dark fermentation: The effect of temperature and initial pH on biohydrogen production and microbial community. *Int. J. Hydrogen Energy* **2019**, *44*, 661–673. [CrossRef]
133. Huang, C.; Wang, W.; Sun, X.; Shen, J.; Wang, L. A novel acetogenic bacteria isolated from waste activated sludge and its potential application for enhancing anaerobic digestion performance. *J. Environ. Manag.* **2020**, *255*, 109842. [CrossRef]
134. Morgan-Sagastume, F.; Hjort, M.; Cirne, D.; Gérardin, F.; Lacroix, S.; Gaval, G.; Karabegovic, L.; Alexandersson, T.; Johansson, P.; Karlsson, A.; et al. Integrated production of polyhydroxyalkanoates (PHAs) with municipal wastewater and sludge treatment at pilot scale. *Bioresour. Technol.* **2015**, *181*, 78–89. [CrossRef]
135. Garcia-Aguirre, J.; Aymerich, E.; de Goñi, J.G.-M.; Esteban-Gutiérrez, M. Selective VFA production potential from organic waste streams: Assessing temperature and pH influence. *Bioresour. Technol.* **2017**, *244*, 1081–1088. [CrossRef]
136. Xiao, B.; Qin, Y.; Wu, J.; Chen, H.; Yu, P.; Liu, J.; Li, Y.Y. Comparison of single-stage and two-stage thermophilic anaerobic digestion of food waste: Performance, energy balance and reaction process. *Energy Convers. Manag.* **2018**, *156*, 215–223. [CrossRef]
137. Zhang, C.; Su, H.; Tan, T. Batch and semi-continuous anaerobic digestion of food waste in a dual solid-liquid system. *Bioresour. Technol.* **2013**, *145*, 10–16. [CrossRef] [PubMed]
138. Zhang, Z.P.; Show, K.Y.; Tay, J.H.; Liang, D.T.; Lee, D.J. Biohydrogen production with anaerobic fluidized bed reactors-A comparison of biofilm-based and granule-based systems. *Int. J. Hydrogen Energy* **2008**, *33*, 1559–1564. [CrossRef]

139. Brar, K.K.; Cortez, A.A.; Pellegrini, V.O.A.; Amulya, K.; Polikarpov, I.; Magdouli, S.; Kumar, M.; Yang, Y.-H.; Bhatia, S.K.; Brar, S.K. An overview on progress, advances, and future outlook for biohydrogen production technology. *Int. J. Hydrogen Energy* 2022, in press. [CrossRef]
140. Castillo-Hernández, A.; Mar-Alvarez, I.; Moreno-Andrade, I. Start-up and operation of continuous stirred-tank reactor for biohydrogen production from restaurant organic solid waste. *Int. J. Hydrogen Energy* **2015**, *2015*, 17239–17245. [CrossRef]
141. Li, Q.; Li, Y. Coproduction of hydrogen and methane in a CSTR-IC two-stage anaerobic digestion system from molasses wastewater. *Water Sci. Technol.* **2019**, *79*, 270–277. [CrossRef]
142. Ri, P.C.; Kim, J.S.; Kim, T.R.; Pang, C.H.; Mun, H.G.; Pak, G.C.; Ren, N.Q. Effect of hydraulic retention time on the hydrogen production in a horizontal and vertical continuous stirred-tank reactor. *Int. J. Hydrogen Energy* **2019**, *44*, 17742–17749. [CrossRef]
143. Tena, M.; Perez, M.; Solera, R. Effect of hydraulic retention time on hydrogen production from sewage sludge and wine vinasse in a thermophilic acidogenic CSTR: A promising approach for hydrogen production within the biorefinery concept. *Int. J. Hydrogen Energy* **2021**, *46*, 7810–7820. [CrossRef]
144. Greses, S.; Tomás-Pejó, E.; Gónzalez-Fernández, C. Agroindustrial waste as a resource for volatile fatty acids production via anaerobic fermentation. *Bioresour. Technol.* **2020**, *297*, 122486. [CrossRef]
145. Jones, R.J.; Fernández-feito, R.; Massanet-nicolau, J.; Dinsdale, R.; Guwy, A. Continuous recovery and enhanced yields of volatile fatty acids from a continually-fed 100 L food waste bioreactor by filtration and electrodialysis. *Waste Manag.* **2021**, *122*, 81–88. [CrossRef]
146. Brindhadevi, K.; Shanmuganathan, R.; Pugazhendhi, A.; Gunasekar, P.; Manigandan, S. Biohydrogen production using horizontal and vertical continuous stirred tank reactor- a numerical optimization. *Int. J. Hydrogen Energy* **2021**, *46*, 11305–11312. [CrossRef]
147. Keskin, T.; Giusti, L.; Azbar, N. Continuous biohydrogen production in immobilized biofilm system versus suspended cell culture. *Int. J. Hydrogen Energy* **2012**, *37*, 1418–1424. [CrossRef]
148. Anjana Anand, A.S.; Adish Kumar, S.; Rajesh Banu, J.; Ginni, G. The performance of fluidized bed solar photo Fenton oxidation in the removal of COD from hospital wastewaters. *Desalin. Water Treat.* **2016**, *57*, 8236–8242. [CrossRef]
149. Chaves, T.C.; Gois, G.N.S.B.; Peiter, F.S.; Vich, D.V.; de Amorim, E.L.C. Biohydrogen production in an AFBR using sugarcane molasses. *Bioprocess. Biosyst. Eng.* **2021**, *44*, 307–316. [CrossRef] [PubMed]
150. Zhang, Z.; Tay, J.; Show, K.; Yan, R.; Teeliang, D.; Lee, D.; Jiang, W. Biohydrogen production in a granular activated carbon anaerobic fluidized bed reactor. *Int. J. Hydrogen Energy* **2007**, *32*, 185–191. [CrossRef]
151. Barros, A.R.; Cavalcante de Amorim, E.L.; Reis, C.M.; Shida, G.M.; Silva, E.L. Biohydrogen production in anaerobic fluidized bed reactors: Effect of support material and hydraulic retention time. *Int. J. Hydrogen Energy* **2010**, *35*, 3379–3388. [CrossRef]
152. Amorim, N.C.S.; Alves, I.; Martins, J.S.; Amorim, E.L.C. Biohydrogen production from cassava wastewater in an anaerobic fluidized bed reactor. *Braz. J. Chem. Eng.* **2014**, *31*, 603–612. [CrossRef]
153. Maaroff, R.M.; Md Jahim, J.; Azahar, A.M.; Abdul, P.M.; Masdar, M.S.; Nordin, D.; Abd Nasir, M.A. Biohydrogen production from palm oil mill effluent (POME) by two stage anaerobic sequencing batch reactor (ASBR) system for better utilization of carbon sources in POME. *Int. J. Hydrogen Energy* **2019**, *44*, 3395–3406. [CrossRef]
154. Santiago, S.G.; Morgan-Sagastume, J.M.; Monroy, O.; Moreno-Andrade, I. Biohydrogen production from organic solid waste in a sequencing batch reactor: An optimization of the hydraulic and solids retention time. *Int. J. Hydrogen Energy* **2020**, *45*, 25681–25688. [CrossRef]
155. Lagoa-Costa, B.; Kennes, C.; Veiga, M.C. Cheese whey fermentation into volatile fatty acids in an anaerobic sequencing batch reactor. *Bioresour. Technol.* **2020**, *308*, 123226. [CrossRef]
156. Show, K.Y.; Lee, D.J.; Tay, J.H.; Lin, C.Y.; Chang, J.S. Biohydrogen production: Current perspectives and the way forward. *Int. J. Hydrogen Energy* **2012**, *37*, 15616–15631. [CrossRef]
157. Anzola-Rojas, M.d.P.; Gonçalves da Fonseca, S.; Canedo da Silva, C.; Maia de Oliveira, V.; Zaiat, M. The use of the carbon/nitrogen ratio and specific organic loading rate as tools for improving biohydrogen production in fixed-bed reactors. *Biotechnol. Rep.* **2015**, *5*, 46–54. [CrossRef] [PubMed]
158. Tomczak, W.; Ferrasse, J.-H.; Giudici-Orticoni, M.-T.; Soric, A. Effect of hydraulic retention time on a continuous biohydrogen production in a packed bed biofilm reactor with recirculation flow of the liquid phase. *Int. J. Hydrogen Energy* **2018**, *43*, 18883–18895. [CrossRef]
159. Muri, P.; Marinšek-Logar, R.; Djinović, P.; Pintar, A. Influence of support materials on continuous hydrogen production in anaerobic packed-bed reactor with immobilized hydrogen producing bacteria at acidic conditions. *Enzyme Microb. Technol.* **2018**, *111*, 87–96. [CrossRef] [PubMed]
160. Latif, M.A.; Ghufran, R.; Wahid, Z.A.; Ahmad, A. Integrated application of upflow anaerobic sludge blanket reactor for the treatment of wastewaters. *Water Res.* **2011**, *45*, 4683–4699. [CrossRef]
161. Castelló, E.; García y Santos, C.; Iglesias, T.; Paolino, G.; Wenzel, J.; Borzacconi, L.; Etchebehere, C. Feasibility of biohydrogen production from cheese whey using a UASB reactor: Links between microbial community and reactor performance. *Int. J. Hydrogen Energy* **2009**, *34*, 5674–5682. [CrossRef]
162. Eregowda, T.; Kokko, M.E.; Rene, E.R.; Rintala, J.; Lens, P.N.L. Volatile fatty acid production from Kraft mill foul condensate in upflow anaerobic sludge blanket reactors. *Environ. Technol.* **2021**, *42*, 2447–2460. [CrossRef]

163. Intanoo, P.; Chaimongkol, P.; Chavadej, S. Hydrogen and methane production from cassava wastewater using two-stage upflow anaerobic sludge blanket reactors (UASB) with an emphasis on maximum hydrogen production. *Int. J. Hydrogen Energy* **2016**, *41*, 6107–6114. [CrossRef]
164. Jung, K.-W.; Kim, D.-H.; Shin, H.-S. Continuous fermentative hydrogen production from coffee drink manufacturing wastewater by applying UASB reactor. *Int. J. Hydrogen Energy* **2010**, *35*, 13370–13378. [CrossRef]
165. Yang, H.; Shao, P.; Lu, T.; Shen, J.; Wang, D.; Xu, Z.; Yuan, X. Continuous bio-hydrogen production from citric acid wastewater via facultative anaerobic bacteria. *Int. J. Hydrogen Energy* **2006**, *31*, 1306–1313. [CrossRef]
166. Estevez, M.M.; Linjordet, R.; Morken, J. Effects of steam explosion and co-digestion in the methane production from Salix by mesophilic batch assays. *Bioresour. Technol.* **2012**, *104*, 749–756. [CrossRef]
167. Argun, H.; Kargi, F.; Kapdan, I.K.; Oztekin, R. Biohydrogen production by dark fermentation of wheat powder solution: Effects of C/N and C/P ratio on hydrogen yield and formation rate. *Int. J. Hydrogen Energy* **2008**, *33*, 1813–1819. [CrossRef]
168. Lin, C.Y.; Lay, C.H. Carbon/nitrogen-ratio effect on fermentative hydrogen production by mixed microflora. *Int. J. Hydrogen Energy* **2004**, *29*, 41–45. [CrossRef]
169. Sun, Y.; He, J.; Yang, G.; Sun, G.; Sage, V. A Review of the Enhancement of Bio-Hydrogen Generation by Chemicals Addition. *Catalysts* **2019**, *9*, 353. [CrossRef]
170. Sun, Y.; Wang, Y.; Yang, G.; Sun, Z. Optimization of biohydrogen production using acid pretreated corn stover hydrolysate followed by nickel nanoparticle addition. *Int. J. Energy Res.* **2020**, *44*, 1843–1857. [CrossRef]
171. Sinharoy, A.; Pakshirajan, K. A novel application of biologically synthesized nanoparticles for enhanced biohydrogen production and carbon monoxide bioconversion. *Renew. Energy* **2020**, *147*, 864–873. [CrossRef]
172. Zhang, H.; Xu, Y.; Tian, Y.; Zheng, L.; Hao, H.; Huang, H. Impact of Fe and Ni Addition on the VFAs' Generation and Process Stability of Anaerobic Fermentation Containing Cd. *Int. J. Environ. Res. Public Health* **2019**, *16*, 4066. [CrossRef]
173. Dahiya, S.; Lakshminarayanan, S.; Venkata Mohan, S. Steering acidogenesis towards selective propionic acid production using co-factors and evaluating environmental sustainability. *Chem. Eng. J.* **2020**, *379*, 122135. [CrossRef]
174. Zhao, L.; Chen, C.; Ren, H.-Y.; Wu, J.-T.; Meng, J.; Nan, J.; Cao, G.-L.; Yang, S.-S.; Ren, N.-Q. Feasibility of enhancing hydrogen production from cornstalk hydrolysate anaerobic fermentation by RCPH-biochar. *Bioresour. Technol.* **2020**, *297*, 122505. [CrossRef]
175. Sugiarto, Y.; Sunyoto, N.M.S.; Zhu, M.; Jones, I.; Zhang, D. Effect of biochar in enhancing hydrogen production by mesophilic anaerobic digestion of food wastes: The role of minerals. *Int. J. Hydrogen Energy* **2021**, *46*, 3695–3703. [CrossRef]
176. Lu, J.-H.; Chen, C.; Huang, C.; Zhuang, H.; Leu, S.-Y.; Lee, D.-J. Dark fermentation production of volatile fatty acids from glucose with biochar amended biological consortium. *Bioresour. Technol.* **2020**, *303*, 122921. [CrossRef]
177. Taheri, E.; Amin, M.M.; Fatehizadeh, A.; Pourzamani, H.; Bina, B.; Spanjers, H. Biohydrogen production under hyper salinity stress by an anaerobic sequencing batch reactor with mixed culture. *J. Environ. Health Sci. Eng.* **2018**, *16*, 159–170. [CrossRef] [PubMed]
178. Sarkar, O.; Kiran Katari, J.; Chatterjee, S.; Venkata Mohan, S. Salinity induced acidogenic fermentation of food waste regulates biohydrogen production and volatile fatty acids profile. *Fuel* **2020**, *276*, 117794. [CrossRef]
179. He, X.; Yin, J.; Liu, J.; Chen, T.; Shen, D. Characteristics of acidogenic fermentation for volatile fatty acid production from food waste at high concentrations of NaCl. *Bioresour. Technol.* **2019**, *271*, 244–250. [CrossRef] [PubMed]
180. Huang, X.; Liu, X.; Chen, F.; Wang, Y.; Li, X.; Wang, D.; Tao, Z.; Xu, D.; Xue, W.; Geng, M.; et al. Clarithromycin affect methane production from anaerobic digestion of waste activated sludge. *J. Clean. Prod.* **2020**, *255*, 120321. [CrossRef]
181. Huang, X.; Xu, Q.; Wu, Y.; Wang, D.; Yang, Q.; Chen, F.; Wu, Y.; Pi, Z.; Chen, Z.; Li, X.; et al. Effect of clarithromycin on the production of volatile fatty acids from waste activated sludge anaerobic fermentation. *Bioresour. Technol.* **2019**, *288*, 121598. [CrossRef]
182. Chen, H.; Zeng, X.; Zhou, Y.; Yang, X.; Lam, S.S.; Wang, D. Influence of roxithromycin as antibiotic residue on volatile fatty acids recovery in anaerobic fermentation of waste activated sludge. *J. Hazard. Mater.* **2020**, *394*, 122570. [CrossRef]
183. Tao, Z.; Yang, Q.; Yao, F.; Huang, X.; Wu, Y.; Du, M.; Chen, S.; Liu, X.; Li, X.; Wang, D. The inhibitory effect of thiosulfinate on volatile fatty acid and hydrogen production from anaerobic co-fermentation of food waste and waste activated sludge. *Bioresour. Technol.* **2020**, *297*, 122428. [CrossRef]
184. Jones, R.J.; Massanet-Nicolau, J.; Mulder, M.J.J.; Premier, G.; Dinsdale, R.; Guwy, A. Increased biohydrogen yields, volatile fatty acid production and substrate utilisation rates via the electrodialysis of a continually fed sucrose fermenter. *Bioresour. Technol.* **2017**, *229*, 46–52. [CrossRef]
185. Hassan, G.K.; Jones, R.J.; Massanet-Nicolau, J.; Dinsdale, R.; Abo-Aly, M.M.; El-Gohary, F.A.; Guwy, A. Increasing 2 -Bio- (H2 and CH4) production from food waste by combining two-stage anaerobic digestion and electrodialysis for continuous volatile fatty acids removal. *Waste Manag.* **2021**, *129*, 20–25. [CrossRef]
186. Bundhoo, M.A.Z.; Mohee, R. Inhibition of dark fermentative bio-hydrogen production: A review. *Int. J. Hydrogen Energy* **2016**, *41*, 6713–6733. [CrossRef]
187. Capson-Tojo, G.; Moscoviz, R.; Astals, S.; Robles, Á.; Steyer, J.P. Unraveling the literature chaos around free ammonia inhibition in anaerobic digestion. *Renew. Sustain. Energy Rev.* **2020**, *117*, 109487. [CrossRef]
188. Rajagopal, R.; Massé, D.I.; Singh, G. A critical review on inhibition of anaerobic digestion process by excess ammonia. *Bioresour. Technol.* **2013**, *143*, 632–641. [CrossRef] [PubMed]

189. Kotoulas, A.; Agathou, D.; Triantaphyllidou, I.; Tatoulis, T.; Akratos, C.; Tekerlekopoulou, A.; Vayenas, D. Zeolite as a Potential Medium for Ammonium Recovery and Second Cheese Whey Treatment. *Water* **2019**, *11*, 136. [CrossRef]
190. Ma, J.; Chen, F.; Xue, S.; Pan, J.; Khoshnevisan, B.; Yang, Y.; Liu, H.; Qiu, L. Improving anaerobic digestion of chicken manure under optimized biochar supplementation strategies. *Bioresour. Technol.* **2021**, *325*, 124697. [CrossRef]
191. Ho, L.; Ho, G. Mitigating ammonia inhibition of thermophilic anaerobic treatment of digested piggery wastewater: Use of pH reduction, zeolite, biomass and humic acid. *Water Res.* **2012**, *46*, 4339–4350. [CrossRef]
192. Quintana-Najera, J.; Blacker, A.J.; Fletcher, L.A.; Ross, A.B. Influence of augmentation of biochar during anaerobic co-digestion of Chlorella vulgaris and cellulose. *Bioresour. Technol.* **2022**, *343*, 126086. [CrossRef]
193. Silva, R.M.; Abreu, A.A.; Salvador, A.F.; Alves, M.M.; Neves, I.C.; Pereira, M.A. Zeolite addition to improve biohydrogen production from dark fermentation of C5/C6-sugars and Sargassum sp. biomass. *Sci. Rep.* **2021**, *11*, 16350. [CrossRef]
194. Hwang, J.-H.; Choi, J.-A.; Oh, Y.-K.; Abou-Shanab, R.A.I.; Song, H.; Min, B.; Cho, Y.; Na, J.-G.; Koo, J.; Jeon, B.-H. Hydrogen production from sulfate- and ferrous-enriched wastewater. *Int. J. Hydrogen Energy* **2011**, *36*, 13984–13990. [CrossRef]
195. Mizuno, O.; Li, Y.Y.; Noike, T. The behavior of sulfate-reducing bacteria in acidogenic phase of anaerobic digestion. *Water Res.* **1998**, *32*, 1626–1634. [CrossRef]
196. Haosagul, S.; Prommeenate, P.; Hobbs, G.; Pisutpaisal, N. Sulfide-oxidizing bacteria community in full-scale bioscrubber treating H2S in biogas from swine anaerobic digester. *Renew. Energy* **2020**, *150*, 973–980. [CrossRef]
197. Kobayashi, T.; Li, Y.-Y.; Kubota, K.; Harada, H.; Maeda, T.; Yu, H.-Q. Characterization of sulfide-oxidizing microbial mats developed inside a full-scale anaerobic digester employing biological desulfurization. *Appl. Microbiol. Biotechnol.* **2012**, *93*, 847–857. [CrossRef] [PubMed]
198. Dhar, B.R.; Elbeshbishy, E.; Nakhla, G. Influence of iron on sulfide inhibition in dark biohydrogen fermentation. *Bioresour. Technol.* **2012**, *126*, 123–130. [CrossRef] [PubMed]
199. Li, J.; Hao, X.; van Loosdrecht, M.C.M.; Liu, R. Relieving the inhibition of humic acid on anaerobic digestion of excess sludge by metal ions. *Water Res.* **2021**, *188*, 116541. [CrossRef] [PubMed]
200. Lin, R.; Cheng, J.; Ding, L.; Song, W.; Zhou, J.; Cen, K. Inhibitory effects of furan derivatives and phenolic compounds on dark hydrogen fermentation. *Bioresour. Technol.* **2015**, *196*, 250–255. [CrossRef] [PubMed]
201. Trujillo-Reyes, Á.; Cubero-Cardoso, J.; Rodríguez-Gutiérrez, G.; García-Martín, J.F.; Rodríguez-Galán, M.; Borja, R.; Serrano, A.; Fermoso, F.G. Extraction of phenolic compounds and production of biomethane from strawberry and raspberry extrudates. *Biochem. Eng. J.* **2019**, *147*, 11–19. [CrossRef]
202. Dong, D.; Wang, R.; Geng, P.; Li, C.; Zhao, Z. Enhancing effects of activated carbon supported nano zero-valent iron on anaerobic digestion of phenol-containing organic wastewater. *J. Environ. Manag.* **2019**, *244*, 1–12. [CrossRef]
203. Kim, J.R.; Karthikeyan, K.G. Effects of severe pretreatment conditions and lignocellulose-derived furan byproducts on anaerobic digestion of dairy manure. *Bioresour. Technol.* **2021**, *340*, 125632. [CrossRef]
204. Wijayasekera, S.C.; Hewage, K.; Siddiqui, O.; Hettiaratchi, P.; Sadiq, R. Waste-to-hydrogen technologies: A critical review of techno-economic and socio-environmental sustainability. *Int. J. Hydrogen Energy* **2022**, *47*, 5842–5870. [CrossRef]
205. Massanet Nicolau, J.; Jones, R.J.; Guwy, A.; Dinsdale, R.; Premier, G.; Mulder, M.J.J. Maximising Biohydrogen Yields via Continuous Electrochemical Hydrogen Removal and Carbon Dioxide Scrubbing. *Bioresour. Technol.* **2016**, *218*, 512–517. [CrossRef]
206. Jones, R.J.; Massanet-Nicolau, J.; Fernandez–Feito, R.; Dinsdale, R.M.; Guwy, A.J. Fermentative volatile fatty acid production and recovery from grass using a novel combination of solids separation, pervaporation, and electrodialysis technologies. *Bioresour. Technol.* **2021**, *342*, 125926. [CrossRef]
207. Jones, R.; Massanet-Nicolau, J.; Guwy, A. A review of carboxylate production and recovery from organic wastes. *Bioresour. Technol. Rep.* **2021**, *16*, 100826.
208. Scoma, A.; Varela-Corredor, F.; Bertin, L.; Gostoli, C.; Bandini, S. Recovery of VFAs from anaerobic digestion of dephenolized Olive Mill Wastewaters by Electrodialysis. *Sep. Purif. Technol.* **2016**, *159*, 81–91. [CrossRef]
209. Dawood, F.; Anda, M.; Shafiullah, G.M. Hydrogen production for energy: An overview. *Int. J. Hydrogen Energy* **2020**, *45*, 3847–3869. [CrossRef]
210. IEA. *The Future of Hydrogen*; IEA: Paris, France, 2019.
211. Maestre, V.M.; Ortiz, A.; Ortiz, I. Challenges and prospects of renewable hydrogen-based strategies for full decarbonization of stationary power applications. *Renew. Sustain. Energy Rev.* **2021**, *152*, 111628. [CrossRef]
212. International Renewable Energy Agency. Renewable Capacity Statistics 2020 Statistiques De Capacité Renouvelable 2020 Estadísticas De Capacidad Renovable 2020. 2020. Available online: https://www.irena.org/-/media/Files/IRENA/Agency/Publication/2020/Mar/IRENA_RE_Capacity_Statistics_2020.pdf (accessed on 8 April 2022).
213. International Renewable Energy Agency. 2050 ENERGY TRANSFORMATION E D I T I O N : 2 0 2 0 GLOBAL RENEWABLES OUTLOOK. 2020. Available online: https://www.irena.org/-/media/Files/IRENA/Agency/Publication/2020/Apr/IRENA_Global_Renewables_Outlook_2020.pdf (accessed on 8 April 2022).
214. Janssen, J.L.L.C.C.; Weeda, M.; Detz, R.J.; van der Zwaan, B. Country-specific cost projections for renewable hydrogen production through off-grid electricity systems. *Appl. Energy* **2022**, *309*, 118398. [CrossRef]
215. Sivaramakrishnan, R.; Shanmugam, S.; Sekar, M.; Mathimani, T.; Incharoensakdi, A.; Kim, S.H.; Parthiban, A.; Edwin Geo, V.; Brindhadevi, K.; Pugazhendhi, A. Insights on biological hydrogen production routes and potential microorganisms for high hydrogen yield. *Fuel* **2021**, *291*, 120136. [CrossRef]

216. Fu, Q.; Wang, D.; Li, X.; Yang, Q.; Xu, Q.; Ni, B.J.; Wang, Q.; Liu, X. Towards hydrogen production from waste activated sludge: Principles, challenges and perspectives. *Renew. Sustain. Energy Rev.* **2021**, *135*, 110283. [CrossRef]
217. Park, J.H.; Chandrasekhar, K.; Jeon, B.H.; Jang, M.; Liu, Y.; Kim, S.H. State-of-the-art technologies for continuous high-rate biohydrogen production. *Bioresour. Technol.* **2021**, *320*, 124304. [CrossRef]
218. Yukesh Kannah, R.; Kavitha, S.; Preethi; Parthiba Karthikeyan, O.; Kumar, G.; Dai-Viet, N.V.; Rajesh Banu, J. Techno-economic assessment of various hydrogen production methods—A review. *Bioresour. Technol.* **2021**, *319*, 124175. [CrossRef]
219. Habashy, M.M.; Ong, E.S.; Abdeldayem, O.M.; Al-Sakkari, E.G.; Rene, E.R. Food Waste: A Promising Source of Sustainable Biohydrogen Fuel. *Trends Biotechnol.* **2021**, *39*, 1274–1288. [CrossRef]
220. Aydin, M.I.; Karaca, A.E.; Qureshy, A.M.M.I.; Dincer, I. A comparative review on clean hydrogen production from wastewaters. *J. Environ. Manag.* **2021**, *279*, 111793. [CrossRef] [PubMed]
221. Ji, M.; Wang, J. Review and comparison of various hydrogen production methods based on costs and life cycle impact assessment indicators. *Int. J. Hydrogen Energy* **2021**, *46*, 38612–38635. [CrossRef]
222. Kumar, R.; Kumar, A.; Pal, A. An overview of conventional and non-conventional hydrogen production methods. *Mater. Today Proc.* **2021**, *46*, 5353–5359. [CrossRef]
223. Lepage, T.; Kammoun, M.; Schmetz, Q.; Richel, A. Biomass-to-hydrogen: A review of main routes production, processes evaluation and techno-economical assessment. *Biomass Bioenergy* **2021**, *144*, 105920. [CrossRef]
224. Abdin, Z.; Zafaranloo, A.; Rafiee, A.; Mérida, W.; Lipiński, W.; Khalilpour, K.R. Hydrogen as an energy vector. *Renew. Sustain. Energy Rev.* **2020**, *120*, 109620. [CrossRef]
225. Duman, A.C.; Güler, Ö. Techno-economic analysis of off-grid PV/wind/fuel cell hybrid system combinations with a comparison of regularly and seasonally occupied households. *Sustain. Cities Soc.* **2018**, *42*, 107–126. [CrossRef]
226. Dawood, F.; Shafiullah, G.M.; Anda, M. Stand-Alone Microgrid with 100% Renewable Energy: A Case Study with Hybrid Solar PV-Battery-Hydrogen. *Sustainability* **2020**, *12*, 2047. [CrossRef]
227. Nascimento da Silva, G.; Rochedo, P.R.R.; Szklo, A. Renewable hydrogen production to deal with wind power surpluses and mitigate carbon dioxide emissions from oil refineries. *Appl. Energy* **2022**, *311*, 118631. [CrossRef]
228. Nazir, H.; Muthuswamy, N.; Louis, C.; Jose, S.; Prakash, J.; Buan, M.E.M.; Flox, C.; Chavan, S.; Shi, X.; Kauranen, P.; et al. Is the H2 economy realizable in the foreseeable future? Part III: H2 usage technologies, applications, and challenges and opportunities. *Int. J. Hydrogen Energy* **2020**, *45*, 28217–28239. [CrossRef]
229. Arora, A.; Zantye, M.S.; Hasan, M.M.F. Sustainable hydrogen manufacturing via renewable-integrated intensified process for refueling stations. *Appl. Energy* **2022**, *311*, 118667. [CrossRef]
230. Symes, D.; Maillard, J.G.; Courtney, J.; Watton, J.; Meadowcroft, A.; Chandan, A.S.; Gurley, L.; Priestly, R.; Serdaroglu, G. Development of a hydrogen fuelling infrastructure in the Northeast U.S.A. *Int. J. Hydrogen Energy* **2014**, *39*, 7460–7466. [CrossRef]
231. Schlund, D.; Schönfisch, M. Analysing the impact of a renewable hydrogen quota on the European electricity and natural gas markets. *Appl. Energy* **2021**, *304*, 117666. [CrossRef]
232. Dolci, F.; Thomas, D.; Hilliard, S.; Guerra, C.F.; Hancke, R.; Ito, H.; Jegoux, M.; Kreeft, G.; Leaver, J.; Newborough, M.; et al. Incentives and legal barriers for power-to-hydrogen pathways: An international snapshot. *Int. J. Hydrogen Energy* **2019**, *44*, 11394–11401. [CrossRef]
233. Veluswamy, G.K.; Shah, K.; Ball, A.S.; Guwy, A.J.; Dinsdale, R.M. A techno-economic case for volatile fatty acid production for increased sustainability in the wastewater treatment industry. *Environ. Sci. Water Res. Technol.* **2021**, *7*, 927–941. [CrossRef]
234. Bahreini, G.; Elbeshbishy, E.; Jimenez, J.; Santoro, D.; Nakhla, G. Integrated fermentation and anaerobic digestion of primary sludges for simultaneous resource and energy recovery: Impact of volatile fatty acids recovery. *Waste Manag.* **2020**, *118*, 341–349. [CrossRef] [PubMed]
235. Moscariello, C.; Matassa, S.; Pirozzi, F.; Esposito, G.; Papirio, S. Valorisation of industrial hemp (*Cannabis sativa* L.) biomass residues through acidogenic fermentation and co-fermentation for volatile fatty acids production. *Bioresour. Technol.* **2022**, *355*, 127289. [CrossRef] [PubMed]
236. Huq, N.A.; Hafenstine, G.R.; Huo, X.; Nguyen, H.; Tifft, S.M.; Conklin, D.R.; Stück, D.; Stunkel, J.; Yang, Z.; Heyne, J.S.; et al. Toward net-zero sustainable aviation fuel with wet waste–derived volatile fatty acids. *Proc. Natl. Acad. Sci. USA* **2021**, *118*, e2023008118. [CrossRef]
237. Patterson, T.; Massanet-Nicolau, J.; Jones, R.; Boldrin, A.; Valentino, F.; Dinsdale, R.; Guwy, A. Utilizing grass for the biological production of polyhydroxyalkanoates (PHAs) via green biorefining: Material and energy flows. *J. Ind. Ecol.* **2020**, *25*, 802–815. [CrossRef]
238. Kumar, G.; Ponnusamy, V.K.; Bhosale, R.R.; Shobana, S.; Yoon, J.-J.; Bhatia, S.K.; Rajesh Banu, J.; Kim, S.-H. A review on the conversion of volatile fatty acids to polyhydroxyalkanoates using dark fermentative effluents from hydrogen production. *Bioresour. Technol.* **2019**, *287*, 121427. [CrossRef]
239. Policastro, G.; Giugliano, M.; Luongo, V.; Napolitano, R.; Fabbricino, M. Enhancing photo fermentative hydrogen production using ethanol rich dark fermentation effluents. *Int. J. Hydrogen Energy* **2022**, *47*, 117–126. [CrossRef]
240. Cheng, J.; Li, H.; Ding, L.; Zhou, J.; Song, W.; Li, Y.-Y.; Lin, R. Improving hydrogen and methane co-generation in cascading dark fermentation and anaerobic digestion: The effect of magnetite nanoparticles on microbial electron transfer and syntrophism. *Chem. Eng. J.* **2020**, *397*, 125394. [CrossRef]

fermentation

MDPI

Article

Shaping an Open Microbiome for Butanol Production through Process Control

Tiago Pinto [1,*], Antonio Grimalt-Alemany [2], Xavier Flores-Alsina [1], Hariklia N. Gavala [2], Krist V. Gernaey [1] and Helena Junicke [1]

[1] Process and Systems Engineering Center (PROSYS), Department of Chemical and Biochemical Engineering, Technical University of Denmark (DTU), Søltofts Plads, Building 227, 2800 Kongens Lyngby, Denmark; xfa@kt.dtu.dk (X.F.-A.); kvg@kt.dtu.dk (K.V.G.); heljun@kt.dtu.dk (H.J.)

[2] Center for Energy Resources Engineering (CERE), Department of Chemical and Biochemical Engineering, Technical University of Denmark (DTU), Søltofts Plads, Building 227, 2800 Kongens Lyngby, Denmark; angral@kt.dtu.dk (A.G.-A.); hnga@kt.dtu.dk (H.N.G.)

* Correspondence: tiapi@kt.dtu.dk

Abstract: The growing awareness of limited resource availability has driven production systems towards greater efficiencies, and motivated the transition of wastewater treatment plants to water resource recovery facilities. Open microbiome fermentation offers a robust platform for resource recovery, due to its higher metabolic versatility, which is capable of dealing with even dilute residual liquid streams. Organic matter, e.g., fatty acids, lost in these streams can potentially be recovered into higher value chemicals such as alcohols. This study aims to shape an open microbiome towards butanol production from butyrate and hydrogen through pH control and continuous hydrogen supply. Two sets of experiments were conducted in Scott bottles (1 L) and a lab-fermenter (3 L). The open microbiome produced up to 4.4 mM butanol in 1 L bottles. More promising conversions were obtained when up-scaling to a lab-fermenter with pH control and an increased hydrogen partial pressure of 2 bar; results included a butanol concentration of 10.9 mM and an average volumetric productivity of 0.68 mmol L^{-1} d^{-1} after 16 days. This corresponds to 2.98- and 4.65-fold increases, respectively, over previously reported values. Thermodynamic calculations revealed that product formation from butyrate was unfeasible, but energetically favorable from bicarbonate present in the inoculum. For the first time, this study provides insights regarding the community structure of an open microbiome producing butanol from butyrate and hydrogen. DNA sequencing combined with 16S rRNA gene amplicon analysis showed high correlation between *Mesotoga* spp. and butanol formation. Microbial diversity can also explain the formation of by-products from non-butyrate carbon sources.

Keywords: butyrate reduction; resource recovery; wastewater remediation; thermodynamic analysis; DNA sequencing

Citation: Pinto, T.; Grimalt-Alemany, A.; Flores-Alsina, X.; Gavala, H.N.; Gernaey, K.V.; Junicke, H. Shaping an Open Microbiome for Butanol Production through Process Control. *Fermentation* **2022**, *8*, 333. https://doi.org/10.3390/fermentation8070333

Academic Editor: Sanjay Nagarajan

Received: 19 June 2022
Accepted: 11 July 2022
Published: 15 July 2022

1. Introduction

Conversion of organic and industrial waste into higher value commodities has gained much attention as an alternative to the use of pure substrates (e.g., glucose) and food-derived feedstocks (e.g., corn and wheat). In addition to lowering processing costs and alleviating competition with the food sector, organic and industrial waste conversions help achieve a more sustainable process and production framework [1]. This extends to the energy sector, where our dependency on fossil fuels is the most critical. In 2017, approximately 80% of global energy consumption was supplied by oil, coal, and natural gas [2].

One of our key energy requirements is a sustainable *drop-in* alternative to current liquid fossil fuels such as gasoline and diesel. For many decades, bioethanol was extolled as a capable alternative to gasoline. However, other than Brazil's massive sugar cane-based bioethanol program in the 1970s and the corn-based program in the United States, few other countries have adapted it so widely, whether with first generation feedstocks or subsequent generations. Moreover, due to limitations in current combustion engines, ethanol can only be blended up to 15% with gasoline. Butanol, an energy-rich alcohol similar to ethanol, is considered to be a superior alternative liquid fuel, despite having been historically overshadowed by its synthetic counterpart [3] and facing some economic challenges [4]. Compared to ethanol, butanol has a higher energy density and is less hygroscopic, less volatile, and less corrosive, making it more compatible with current infrastructures for gasoline storage and transportation [5]. Furthermore, butanol can be produced from a wide range of organic and industrial wastes; while much research is focused on solid residues, wastewater is also a prime candidate for butanol production [6,7].

Wastewater generated in agriculture, food, and fermentation-based biotechnology sectors is commonly treated with anaerobic digestion (AD) [8–10]. As a result of AD, soluble metabolites such as volatile fatty acids (VFAs) are present in fermentation processed wastewater. These metabolites (e.g., butyrate and acetate) are building blocks in alcohol synthesis through acetone–butanol–ethanol (ABE) fermentation by *Clostridium* species [11,12]. Butyrate is the most promising VFA for butanol production. However, conversion of butyrate in an anaerobic digester results in acetate production (Equation (1)) rather than butanol production. Under mild acidic conditions, butyrate-oxidizing bacteria convert one mole of butyrate to two moles of acetate and two moles of hydrogen (H_2).

$$C_4H_7O_2^- + 2\,H_2O \rightarrow 2\,C_2H_3O_2^- + H^+ + 2\,H_2 \tag{1}$$

Conversely, if the hydrogen partial pressure, p_{H_2}, increases substantially, anaerobic conversion of butyrate is inhibited. This is even more interesting when considering that the reduction of butyrate to butanol requires hydrogen (Equation (2)). A high proton concentration (i.e., low pH) renders Equation (1) less favorable and Equation (2) more thermodynamically feasible.

$$C_4H_7O_2^- + H^+ + 2\,H_2 \rightarrow C_4H_9OH + H_2O \tag{2}$$

Previous works have already demonstrated how p_{H_2} can be used as a control parameter to direct butanol production from butyrate and hydrogen [13,14]. Steinbusch et al. (2008) [13] reported the capability of undefined microbial cultures to mediate the formation of butanol from butyrate and hydrogen, achieving a final butanol concentration of 3.66 mM after 21 days of batch fermentation in serum bottles (37.5 mL working volume). To drive butyrate reduction, the headspace was flushed with pure hydrogen to a final total pressure of 1.5 bar. Methane was the main by-product of this fermentation due to an increase in pH (pH 5 to pH 5.7); less than 10% of the initial butyrate concentration (50 mM) was consumed. Junicke et al. (2016) [14] perturbed a microbial culture enriched with butyrate and ethanol to find that an increase in p_{H_2} was correlated with an increase in butanol production from butyrate. However, the maximum p_{H_2} (0.0012 bar) was much inferior to that reported by Steinbusch et al. (2008). Although thermodynamic calculations show that direct conversion of butyrate to butanol is more favorable at elevated hydrogen partial pressures, Junicke et al.'s results (2016) hint at the possibility of a more flexible fermentation culture with lower hydrogen overpressure requirements.

The present study aims to direct an open microbiome, often referred to as an undefined microbial culture or mixed microbial culture, towards the anaerobic production of butanol solely from butyrate and hydrogen through process control. The study goes beyond the state-of-the-art by reporting on the use of a nonconventional carbon source (butyrate) contained in waste streams, the effects of operational parameters (e.g., pH, p_{H_2}), and the importance of process control over reaction feasibility. In addition, a detailed thermodynamic assessment based on actual experimental results provides further insights regarding the feasibility of catabolic reactions. The effects of ecological control on the structure of the microbiome are also analyzed via Illumina sequencing of 16S rRNA genes; the dynamics of the main microbial populations are linked to butanol and by-product formation.

2. Methods

2.1. Schott Bottle Fermentation

To start, 1-L Schott (Duran) bottles were inoculated with non-enriched granular sludge (Novozymes, Kalundborg, Denmark) from an anaerobic industrial effluent treatment BIOPAQ®IC reactor (Paques BV, Tjalke de Boerstrjitte, The Netherlands), with either a 15% or 50% volume of sludge (Table 1) in a total working volume of 400 mL. Each condition was performed in duplicate. The biomass elemental composition was taken from Junicke et al. (2016). A medium was designed to fulfill minimum element requirements for microbial growth: butyrate (4405 mg L^{-1}), KH_2PO_4 (3.7 mg L^{-1}), H_3PO_4 (7.5 mg L^{-1}), NH_4Cl (57.1 mg L^{-1}), NaCl (2.6 mg L^{-1}), $CaCl_2 \cdot 2H_2O$ (1.9 mg L^{-1}), $MgSO_4 \cdot 7H_2O$ (1.5 mg L^{-1}), $MgCl_2 \cdot 6H_2O$ (3.1 mg L^{-1}), $FeCl_3 \cdot 6H_2O$ (1.0 mg L^{-1}), $ZnSO \cdot 7H_2O$ (0.2 mg L^{-1}), $MnCl_2 \cdot 4H_2O$ (0.1 mg L^{-1}), H_3BO_3 (0.2 mg L^{-1}), $CoCl_2 \cdot 6H_2O$ (0.2 mg L^{-1}), $CuCl_2 \cdot 2H_2O$ (0.1 mg L^{-1}), $NiCl_2 \cdot 6H_2O$ (0.2 mg L^{-1}), $Na_2MoO_4 \cdot 2H_2O$ (0.1 mg L^{-1}), $Na_2SeO_3 \cdot 5H_2O$ (0.1 mg L^{-1}), thiamine (10 mg L^{-1}), P-aminobenzoic acid (10 mg L^{-1}), Ca-D-pantothenate (10 mg L^{-1}), and biotin (1 mg L^{-1}). Each bottle was buffered (100 mM potassium phosphate, pH 5.5) and the control was inoculated without the presence of butyrate. Prior to inoculation, the bottles and media were sparged with nitrogen to ensure oxygen removal. After inoculation, the headspace was flushed for 10 min at a high flow rate with hydrogen gas to ensure a full hydrogen atmosphere; a final p_{H_2} of 1.5 bar was built with precise gas injection using a mass flow controller (MFC) (red-y Smart Series, Vögtlin Instruments GmbH, Switzerland) and a hydrogen generator (Precision Hydrogen 100, PEAK® Scientific, UK). The Schott bottles were incubated (Ecotron, Infors HT, Switzerland) at 35 °C and 150 rpm. The pH was adjusted to 5.5 with 2 M HCl prior to sealing the bottles with GL 45 bromobutyl rubber septa (Duran, USA). The experiments were carried out for 10 days, with daily sampling and reflushing of the headspace with hydrogen gas for 10 min to a final p_{H_2} of 1.5 bar.

Table 1. Experimental conditions for the 1 L Schott bottle experiments.

	Control	Experiment 50	Experiment 15
Inoculum size (*v/v*)	50%	50%	15%
Total suspended solids (g L^{-1})	23.8	23.8	7.1
Initial butyrate concentration (mM)	0	50	50
Medium	Yes	Yes	Yes
Hydrogen partial pressure (bar)	1.5	1.5	1.5

2.2. Bioreactor Fermentation

A modified (3 L total volume stainless steel vessel) continuous stirred-tank reactor (CSTR) system (ez-Control, Applikon, The Netherlands) was inoculated (50% *v/v*) with non-enriched granular sludge (Novozymes, Kalundborg, Denmark) from an anaerobic industrial effluent treatment BIOPAQ®IC reactor (Paques BV, The Netherlands) to a final volume of 2 L. The same medium composition was used as for the Schott bottle fermentations (see Section 2.1) except for the addition of the potassium phosphate buffer. Anaerobic conditions in the reactor were maintained by continuously sparging with hydrogen gas (0.050 L_N min^{-1}) derived from a hydrogen generator (Precision Hydrogen 100, PEAK® Scientific, Inchinnan, UK). A total pressure of 2 bar was maintained using a low-pressure proportional relief valve (SS-RL3S4, Swagelok, Solon, OH, USA) coupled to a manometer (WIKA, Klingenberg am Main, Germany). The reactor was operated in batch mode with respect to the liquid phase, at 35 °C, p_{H_2} of 2 bar, 400 rpm, and the pH was controlled at 5.5 ± 0.1 using 2 M of NaOH and 2 M of HCl. The full experimental set-up is depicted in Figure 1.

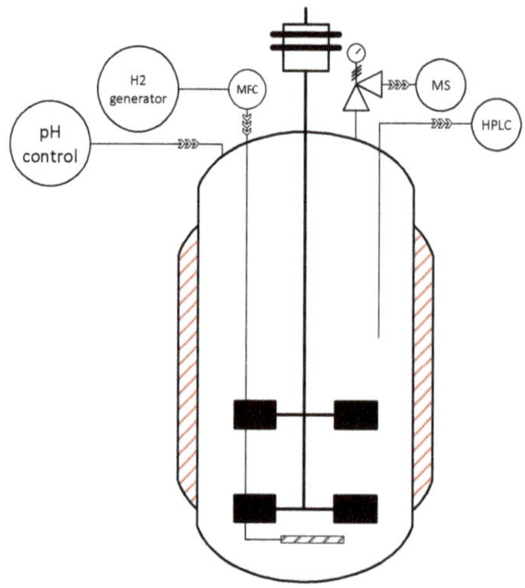

Figure 1. Bioreactor set-up for biobutanol production.

2.3. Analytical Methods

Liquid samples were analyzed for VFAs and alcohols using high-performance liquid chromatography (HPLC) with a Dionex Ultimate 3000 (Thermofisher Scientific, Waltham, MA, USA), equipped with a refractive index detector and an Aminex HPX-87H column (Bio-Rad, Hercules, CA, USA) after filtration through a 0.2 μm pore size cellulose acetate filter (Sartorius, Göttingen, Germany). The RI detector temperature was 50 °C, the column temperature was 20 °C, and the mobile phase (5 mM H_2SO_4) flow rate was maintained at 0.6 mL min^{-1}. The measurement error for HPLC measurements is less than 5%.

The HPR-20 R&D mass spectrometer (MS) (Hiden Analytical, Warrington, UK) was used for online analysis of the bioreactor off-gas stream for hydrogen, water, methane (CH_4), and carbon dioxide (CO_2). Cumulative gas productions were calculated based on the daily net production rate of each gas, corrected for the total gas outflow rate and the mole fraction of the respective gas. The measurement error for MS measurements is less than 2%.

2.4. Thermodynamic Calculations

The actual Gibbs energy change, ΔG^1, for reactions discussed in this study was calculated according to Equation (3):

$$\Delta G^1 = \Delta G^0 + RT \sum y_i \cdot \ln\ c_i \quad (3)$$

where ΔG^0 denotes the standard Gibbs energy change, R is the gas constant (8.314 J K^{-1} mol^{-1}), T is the temperature in Kelvin, y_i is the stoichiometric coefficient of compound i, and c_i is the concentration of compound i. The correction for the pH dependency of alcohol formation on its corresponding VFA can be described by Equation (3) (the derivation can be found in the Supplementary Material):

$$\Delta G^1 = \Delta G^0 + RT \cdot \ln \frac{[\text{Alcohol}]}{[\text{VFA}_t] \cdot p_{H_2}^2} + RT \cdot \ln \frac{K_a + [H^+]}{K_a \cdot [H^+]} \quad (4)$$

where K_a denotes the acid dissociation constant. The Gibbs–Helmholtz equation was used for ΔG^0 temperature correction [15]. The derivation of pH-dependent change in the CO_2 partial pressure for the actual Gibbs energy change in participating catabolic reactions can also be found in the Supplementary Material. The standard Gibbs energy of formation for each compound was found in Kleerebezem and van Loosdrecht (2010).

2.5. Carbon and Electron Balances

At each sampling point, carbon and electron balances were determined. The total carbon amount (C-mol) was obtained by multiplying all measured compounds by their number of carbon atoms. The total electron amount (e-mol) was obtained by multiplying all measured compounds by their respective degree of reduction (e-mol/mol-compound). Both C-mol and e-mol gaps in percent were obtained from the difference between the total amount of carbon/electron at each sampling point and the initial total amount of carbon/electron, divided by the initial total amount of carbon/electron.

2.6. DNA Isolation and Amplicon Sequencing

A total of 5 samples of 2 ml each were selected from the bioreactor fermentation for microbial composition analysis. The samples selected for analysis were collected on days 0 (at inoculation), 2, 8, 10, and 20. Microbial genomic DNA was isolated from all samples using the DNeasy Powersoil Kit (Qiagen, Vedbæk, Denmark) following the manufacturer's recommendations. DNA samples were shipped to Macrogen Inc. (Seoul, Korea) for 16S rRNA amplicon library preparation and sequencing using the Illumina Miseq instrument (300 bp paired-end sequencing). The libraries were constructed according to the 16S Metagenomic Sequencing Library Preparation Protocol (Part #15044223, Rev. B) using Herculase II Fusion DNA Polymerase Nextera XT Index Kit V2. Regions V3 and V4 of the 16S rRNA gene were amplified with primers Pro341F (5′-CCTACGGGNBGCASCAG-3′) and Pro805R (5′-GACTACNVGGGTATCTAATCC-3′) [16]. Raw sequences were uploaded to the NCBI SRA database with BioProject ID PRJNA741687 and BioSample accession SAMN19895498.

2.7. Analysis of 16S rRNA Gene Amplicons

Raw reads were primer-trimmed with cutadapt, discarding all untrimmed reads [17]. Next, low quality tails were trimmed by a fixed length of 15 bases in forward reads and 50 bases in reverse reads. Paired reads were merged using usearch-fastq_mergepairs allowing for 2 mismatches in the alignment, and were quality-filtered using usearch-fastq_filter with a maximum expected error threshold of 1.0 [18]. Unique reads were obtained by dereplicating quality filtered reads using vsearch-derep_fulllength [19]. Generation of amplicon sequence variants (ASVs) (or zero-radius operational taxonomic units (zOTUs)) and mapping of merged reads to ASVs were performed using the UNOISE algorithm [20]; unique sequences with a minimum count of 8 and at least 99% identity were considered in the ASV counts. Taxonomic assignment to ASVs was accomplished using Qiime2 and the SILVAv132 database using classify-consensus-vsearch [19,21]. Downstream analyses, including canonical correspondence analysis (CCA) and statistical correlations, were performed using the Phyloseq, Vegan, ggpubr, and R packages (Phyloseq version 1.28.0, Vegan version 2.5.6, ggpubr version 0.4.0, and R version 3.6.0) [22–24].

3. Results and Discussion

3.1. Butanol Production in 1 L Schott Bottles

The non-enriched microbiome was capable of butyrate reduction to butanol (Table 1) at an elevated p_{H_2} of 1.5 bar. Table 2 shows the measured metabolite concentrations for each condition in the bottle trials. The highest butanol concentration of 4.40 mM was achieved using 50% inoculum; the 15% inoculum was capable of reaching 1.33 mM. The former was similar to the maximum butanol concentration of 3.66 mM reported by Steinbusch et al. (2008) [13], but represented a 3-fold increase in daily average productivity from 0.15 mM d^{-1} to 0.44 mM d^{-1}. Despite this process improvement, and similar to the findings (<10%) of Steinbusch et al. (2008) [13], substrate consumption was not complete; it amounted to less than 20% for the 50% inoculum and 10% for the 15% inoculum. This can potentially be attributed to the lack of pH control during the bottle trial, as an increase in pH renders butanol formation thermodynamically less feasible. By-product formation recurred in all experiments, with acetate being the most predominant by-product, along with the formation of iso-butyrate, propionate, and iso-valerate.

Table 2. Measured substrate concentration (butyrate) and maximum product concentrations after 10 days of fermentation in 1 L Schott bottles.

	Control	Experiment 50	Experiment 15
Butyrate (mM)	0.52	41.70	47.36
Products (mM)			
Acetate	14.33	19.58	5.19
Butanol	0.00	4.40	1.33
i-Butyrate	0.67	1.02	0.26
Propionate	0.82	1.81	0.34
i-Valerate	1.02	2.27	0.57

3.2. Butanol Production under High Hydrogen Partial Pressure

Thermodynamic calculations for butanol formation confirm the experimental observations presented in Table 2. Figure 2 shows how a 10-fold increase in p_{H_2}, from 0.01 bar to 0.1 bar, brings the reaction closer to the minimum biological energy quantum of -20 kJ mol^{-1} necessary for ATP synthesis and, thus, cell growth [25]. A further 10-fold increase in p_{H_2}, to 1 bar, theoretically generates enough excess energy for a thriving microbial community. In practice, microorganisms in natural ecosystems can be metabolically active at lower Gibbs energy changes between -9 to -12 kJ mol^{-1} [26], ensuring some flexibility to the butanol production system in this study.

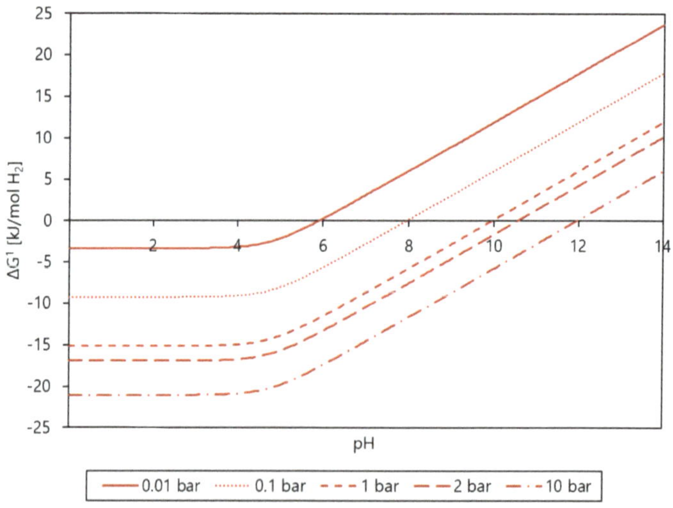

Figure 2. Actual Gibbs energy changes for butanol formation from butyrate and hydrogen (Equation (2)) as a function of pH at different H_2 partial pressures of 0.01 bar (____), 0.1 bar (. . . . ..), 1 bar (—), 2 bar (- - -), and 10 bar (_ . _ . _) at 50 mM of butyrate and 10 mM of butanol.

Additional calculations depicted in Figure 3 reveal that at the experimental conditions of pH 5.5 and p_{H_2} of 1.5 bar, anaerobic butyrate conversion to acetate is endergonic ($\Delta G^1 > 0$). However, acetate production was significant in the control and in the 50% inoculum experiment, as compared to the 15% inoculum experiment. Granular sludge originating from anaerobic wastewater digesters is known to contain calcium carbonate precipitates [27,28], which can function as building blocks in homoacetogenesis and hydrogenotrophic methanogenesis (Equations (5) and (6)).

$$2\,CO_2 + 4\,H_2 \rightarrow C_2H_3O_2^- + H^+ + 2\,H_2O \quad (5)$$

$$CO_2 + 4\,H_2 \rightarrow CH_4 + 2\,H_2O \quad (6)$$

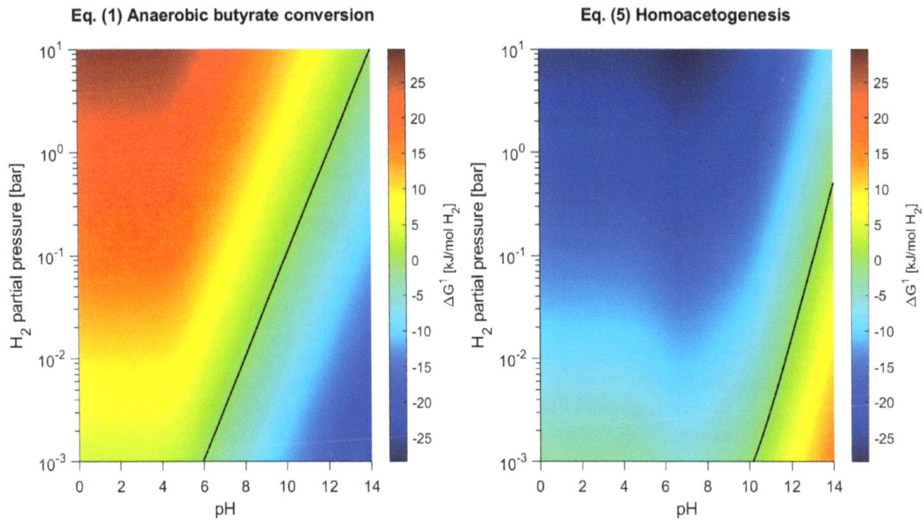

Figure 3. Actual Gibbs energy changes for acetate formation from butyrate (Equation (1)) and homoacetogenesis (Equation (5)) at 50 mM butyrate and 10 mM acetate. For the homoacetogenic reaction, the partial pressure of carbon dioxide was calculated according to Equations (S1)–(S20) in the Supplementary Material. Black line represents 0 kJ mol^{-1} H_2 limit.

With the present anaerobic granular sludge, a HCO_3^- concentration of 50 mM is typically common in the digester of origin (data not shown), and might contribute positively to the formation of acetate. Moreover, homoacetogenesis is a thermodynamically favorable reaction under the applied experimental conditions (Figure 3), hence supporting acetate formation through homoacetogenesis. This is further supported by the lack of an exogenous carbon source present in the control experiment, leaving carbon dioxide as the sole carbon precursor for acetate formation.

3.3. Improved Butanol Formation Using a pH Controlled Bioreactor

A CSTR was used to ensure adequate control of the pH and the hydrogen partial pressure. A pH of 5.5 was selected based on previous finding for the anaerobic sludge used in this work [29]. Figure 4 shows the measured changes in metabolite concentrations in the controlled bioreactor. Butanol formation significantly improved, with the highest butanol concentration of 10.9 mM and an average volumetric productivity of 0.68 mmol L^{-1} d^{-1} after 16 days (Figure 4). Compared to previous work by Steinbusch et al. (2008) [13], this corresponds to 2.98- and 4.65-fold increases, respectively, and 2.47- and 1.55-fold improvements over the 1 L Schott bottle trials, respectively (see Experiment 50 in Table 2 for comparison).

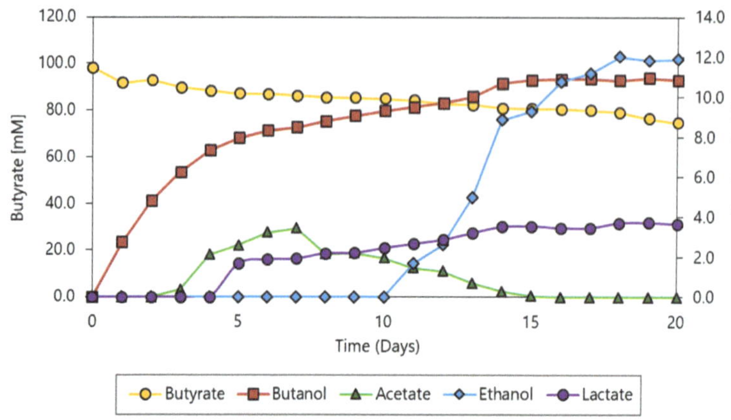

Figure 4. Measured concentrations of butyrate as substrate (primary *y*-axis) and products (secondary *y*-axis) with time in the controlled bioreactor.

By-product formation was largely directed towards ethanol production at the end of the fermentation. Whereas the lack of pH control in the Schott bottles resulted in acetate accumulation, in the bioreactor experiment the acetate produced was further reduced to ethanol (Equation (7)) to a final concentration of 11.9 mM.

$$C_2H_3O_2^- + H^+ + 2\,H_2 \rightarrow C_2H_5OH + 2\,H_2O \quad (7)$$

Ethanol formation from acetate and hydrogen was limited by the availability of reducible acetate in the fermentation broth (Figure 4), further evidencing the uncoupling of by-product formation from butyrate consumption. Methane and carbon dioxide formation reached 18.9 mM and 11.8 mM, respectively (Figure 5); lactate production (3.7 mM) was also found.

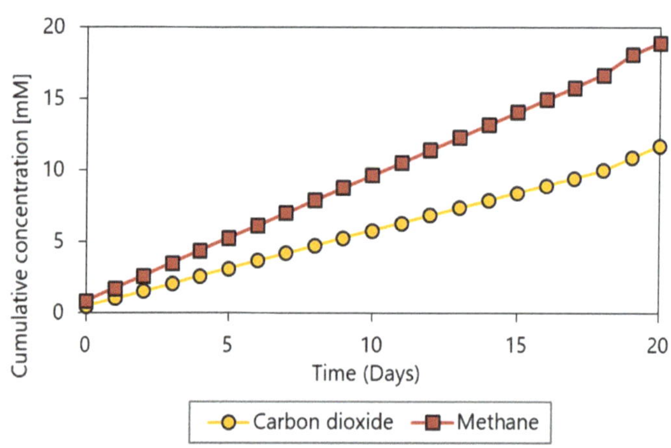

Figure 5. Cumulative concentrations of carbon dioxide and methane with time for the bioreactor run.

Carbon and electron balances show a gap of less than 7% and 10%, respectively, in the course of the experiment, mostly justified by the formation of by-products (Figure 6). As previously discussed, calcium carbonate precipitates are expected in anaerobic granular sludge and can contribute to the formation of by-products; however, they were not included in the balances due to an inherent difficulty in measuring the precipitates' c-mol contribution. The same is true for dry weight determination of granular sludge, where the associated sampling/measurement error is higher than the biomass contribution (one carbon atom) to the c-mol balance. In turn, this leads to an underestimation of total c-mol and a consequent overestimation of balance gaps. Nevertheless, a relative analysis shows that lactate and CO_2 contributed the least to balance gaps, with ethanol being the most predominant by-product c-mol contributor.

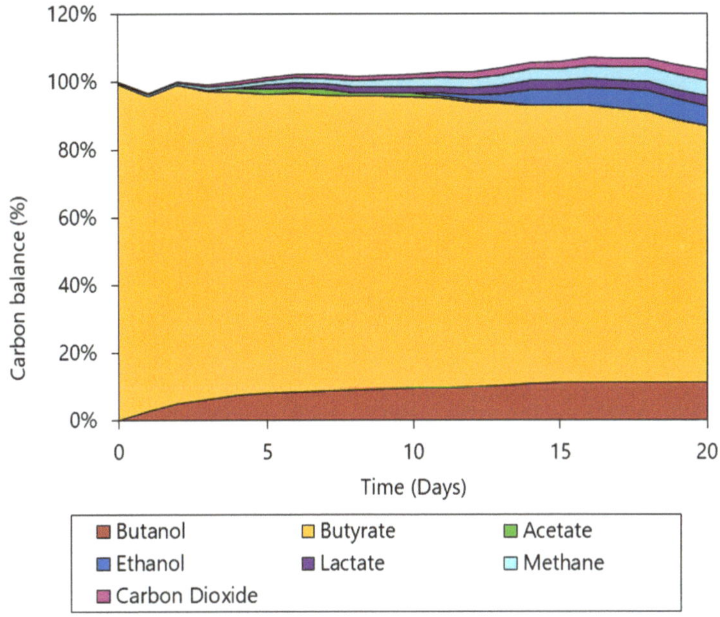

Figure 6. Carbon balance during the course of the bioreactor experiment. By-products include acetate, ethanol, lactate, and methane.

3.4. Product Formation Controlled by Thermodynamics

Figure 7 shows the ΔG^1 of catabolic reactions outlined in Figure 8 for the bioreactor experiment. Similar to the 1-L bottle trials, acetate formation from butyrate (Equation (1)) is endergonic ($\Delta G^1 > +15$ kJ mol H_2^{-1}) and should be considered a result of homoacetogenesis (Equation (5)) according to the discussion in Section 3.2. Butyrate reduction to butanol (Equation (2)) remained exergonic ($\Delta G^1 \approx -16$ kJ mol H_2^{-1}) throughout the entire experiment. Interestingly, butanol formation from butyrate and H_2 seems to occur below the minimum energy quantum of approximately -20 kJ mol^{-1} postulated by Schink (1977) [25]. The relatively constant ΔG^1 for butanol formation, together with production up to day 14, strongly indicates conversion was restricted by other than thermodynamic limitations. However, the final butanol concentration was considerably higher than in the previous bottle experiments (10.9 mM compared to 4.4 mM, respectively), highlighting again the relevance of pH control. Ethanol formation from acetate and H_2 (Equation (7)) is thermodynamically feasible as long as acetate is present in the fermentation broth. The ΔG^1 of the ethanol-forming reaction (Equation (7)) increases from -6 to +0.1 kJ mol H_2^{-1}, which can be mainly attributed to acetate limitation and ethanol accumulation. Again, ethanol formation from acetate and H_2 seems to occur well below -20 kJ mol^{-1}, and much closer to 0 kJ mol^{-1}. These observations give rise to the capability of microbes to survive at life-threatening energy limits.

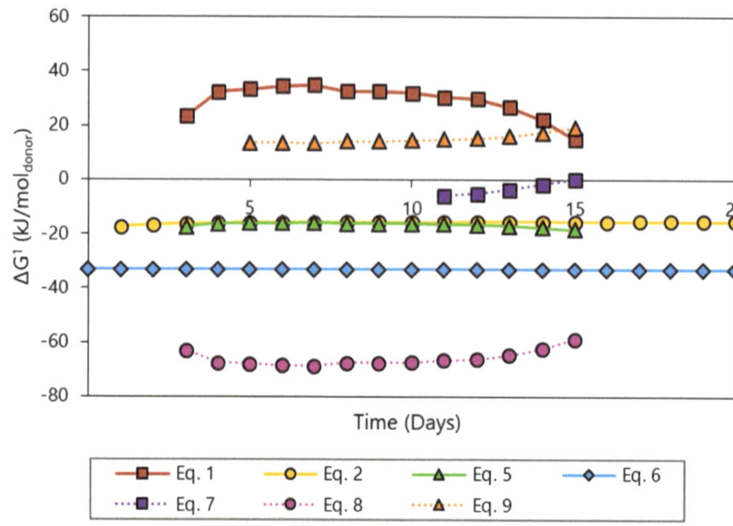

Figure 7. Actual Gibbs energy changes in kJ per mol of electron donor for (Equation (1)) butyrate oxidation to acetate, (Equation (2)) butyrate reduction to butanol, (Equation (5)) homoacetogenesis, (Equation (6)) hydrogenotrophic methanogenesis, (Equation (7)) acetate reduction to ethanol, (Equation (8)) acetoclastic methanogenesis, and (Equation (9)) lactate formation from acetate and CO_2, with time according to experimental data in the bioreactor experiment.

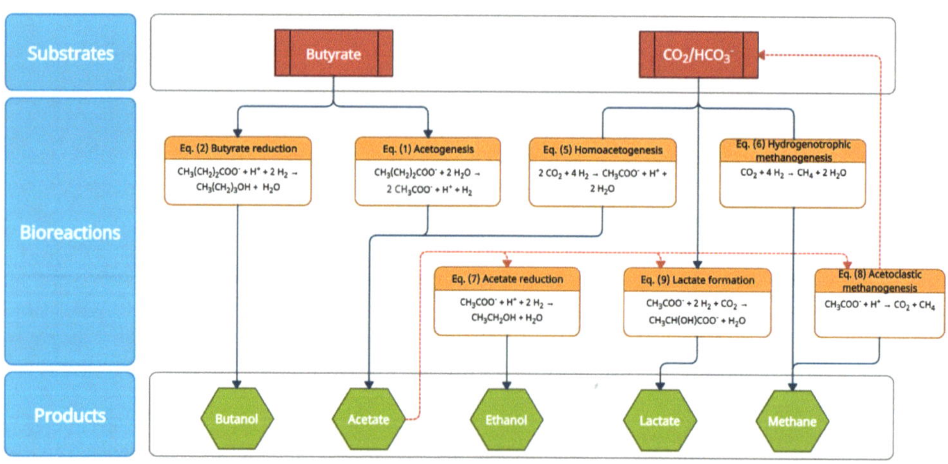

Figure 8. Catabolic reactions used for the analysis of the thermodynamic system state.

Figure 8 shows the catabolic reactions used to analyze the thermodynamic system state. Notably, acetate seems to play a key role in the formation of multiple by-products. CH_4 formation from acetate via acetoclastic methanogenesis (Equation (8)) and CH_4 formation from hydrogen and carbon dioxide via hydrogenotrophic methanogenesis (Equation (6)) are thermodynamically feasible. However, when acetoclastic methanogenesis is normalized to a hydrogen equivalent (i.e., two electrons), ΔG^1 for the reaction raises to approximately -17 kJ mol^{-1}. This, combined with acetoclastic methanogens' inhibition below pH 6 [30], indicates hydrogenotrophic methanogenesis is the more likely reaction leading to CH_4 formation.

$$C_2H_3O_2^- + H^+ \rightarrow CO_2 + CH_4 \tag{8}$$

Lactate formation from acetate and CO_2 (Equation (9)) is not thermodynamically feasible (Figure 7), despite being detected in the bioreactor. Further investigation is required to determine which catabolic reaction could, in fact, lead to lactate formation, with one possible explanation being the presence of non-measured carbohydrates or proteinaceous material.

$$C_2H_3O_2^- + 2\,H_2 + CO_2 \rightarrow C_3H_6O_3 + H_2O \quad (9)$$

Although reduction of by-product formation was the end goal of the controlled fermentation, the current production platform might still be of interest if downstream processing is able to provide feasible separation processes for all produced metabolites.

3.5. Open Microbiome Analysis

Analysis of the microbiome composition during the bioreactor fermentation showed a high microbial diversity and even composition, with representation of a variety of phyla including Thermotogae, Proteobacteria, and Bacteroidetes, among others. The most abundant families identified corresponded to *Kosmotogaceae* (12.4–23.6% of reads mapping to their corresponding ASVs), *Geobacteraceae* (3.5–16.6% of reads mapped), *Synergistaceae* (8.5–13.4% of reads mapped), *Bacteroidaceae* (7.9–11.7% of reads mapped) and *Methanosaetaceae* (6.1–8.9% of reads mapped). However, the percentage of reads mapped to several of these families did not present an increasing trend during the fermentation; this suggests that their presence in the microbial community was due to their high abundance in the granular sludge used as inoculum and the large inoculum size (50% *v/v*), rather than corresponding to an actual active role during the conversion of butyrate. This is likely the case for the putative *Methanosaetaceae*, *Geobacteraceae*, and *Bacteroidaceae* spp. identified, and several other families with minor representation in the microbial community (Figure 9). A preliminary analysis of the dynamics of the reads mapped suggests that the family *Kosmotogaceae*, represented exclusively by putative *Mesotoga* spp., was most probably involved in the conversion of butyrate into butanol, as the percentage of reads mapping to this family increases up to 23.6% during the fermentation (Figure 9).

The results of the microbiome analysis were generally consistent with the product profile obtained experimentally. As mentioned above, the main products of the fermentation were butanol, ethanol, acetate, lactate, and methane. This indicated the presence of several functional groups in the microbial community, namely (i) a variety of fermentative bacteria likely performing the catabolic activities leading to acids and alcohols production, (ii) methanogenic archaea producing methane, and (iii) probably autotrophic bacteria contributing to the production of acetate from CO_2 using H_2 as an electron donor. The composition of the microbiome was consistent with these observations, as a significant fraction of reads were mapped to several fermentative bacteria corresponding to putative *Mesotoga* spp. [31], *Anaerolineaceae* spp. [32], *Clostridium* spp. [33], and *Lactobacillus* spp. [34], all of which increased during the fermentation (Figure 9). Reads mapping to methanogenic archaea other than *Methanosaetaceae* spp. were also identified in small amounts (in line with the limited methane production during the fermentation) and corresponded to *Methanobacteriaceae* spp. and *Methanospirillaceae* spp., both likely growing hydrogenotrophically [35,36]. Nevertheless, it was not possible to confirm the presence of autotrophic bacteria, despite the transient accumulation of significant amounts of acetate during the fermentation. Other putative species that might have contributed to the production of acetate and lactate include the aforementioned *Anaerolineaceae* spp. and *Synergistaceae* spp., as both families present an increasing percentage of reads mapped along the fermentation and count members that were previously reported to convert amino acids into carboxylic acids anaerobically [32,37] (Figure 9).

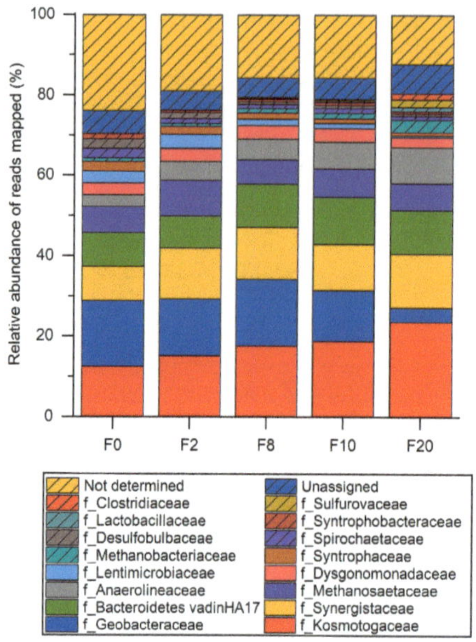

Figure 9. Relative abundance of reads mapped to ASVs at family taxonomic level for bioreactor fermentation samples. Samples were collected on days 0, 2, 8, 10, and 20 of the fermentation. The "Unassigned" category corresponds to reads mapped to ASVs assigned to families with a percentage of reads below 1% in any of the samples.

Keeping in mind the limitations of the microbial community analysis strictly based on the abundance of 16S rRNA gene amplicon reads, without considering the microbial load [38], the population dynamics of the microbiome were further investigated through a canonical correspondence analysis (CCA) (Figure 10) and Pearson correlation for selected genera (Figure 11) to infer their potential roles in the fermentation. The CCA shows that the fermentation samples (F0–F20) were ordinated according to the pattern of activity observed along the fermentation, which was characterized by an initial consumption/production of butyrate/butanol, followed by a transient production of acetate and its further reduction into ethanol at the end of the fermentation (Figure 4). The initial fermentation samples (F0 and F2) are located close to and move along the butyrate/butanol vector, followed by the proximity of samples F8 and F10 to the acetate vector; finally, F20 is aligned with the ethanol vector. In turn, the genera scores resulted in a mainly horizontal distribution aligned with the butyrate vector, with few exceptions, such as *Clostridium* spp. (Figure 10). This indicated that changes in the percentage of reads mapped to these genera are closely related to the changes in butyrate, butanol, and lactate concentrations in the broth, as well as CH_4 evolution. However, among those aligned with the butyrate vector, the few genera located in the negative side of the vector, e.g., *Mesotoga* spp. (family *Kosmotogaceae*) and *Syner*-01 spp. (family *Synergistaceae*), are those that presented an increase in relative abundance as butyrate was converted (Figure 10). This implies that these genera most likely had an active role during the fermentation. As shown in Figure 11A,B, the changes in relative abundance of the putative *Mesotoga* spp. have a significant correlation with the changes in butyrate and butanol concentration in the broth, which suggests that this genus was responsible for the reduction of butyrate into butanol. *Mesotoga* spp. were previously reported to produce a variety of acids including butyrate [31], for which the re-assimilation and further reduction of butyrate into butanol is possible. Other genera likely responsible for the reduction of acetate into ethanol and the conversion of H_2/CO_2 into methane are *Clostridium* and *Methanobacterium*, both of which presented a significant correlation with the evolution of ethanol and methane, respectively (Figure 11C,D).

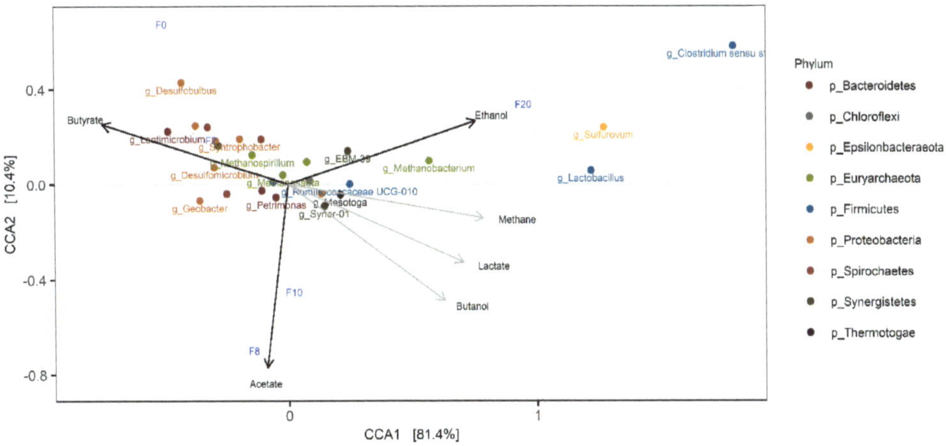

Figure 10. Canonical correspondence analysis (CCA) ordination triplot of microbial species at genus level, fermentation samples, and substrate/product concentrations using Bray–Curtis distances. The constrained ordination explains 97.4% of the variation; the corresponding Eigen values for CCA1 and CCA2 are 0.103 and 0.013, respectively. The overall solution has a *p*-value of 0.0083. ASVs mapped to genera were color-coded according to phyla (given in the legend); fermentation samples are depicted as labels in blue; and substrate/products are shown as arrows. Methane, lactate, and butanol (shown in grey) were not used as ordination constraints due to high collinearity with butyrate. Sample scores and product concentration scores were scaled by a factor of 0.5 and 0.8, respectively, to enhance visualization of the data.

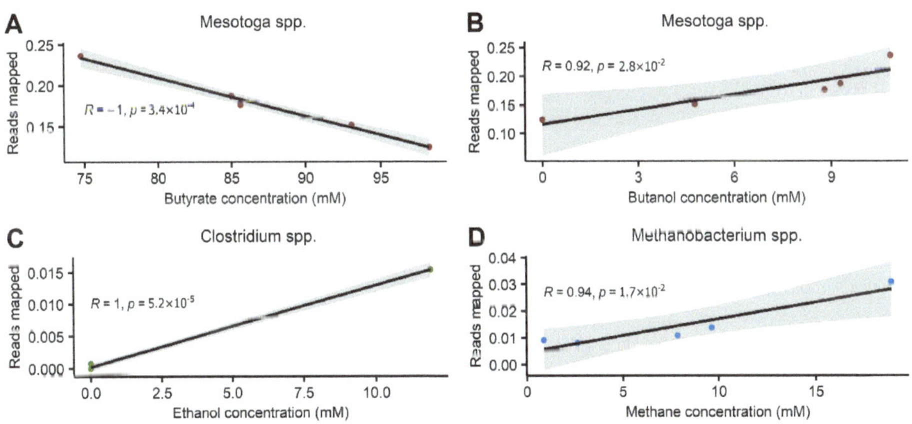

Figure 11. (**A–D**) Comparison of the relative abundance of selected genera to substrate/product concentration along the bioreactor fermentation. "R" corresponds to the Pearson correlation coefficient and "p" is the *p*-value of the correlation.

Overall, the results of the microbiome analysis support the fact that butyrate was exclusively converted to butanol, while the synthesis of other products found at the end of the fermentation originated from other carbon sources present in the inoculated sludge, such as carbon dioxide and proteinaceous material.

4. Conclusions

- Schott bottle experiments showed butanol production from butyrate and hydrogen to a highest titer of 4.4 mM and volumetric productivity of 0.44 mmol L^{-1} d^{-1} of butanol. The use of a large inoculum size of anaerobic granular sludge (50% *v/v*) and lack of pH control contributed largely to by-product formation, with acetate as the most predominant measured by-product.
- A bioreactor operated at pH 5.5 and a p_{H_2} of 2 bar showed an increase in butanol titer (10.9 mM) and volumetric productivity (0.68 mmol L^{-1} d^{-1}); 2.98- and 4.65-fold increases from previously reported values, respectively. By-product formation from granular sludge was still prevalent, but directed towards ethanol production.
- Butyrate conversion is solely directed at butanol formation according to thermodynamics. Calculations of the actual Gibbs energy changes for the proposed catabolic reactions support the thermodynamic feasibility of by-product formation from bicarbonate in granular sludge, with the exception of lactate formation.
- Open microbiome analysis further supports exclusive butyrate conversion to butanol, probably by *Mesotoga* spp., and formation of by-products from residual carbon sources present in the inoculum. Reduced by-products such as ethanol and methane are most likely produced by *Clostridium* spp. and *Methanobacterium* spp., respectively.

Supplementary Materials: The following supporting information can be downloaded at: https://www.mdpi.com/article/10.3390/fermentation8070333/s1. File S1: Derivations for Gibbs energy change.

Author Contributions: Conceptualization, T.P., H.J. and X.F.-A.; methodology, T.P., A.G.-A. and H.J.; software, T.P., A.G.-A. and H.J.; validation, T.P., A.G.-A. and X.F.-A.; formal analysis, T.P., A.G.-A., H.N.G. and H.J.; writing—original draft preparation, T.P., A.G.-A., H.N.G. and X.F.-A.; writing—review and editing, all authors; visualization, T.P., A.G.-A. and H.J.; supervision, K.V.G., X.F.-A. and H.J.; project administration, K.V.G. and H.J.; funding acquisition, K.V.G. and H.J. All authors have read and agreed to the published version of the manuscript.

Funding: This research was funded by the European Union's Horizon 2020 research and innovation program under the Marie Sklodowska-Curie grant agreement number 713683 (COFUNDfellowsDTU), the Danish Council for Independent Research in the frame of the DFF FTP research project GREENLOGIC (grant agreement No. 7017-00175A) and the DFF-International Postdoc (grant no. 0170-00014B), and the Novo Nordisk Fonden in the frame of the Fermentation-Based Biomanufacturing education initiative (grant agreement No. NNF17SA0031362).

Institutional Review Board Statement: Not applicable.

Informed Consent Statement: Not applicable.

Data Availability Statement: Not applicable.

Acknowledgments: The authors acknowledge support obtained from the European Union's Horizon 2020 research and innovation program under the Marie Sklodowska-Curie grant agreement number 713683 (COFUNDfellowsDTU), from the Danish Council for Independent Research in the frame of the DFF FTP research project GREENLOGIC (grant agreement No. 7017-00175A) and DFF-International Postdoc (grant no. 0170-00014B), and from the Novo Nordisk Fonden in the frame of the Fermentation-Based Biomanufacturing education initiative (grant agreement No. NNF17SA0031362).

Conflicts of Interest: The authors declare that they have no known competing conflict of interest or personal relationship that could have appeared to influence the work reported in this paper and are aware of the ethics regarding a journal publication.

Abbreviations

AD	anaerobic digestion
VFA	volatile fatty acid
ABE	acetone, butanol, ethanol
HPLC	high-performance liquid chromatography
MS	mass spectrometer
MFC	mass flow controller
p_{H_2}	hydrogen partial pressure
ΔG^1	actual Gibbs energy change
ΔG^0	standard Gibbs energy change
R	gas constant
T	temperature
y_i	stoichiometric coefficient of compound *i*
c_i	concentration of compound *i*
K_a	acid dissociation constant
H_2	hydrogen
HCO_3^-	bicarbonate
CO_2	carbon dioxide
CH_4	methane
ASV	amplicon sequence variant
CCA	canonical correspondence analysis

References

1. Mansouri, S.S.; Udugama, I.A.; Cignitti, S.; Mitic, A.; Flores-Alsina, X.; Gernaey, K.V. Resource recovery from bio-based production processes: A future necessity? *Curr. Opin. Chem. Eng.* **2017**, *18*, 1–9. [CrossRef]
2. International Energy Agency. *Key World Energy Statistics*; International Energy Agency: Paris, France, 2019.
3. Dürre, P. Fermentative butanol production: Bulk chemical and biofuel. *Ann. N. Y. Acad. Sci.* **2008**, *1125*, 353–362. [CrossRef] [PubMed]
4. Pugazhendhi, A.; Mathimani, T.; Varjani, S.; Rene, E.R.; Kumar, G.; Kim, S.H.; Ponnusamy, V.K.; Yoon, J.J. Biobutanol as a promising liquid fuel for the future—Recent updates and perspectives. *Fuel* **2019**, *253*, 637–646. [CrossRef]
5. Harvey, B.G.; Meylemans, H.A. The role of butanol in the development of sustainable fuel technologies. *J. Chem. Technol. Biotechnol.* **2011**, *86*, 2–9. [CrossRef]
6. Pinto, T.; Flores-Alsina, X.; Gernaey, K.V.; Junicke, H. Alone or together? A review on pure and mixed microbial cultures for butanol production. *Renew. Sustain. Energy Rev.* **2021**, *147*, 111244. [CrossRef]
7. Udugama, I.A.; Petersen, L.A.H.; Falco, F.C.; Junicke, H.; Mitic, A.; Alsina, X.F.; Mansouri, S.S.; Gernaey, K.V. Resource recovery from waste streams in a water-energy-food nexus perspective: Toward more sustainable food processing. *Food Bioprod. Process.* **2020**, *119*, 133–147. [CrossRef]
8. Chen, H.; Hao, S.; Chen, Z.; O-Thong, S.; Fan, J.; Clark, J.; Luo, G.; Zhang, S. Mesophilic and thermophilic anaerobic digestion of aqueous phase generated from hydrothermal liquefaction of cornstalk: Molecular and metabolic insights. *Water Res.* **2020**, *168*, 115199. [CrossRef]
9. Amha, Y.M.; Sinha, P.; Lagman, J.; Gregori, M.; Smith, A.L. Elucidating microbial community adaptation to anaerobic co-digestion of fats, oils, and grease and food waste. *Water Res.* **2017**, *123*, 277–289. [CrossRef]
10. Feldman, H.; Flores-Alsina, X.; Ramin, P.; Kjellberg, K.; Jeppsson, U.; Batstone, D.J.; Gernaey, K.V. Modelling an industrial anaerobic granular reactor using a multi-scale approach. *Water Res.* **2017**, *126*, 488–500. [CrossRef]
11. Jones, D.T.; Woods, D.R. Acetone-butanol fermentation revisited. *Microbiol. Rev.* **1986**, *50*, 484–524. Available online: http://www.ncbi.nlm.nih.gov/pubmed/3540574 (accessed on 31 January 2018).
12. Jankowska, E.; Duber, A.; Chwialkowska, J.; Stodolny, M.; Oleskowicz-Popiel, P. Conversion of organic waste into volatile fatty acids—The influence of process operating parameters. *Chem. Eng. J.* **2018**, *345*, 395–403. [CrossRef]
13. Steinbusch, K.J.J.; Hamelers, H.V.M.; Buisman, C.J.N. Alcohol production through volatile fatty acids reduction with hydrogen as electron donor by mixed cultures. *Water Res.* **2008**, *42*, 4059–4066. [CrossRef] [PubMed]
14. Junicke, H.; van Loosdrecht, M.C.M.; Kleerebezem, R. Kinetic and thermodynamic control of butyrate conversion in non-defined methanogenic communities. *Appl. Microbiol. Biotechnol.* **2016**, *100*, 915–925. [CrossRef]
15. Kleerebezem, R.; van Loosdrecht, M.C.M. A Generalized Method for Thermodynamic State Analysis of Environmental Systems. *Crit. Rev. Environ. Sci. Technol.* **2010**, *40*, 1–54. [CrossRef]
16. Takahashi, S.; Tomita, J.; Nishioka, K.; Hisada, T.; Nishijima, M. Development of a Prokaryotic Universal Primer for Simultaneous Analysis of Bacteria and Archaea Using Next-Generation Sequencing. *PLoS ONE* **2014**, *9*, e105592. [CrossRef]
17. Martin, M. Cutadapt removes adapter sequences from high-thoughput sequencing reads. *EMBnet. J.* **2011**, *17*, 10–12. [CrossRef]

18. Edgar, R.C. UPARSE: Highly accurate OTU sequences from microbial amplicon reads. *Nat. Methods* **2013**, *10*, 996–998. [CrossRef]
19. Rognes, T.; Flouri, T.; Nichols, B.; Quince, C.; Mahé, F. VSEARCH: A versatile open source tool for metagenomics. *PeerJ* **2016**, *4*, e2584. [CrossRef]
20. Edgar, R.C. UNOISE2: Improved error-correction for Illumina 16S and ITS amplicon sequencing. *BioRxiv* **2016**. [CrossRef]
21. Bolyen, E.; Rideout, J.R.; Dillon, M.R.; Bokulich, N.A.; Abnet, C.C.; Al-Ghalith, G.A.; Alexander, H.; Alm, E.J.; Arumugam, M.; Asnicar, F.; et al. Reproducible, interactive, scalable and extensible microbiome data science using QIIME 2. *Nat. Biotechnol.* **2019**, *37*, 852–857. [CrossRef]
22. R Core Team. *R: A Language and Environment for Statistical Computing*; R Foundation for Statistical Computing: Vienna, Austria, 2013; Available online: http://www.R-project.org/ (accessed on 21 January 2021).
23. McMurdie, P.J.; Holmes, S. Phyloseq: An R Package for Reproducible Interactive Analysis and Graphics of Microbiome Census Data. *PLoS ONE* **2013**, *8*, e61217. [CrossRef] [PubMed]
24. Oksanen, J.; Blanchet, G.F.; Friendly, M.; Kindt, R.; Legendre, P.; McGlinn, D.; Minchin, P.R.; O'Hara, R.B.; Simpson, G.L.; Solymos, P.; et al. Vegan: Community Ecology Package. R package Version 2.5–6. 2019. Available online: https://CRAN.R-project.org/package=vegan (accessed on 21 January 2021).
25. Schink, B. Energetics of syntrophic cooperation in methanogenic degradation. *Microbiol. Mol. Biol. Rev.* **1997**, *61*, 262–280. [CrossRef] [PubMed]
26. Hoehler, T.M. Biological energy requirements as quantitative boundary conditions for life in the subsurface. *Geobiology* **2004**, *2*, 205–215. [CrossRef]
27. Feldman, H.; Flores-Alsina, X.; Ramin, P.; Kjellberg, K.; Jeppsson, U.; Batstone, D.J.; Gernaey, K.V. Assessing the effects of intra-granule precipitation in a full-scale industrial anaerobic digester. *Water Sci. Technol.* **2019**, *79*, 1327–1337. [CrossRef]
28. Mañas, A.; Pocquet, M.; Biscans, B.; Sperandio, M. Parameters influencing calcium phosphate precipitation in granular sludge sequencing batch reactor. *Chem. Eng. Sci.* **2012**, *77*, 165–175. [CrossRef]
29. El-Zouheiri, F.A. *Screening of Optimum Process Conditions for Butanol Production in Mixed Microbial Cultures*; Danmarks Tekniske Universitet: Copenhagen, Denmark, 2019; Available online: https://findit.dtu.dk/en/catalog/2453508175 (accessed on 15 June 2021).
30. Batstone, D.J.; Keller, J.; Angelidaki, I.; Kalyuzhnyi, S.V.; Pavlostathis, S.G.; Rozzi, A.; Sanders, W.T.M.; Siegrist, H.; Vavilin, V.A. The IWA Anaerobic Digestion Model No 1 (ADM1). *Water Sci. Technol.* **2002**, *45*, 65–73. [CrossRef]
31. Cappelletti, M.; Zannoni, D.; Postec, A.; Ollivier, B. Members of the Order Thermotogales: From Microbiology to Hydrogen Production. In *Microbial Bioenergy: Hydrogen Production*; Springer: Dordrecht, The Netherlands, 2014; pp. 197–224. [CrossRef]
32. Yamada, T.; Sekiguchi, Y. A naerolineaceae. In *Bergey's Manual of Systematics of Archaea and Bacteria*; John Wiley & Sons, Inc.: New York, NY, USA, 2018; pp. 1–5. [CrossRef]
33. Xu, P.-X.; Chai, L.-J.; Qiu, T.; Zhang, X.-J.; Lu, Z.-M.; Xiao, C.; Wang, S.-T.; Shen, C.-H.; Shi, J.-S.; Xu, Z.-H. *Clostridium fermenticellae* sp. nov., isolated from the mud in a fermentation cellar for the production of the Chinese liquor, baijiu. *Int. J. Syst. Evol. Microbiol.* **2019**, *69*, 859–865. [CrossRef]
34. Oki, K.; Kudo, Y.; Watanabe, K. *Lactobacillus saniviri* sp. nov. and *Lactobacillus senioris* sp. nov., isolated from human faeces. *Int. J. Syst. Evol. Microbiol.* **2012**, *62*, 601–607. [CrossRef]
35. Ma, K.; Liu, X.; Dong, X. *Methanobacterium beijingense* sp. nov., a novel methanogen isolated from anaerobic digesters. *Int. J. Syst. Evol. Microbiol.* **2005**, *55*, 325–329. [CrossRef]
36. Fournier, G.P.; Gogarten, J.P. Evolution of Acetoclastic Methanogenesis in Methanosarcina via Horizontal Gene Transfer from Cellulolytic Clostridia. *J. Bacteriol.* **2008**, *190*, 1124–1127. [CrossRef]
37. Honda, T.; Fujita, T.; Tonouchi, A. *Aminivibrio pyruvatiphilus* gen. nov., sp. nov., an anaerobic, amino-acid-degrading bacterium from soil of a Japanese rice field. *Int. J. Syst. Evol. Microbiol.* **2013**, *63*, 3679–3686. [CrossRef] [PubMed]
38. Props, R.; Kerckhof, F.M.; Rubbens, P.; de Vrieze, J.; Sanabria, E.H.; Waegeman, W.; Monsieurs, P.; Hammes, F.; Boon, N. Absolute quantification of microbial taxon abundances. *ISME J.* **2017**, *11*, 584–587. [CrossRef] [PubMed]

fermentation

Article

Biohydrogen and Methane Production from Sugarcane Leaves Pretreated by Deep Eutectic Solvents and Enzymatic Hydrolysis by Cellulolytic Consortia

Apik Khautsart Miftah [1], Sureewan Sittijunda [2], Tsuyoshi Imai [3], Apilak Salakkam [1] and Alissara Reungsang [1,4,5,*]

1 Department of Biotechnology, Faculty of Technology, Khon Kaen University, Khon Kaen 40002, Thailand
2 Faculty of Environment and Resource Studies, Mahidol University, Nakhon Pathom 73170, Thailand
3 Graduate School of Sciences and Technology for Innovation, Yamaguchi University, Yamaguchi 755-8611, Japan
4 Academy of Science, Royal Society of Thailand, Bangkok 10300, Thailand
5 Research Group for Development of Microbial Hydrogen Production Process, Khon Kaen University, Khon Kaen 40002, Thailand
* Correspondence: alissara@kku.ac.th

Abstract: This study determined the optimal conditions for the deep eutectic solvent (DES) pretreatment of sugarcane leaves and the best fermentation mode for hydrogen and methane production from DES-pretreated sugarcane leaves. Choline chloride (ChCl):monoethanolamine (MEA) is the most effective solvent for removing lignin from sugarcane leaves. The optimum conditions were a ChCl: MEA molar ratio of 1:6, 120 °C, 3 h, and substrate-to-DES solution ratio of 1:12. Under these conditions, 86.37 ± 0.36% lignin removal and 73.98 ± 0.42% hemicellulose removal were achieved, whereas 84.13 ± 0.77% cellulose was recovered. At a substrate loading of 4 g volatile solids (VS), the simultaneous saccharification and fermentation (SSF) and separate hydrolysis and fermentation (SHF) processes yielded maximum hydrogen productions of 3187 ± 202 and 2135 ± 315 mL H_2/L, respectively. In the second stage, methane was produced using the hydrogenic effluent. SSF produced 5923 ± 251 mL CH_4/L, whereas SHF produced 3583 ± 128 mL CH_4/L. In a one-stage methane production process, a maximum methane production of 4067 ± 320 mL CH_4/L with a substrate loading of 4 g VS was achieved from the SSF process. SSF proved to be more efficient than SHF for producing hydrogen from DES-pretreated sugarcane leaves in a two-stage hydrogen and methane production process as well as a one-stage methane production process.

Keywords: ionic liquid; anaerobic digestion; lignocellulosic biomass; pretreatment; clean energy

Citation: Miftah, A.K.; Sittijunda, S.; Imai, T.; Salakkam, A.; Reungsang, A. Biohydrogen and Methane Production from Sugarcane Leaves Pretreated by Deep Eutectic Solvents and Enzymatic Hydrolysis by Cellulolytic Consortia. *Fermentation* **2022**, *8*, 396. https://doi.org/10.3390/fermentation8080396

Academic Editor: Sanjay Nagarajan

Received: 23 July 2022
Accepted: 12 August 2022
Published: 16 August 2022

1. Introduction

Biofuels are any liquid, gas, or solid fuel produced from renewable biomass. Examples of biofuels include ethanol, methanol, synthetic gas (syngas), biodiesel, biogas (methane), biochar, bio-oil, and biohydrogen [1]. Biogas, typically a mixture of methane (CH_4) and carbon dioxide (CO_2), is produced worldwide from agricultural, municipal, and industrial wastes via anaerobic digestion (AD) [2].

AD is a well-established process for converting different organic waste into renewable energy, such as methane, with limited environmental impact [3]. AD involves the biological degradation of organic matter. Digestion is driven by anaerobic microorganisms and involves a series of steps, including hydrolysis, acidogenesis, acetogenesis, and methanogenesis [4]. At the beginning of the process, in the hydrolysis step, complex organic polymers are decomposed into monomers such as amino acids, fatty acids, and monosaccharides. Next, acidogenic bacteria convert these monomers into a mixture of short-chain volatile fatty acids during acidogenesis. Acidogenic bacteria are either facultative or strictly anaerobic bacteria belonging to the family *Enterobacteriaceae* [5]. Next, acetogenic bacteria or acetogens convert volatile fatty acids to acetate, carbon dioxide, and hydrogen in the acetogenesis step which are further used as substrates to produce methane in the methanogenesis step [6]. Therefore, a feedstock used for methane production by AD should be readily biodegradable and free of toxic components that would cause adverse effects on bacteria [7]. AD feedstocks are divided into three main categories: edible food crop resources (first-generation feedstocks), lignocellulosic materials and different organic waste materials (second-generation feedstocks), and algal biomass (third-generation feedstocks) [8]. Despite their high biogas production rates, first-generation feedstocks compete with food production, making them an undesirable biomass source. Therefore, lignocellulosic materials have received attention as alternative feedstocks for biogas production in recent decades owing to their abundance and cost. In this study, sugarcane leaves were used as lignocellulosic material for methane production.

Sugarcane leaves are usually either burnt to enable manual harvest, adding to environmental pollution and greenhouse gases, or left in the field as part of fertilizer, providing soil nutrients. [9]. It is usually left in abundance for up to 600–800 g/m^2 of sugarcane crops [10]. Dry leaves possess the energy equivalent of 1000 g/m^2, which is a significant advantage of this feedstock [11]. Sugarcane leaves are primarily composed of lignin (15–20%), hemicellulose (20–35%), and cellulose (35–50%) [12]. Biochemical or thermal methods can convert them into different bioenergy forms and other marketable co-products. Thus, using sugarcane leaves as feedstock to produce methane is necessary to obtain biofuel and mitigate environmental problems. However, the complexity of the structure of lignocellulosic biomasses, such as sugarcane leaves, is a major challenge, making them highly recalcitrant to AD and ultimately resulting in low methane yield [6]. Lignin is a physical barrier to lignocellulosic biomass [13–15], preventing enzymes from accessing cellulose [16–19]. Therefore, a pretreatment step is required to overcome this problem.

Pretreatment steps are essential to make cellulose more accessible for enzymatic hydrolysis by changing the physical and chemical structure of the lignocellulosic biomass and facilitating the conversion of polysaccharides into fermentable sugars [20]. Consequently, methane production from the lignocellulosic biomass is enhanced. One of the pretreatment methods that has received attention in the last decade is ionic liquid (IL) pretreatment. Lignocellulosic biomass pretreatment using ILs offers several attractive features compared to conventional methods [21]. As a molten salt, ILs are composed only of ions, usually a combination of larger organic cations and smaller anions. These features enable ILs ion properties and chemistry to be designed and tailored for desired applications. The dissolution of cellulose in ILs can be increased by introducing higher hydrogen bond (H-bond) basicity and polarity on the anion side and reducing the alkyl chain length on the cation side. This shows that the anion and cation sides of ionic liquids can be changed and tuned to enhance lignocellulosic biomass pretreatment [21]. ILs are considered green solvents due to their characteristics, such as nonvolatility, which makes the application of ILs reduce the air pollution caused by solvent evaporation, having a low toxicity to human health and the environment, biodegradability, being easy to recycle and reuse, and non-corrosive [21]. However, ILs have some disadvantages, including that they are not always "green" and are generally costly. In addition, they can absorb water from the air and evaporate at moderate temperatures. IL synthesis, separation, and purification require many solvents and energy [22–25]. Recently, a solvent with similar qualities and fewer limitations than IL has been proposed. They are referred to as deep eutectic solvents (DESs). DESs have a high dissolving capacity and a low melting point but are simpler to prepare and less expensive in terms of raw materials [25]. DESs are produced by combining multiple H-bond donors (HBD) and acceptors (HBA) [21]. The number of DES components can be increased to three or more. A DES has a substantially lower melting point than its components (HBD and HBA) due to the strong hydrogen-bonding interaction between HBD and HBA [21]. The capacity of a DES to preferentially cleave the ether bonds between phenylpropane units in lignin heteropolymers [26] is the mechanism by which it delignifies biomass. Although a DES can remove lignin from lignocellulosic biomass, its use as a pretreatment method for feedstock for biogas production remains limited. In this work, DES conditions were optimized to pretreat sugarcane leaves before their usage as an AD feedstock for biogas production. Despite the fact that lignocellulosic biomass pretreatment can improve biogas production, lignocellulosic biomass hydrolysis in AD is still a bottleneck because of the low rate of hydrolysis products slows down the entire AD process. To solve these problems, an enzymatic process was introduced to aid hydrolysis. This can be achieved using enzymes or cellulolytic microorganisms. Cellulolytic microorganisms are effective in recalcitrant cellulosic biomass hydrolysis. These microorganisms have been enriched and isolated from various ecological niches. Consortia are more robust to environmental fluctuations because they are naturally occurring. In this study, cellulolytic microorganisms enriched from rice straw compost (RSC), termite intestines (TI), and the soil around goat and sheep stalls (SGS) were used to construct a cellulolytic consortium. The consortium was further used to hydrolyze the cellulose fraction of sugarcane leaves via separate hydrolysis fermentation (SHF) and simultaneous saccharification and fermentation (SSF). SHF is a method in which enzymatic hydrolysis and fermentation are performed sequentially. This process begins with the enzymatic hydrolysis of biomass or pretreated lignocellulosic biomass at the appropriate temperature for the hydrolyzing enzyme. Subsequently, the fermentation conditions were optimized. However, these two separate processes increase the capital cost of the SHF process. Unlike SHF, SSF involves simultaneous enzymatic hydrolysis and fermentation in the same reactor. SSF can eliminate substrate inhibition because the sugars from the hydrolyzed biomass are directly transformed into biogas by AD [26]. Hence, this study employed SHF and SSF as fermentation modes to hydrolyze sugarcane leaves and produce methane.

There are two ADs for producing methane. The first was a one-stage methane production process. This process is the simplest and most conventional system to perform AD, one in which all AD steps occur in one reactor. One-stage AD has the advantages of low installation and operating costs and a short processing time [27]. The second step is two-stage AD, which separates the first step (acidogenesis and acetogenesis) and the second step (methanogenesis) in two different reactors, allowing the recovery of both hydrogen and methane from this process. By implementing two-stage AD, the system can achieve a more stable operation, higher organic loading capacity, and higher resistance to toxicants and inhibiting substances [28]. To produce methane from DES-treated sugarcane leaves, this study used a one-stage methane production process and a two-stage hydrogen and methane production process. This study aimed to find the best conditions for the DES pretreatment of sugarcane leaves and the best fermentation mode for hydrogen and methane production from DES-pretreated sugarcane leaves.

2. Materials and Methods

2.1. Sugarcane Leaves Preparation

Sugarcane leaves were collected from sugarcane plantations in the Khon Kaen Province, Thailand. First, the sugarcane leaves were dried under the sunlight until less than 10% of the moisture content remained. Next, the dried sugarcane leaves were cut, ground, and sieved using sieve no. 18 (mesh size 1.0 mm) (Central World Intertrade Co., Ltd., Ladkrabang, Bangkok, Thailand) to obtain particles smaller than 1.0 mm. Finally, the sugarcane leaves were stored in a dry plastic box at room temperature (32 ± 2 °C) for further use. The compositions of the sugarcane leaves (all in % (*w*/*w*) dry weight) are 36.18 cellulose, 25.23 hemicellulose, 27.68 lignin, and 10.91 ash.

2.2. Deep Eutectic Solvents Preparations

DESs were prepared using different types and molar ratios of HBA and HBD. The DESs used in this study were ChCl:glycerol (G), ChCl:G:aluminum chloride ($AlCl_3$), and ChCl:MEA. ChCl acts as an HBD, whereas G, a mixture of G, aluminum chloride ($AlCl_3$), and MEA, acts as HBA. DESs were prepared by varying the molar ratio of each mixture. The ChCl:G mixtures were at 1:2, 1:4, and 1:6 molar ratios, the ChCl:G:$AlCl_3$ mixtures were at

1:2:0.33, 1:4:0.33, and 1:6:0.33 molar ratios, and the ChCl:MEA mixtures were at 1:6, 1:8, and 1:10, respectively. All analytical grade chemicals were purchased from Elago Enterprises Pty., Ltd. (Elago Enterprises, 5 The Cloisters, Cherrybrook, Sydney, NSW, Australia)

2.3. Cellulolytic Consortium

RSC, TI, and SGS from the Department of Animal Science, Faculty of Agriculture, Khon Kaen University, Thailand, were used as cellulolytic consortium sources. Peptone-cellulose solution (PCS) medium (all in g/L; yeast extract 1, peptones 5, $CaCO_3$ 2, and NaCl 5) was used to enrich the cellulolytic consortium from RSC, TI, and SGS. Initially, one gram each of RSC, TI, and SGS was added to 30 mL of PCS medium containing 0.15 g of filter paper (Whatman No.1) as the carbon source and an indicator for cellulase activity. Enrichment was performed at 37 °C. Once the filter paper strip was completely degraded, 10% (*v*/*v*) of the culture was transferred into the fresh PCS media containing 0.15 g of filter paper Whatman No.1 with 3.0 × 5.5 cm. This process was repeated at least 10 times or until the filter paper weight loss was stable (relative standard deviation < 0.05). Finally, the enriched culture was collected to analyze enzyme activity, including carboxymethyl cellulose degradation (CMCase), filter paper degradation (FPUase), and xylanase, before hydrolysis. The enriched cellulolytic consortium showed CMCase, FPUase, and xylanase activities of 0.38, 8.05, and 1.21 IU/mL. For long-term storage, the cellulolytic consortium was kept in a PCS medium containing 20% glycerol without cellulosic substrates at −80 °C.

2.4. *Clostridium butyricum* TISTR 1032 Prepartions

The hydrogen producer in this study was *C. butyricum* TISTR1032. It was purchased from the Thailand Institute of Scientific and Technological Research (TISTR). Strain TISTR1032 was regenerated in a serum bottle containing a cooked meat medium (CMM) (Himedia, Analytical grade, Thane, India). The serum bottle was flushed with nitrogen gas to create anaerobic conditions and incubated at 37 °C for 10 h. Then, 1 mL of the stock culture was added to a serum bottle containing 10 mL of tryptone sucrose yeast extract (TSY) (Himedia, Analytical grade, Thane, India) medium under anaerobic conditions at 150 rpm for 10 h at 37 °C. This process was repeated until an initial cell concentration of 10^7 cells/mL was obtained. The composition of TSY (all in g/L) was tryptone 5, sucrose 3, yeast extract 5, and K_2HPO_4 1 [29].

2.5. Anaerobic Sludge Preparations

Anaerobic sludge from the anaerobic digester of SF Khon Kaen Co., Ltd., (SF Khon Kaen, Nai Mueang, Thailand) was used as the inoculum source for the methane production process. The anaerobic digester produced biogas through the Napier silage and chicken manure co-digestion. To prepare methane producers, 5 L of anaerobic sludge was cultivated in a 10 L closed container containing 10 g/L of sugarcane leaves as the carbon source. The container was purged with nitrogen gas for 15 min to create anaerobic conditions and incubated for 2 weeks at room temperature (30 ± 5 °C) for degassing. The acclimatized inoculum was filtered through a filter cloth to remove non-degradable materials before being used for methane production. The anaerobic sludge compositions were 5.62 ± 0.27% of total solids (TS), 3.35 ± 0.08% of VS, 1.46 ± 0.12% of total suspended solids (TSS), and 2.11 ± 0.11% of volatile suspended solids (VSS.)

2.6. The Optimization Factors Affecting the Pretreatment of Sugarcane Leaves

The optimization of the different DESs, the molar ratio of HBA and HBD (ChCl: MEA (1:6, 1:8, and 1:10 molar ratio), ChCl:G (1:2, 1:4, and 1:6 molar ratio), and ChCl:G:$AlCl_3$ (1:2:0.33, 1:4:0.33, and 1:6:0.33 molar ratio), pretreatment temperature (80 °C, 100 °C, and 120 °C) and time (3 h, 6 h, and 9 h), and substrate-to-DES solution ratio for the pretreatment of sugarcane leaves were examined in the batch test. Next, the substrate-to-DES solution ratios were varied at 1:8, 1:12, 1:16, 1:20, and 1:24 (*w*/*v*). All treatments were performed in triplicate. At the end of the heating process, the pretreated solids were washed with hot distilled water several times until a pH of 7.0 was obtained. The pretreated solids were then dried at 60 °C in a hot-air oven until less than 10% of the moisture content remained. Finally, the pretreated solids were stored in sealed plastic bags at 31 ± 2 °C before analysis. The pretreated solids were used as substrates in the SHF and SSF processes for the two-stage hydrogen and methane production and one-stage methane production.

2.7. SHF for Two-Stage Hydrogen and Methane Production and One-Stage Methane Production from Pretreated Sugarcane Leaves

The SHF for the two-stage hydrogen and methane production and one-stage methane production was carried out in batch experiments. The hydrolysis step was investigated separately from that of the fermentation process. The two-stage hydrogen and methane production and one-stage methane production processes were initiated once hydrolysis was completed.

In the hydrolysis step, the experiment was conducted in 120 mL serum bottles with variations of pretreated sugarcane leaf loading of 1, 2, 3, and 4 g VS, respectively. Non-pretreated sugarcane leaves at loadings of 1, 2, 3, and 4 g VS were used as controls. The pretreated and untreated sugarcane leaves at the various loadings were added to the serum bottles containing 50 mL of fermentation media and 10% (*v*/*v*) of cellulolytic consortium. The enzyme activities of the cellulolytic consortium, including CMCase, FPUase, and xylanase, were analyzed. The fermentation media was comprised of (all in mg/L) K_2HPO_4 125, $MgCl_2{\cdot}6H_2O$ 15, $FeSO_4{\cdot}7H_2O$ 25, $CuSO_4{\cdot}5H_2O$ 5, $CoCl_2{\cdot}5H_2O$ 0.125, NH_4HCO_3 5240, and $NaHCO_3$ 6720 [30]. The initial fermentation broth pH was adjusted to 6.5 using either 5 M NaOH or 5 M HCl. The serum bottles were closed tightly using a rubber stopper and aluminum caps and incubated at 37 °C for 7 d [30]. Every 24 h, 1 mL of the hydrolysate was collected from each serum bottle to measure the pH and reduce sugar concentration. At the end of the hydrolysis step, the hydrolysate and solid residue in each serum bottle were used as substrates for the two-stage hydrogen and methane production and one-stage methane production processes.

For the two-stage hydrogen and methane production process, all serum bottles from the hydrolysis step were uncapped, and the pH was adjusted to 6.5 using 5 M NaOH or 5 M HCl. Then, 10% of *C. butyricum* TISTR 1032 (10^7 cells/mL) was added to the serum bottles as the inoculum for hydrogen production. The serum bottles were recapped using a rubber stopper and aluminum caps and then purged with nitrogen gas for 15 min to ensure anaerobic conditions. The serum bottles were then incubated at 37 °C. The hydrogen production stage continued until biogas production ceased. After the first hydrogen fermentation stage, the hydrogenic effluent and solid residues were used as substrates for methane production in the second stage. The serum bottles were uncapped, and 20 g VS/L of acclimatized anaerobic sludge was added to produce methane in the methane stage. The initial fermentation broth pH was adjusted to 7.0 using 5 M NaOH or 5 M HCl.

The hydrolysate and solid residue from the hydrolysis step were used directly to produce methane for the one-stage methane production. The experiment was conducted in a serum bottle containing 20 g VS/L anaerobic sludge as the inoculum. The fermentation broth pH was adjusted to 7.0 using 5M NaOH or 5M HCl. The biogas volume was measured using a wetted glass syringe [31], and the biogas content was analyzed using gas chromatography (GC). The measurement of hydrogen and methane production continued until biogas production ceased. All treatments were performed in triplicate.

2.8. SSF for Two-Stage Hydrogen and Methane Production and One-Stage Methane Production from Pretreated Sugarcane Leaves

SSF for the two-stage hydrogen and methane production and one-stage methane production of pretreated sugarcane leaves was carried out in batch tests. Hydrolysis and fermentation occur simultaneously in this process.

The SSF for the two-stage hydrogen and methane production process was conducted using various loadings of pretreated sugarcane leaves (1, 2, 3, and 4 g VS). The control comprised untreated sugarcane leaves at different loadings as the DES-pretreated sugarcane leaves. The pretreated and untreated sugarcane leaves at the different substrate loadings were added to the serum bottles containing 50 mL of fermentation media, 10% (*v*/*v*) of the cellulolytic consortium, and 10% (*v*/*v*) of *C. butyricum* TISTR1032. The pH of the fermentation broth was adjusted to 6.5. The hydrogen production stage continued until biogas production ceased. After the first hydrogen fermentation stage, the hydrogenic effluent and solid residues were used as substrates for methane production in the second stage. At the end of the hydrogen production stage, the serum bottles were uncapped and 20 g VS/L of anaerobic sludge was added to produce methane. The fermentation broth pH was adjusted to 7.0.

Each serum bottle contained 50 mL of fermentation medium, 10% *v*/*v* of the cellulolytic consortium, 20 g-VS/L of anaerobic sludge as the inoculum for the one-stage methane production process, and different pretreated sugarcane leaves at various loadings of 1, 2, 3, and 4 g-VS. An experimental setup using untreated sugarcane leaves was used as a control. The solution pH was adjusted to 7.0 by 5 M NaOH or 5 M HCl.

Fermentation was performed in 120 mL serum bottles with a 50 mL working volume. Each serum bottle was capped using a rubber stopper and aluminum caps and closed tightly using a rubber stopper and aluminum caps. The serum bottles were incubated at 37 °C after being purged with nitrogen gas for 15 min to create anaerobic conditions. All procedures were carried out in triplicate. The measurement of hydrogen and methane production continued until biogas production ceased.

2.9. Enzyme acTIVITY Assay

The enriched culture from Section 2.3 was centrifuged at 10,000 rpm for 5 min to remove the cells and residues of the filter paper. The supernatant was used to analyze the enzyme activity. The activity of CMCase, FPUase, and xylanase was determined according to the IUPAC Commission of Biotechnology [32]. Enzyme activity assays were performed using different substrates for each enzyme. As substrates, the CMCase used 0.5 mL of 2% of carboxymethyl cellulose in sodium citrate buffer (0.05 M, pH 4.8). The FPUase used Whatman No.1 filter paper strip (1.0 × 6.0 cm) as the substrate, and xylanase used 0.9 mL of 1% (*w*/*v*) birchwood xylan in citrate-phosphate buffer (pH 5.0) as the substrate. The enzymatic assays were performed as described below.

CMCase and FPUase were determined by adding 0.5 mL supernatant to 1.0 mL sodium citrate buffer (0.05 M, pH 4.8). The substrate for each enzyme was then added to the reaction mixture. The mixture was then incubated at 50 °C for 60 min. To stop the reaction, 3.0 mL 3.5-dinitrosalicylic acid (DNS) reagent solution was added to the reaction mixture. The mixture was added to 20 mL of deionized water (DI water) and then left at room temperature for 20 min before measuring the absorbance at 540 nm using a spectrophotometer. Glucose was used as the standard.

The xylanase was determined by adding 1.0 mL of the supernatant to the 0.9 mL of 1% (*w*/*v*) birchwood xylan in citrate-phosphate buffer (pH 5.0). The mixture was incubated at 50 °C for 10 min and then cooled to room temperature. Subsequently, 20 mL of DI water was added to the cooled mixture before measuring the absorbance at 540 nm. Xylose was used as the standard.

A blank containing only water and the reaction mixtures was included to compensate for the effects of enzymes and substrates on the enzyme reaction. The control was the reaction mixture without an enzyme or substrate. CMCase, FPUase, and xylanase were measured at 540 nm using an EMC-11D-V spectrophotometer (EMCLab, Duisburg, Germany).

One enzyme unit (IU) was defined as the amount of enzyme hydrolyzing cellulose to release one microgram of glucose (FPUase and CMCase) or xylose (xylanase) per minute under assay conditions.

2.10. Analytical Methods

According to standard methods, the TS, TSS, VS, and VSS were determined [33]. The pH was measured using a pH meter (pH-500, Queen, New York, USA). The chemical compositions of untreated and pretreated sugarcane leaves, including cellulose, hemicellulose, and lignin, were analyzed according to the laboratory analytical procedures (LAP) of the National Renewable Energy Laboratory (NREL) to determine the structural carbohydrates and lignin in biomass [34]. The reducing sugars were analyzed using the 3,5-dinitrosalicylic acid (DNS) method [35]. The biogas composition was analyzed using a GC (GC-104, Shimadzu, Kyoto, Japan) with a thermal conductivity detector and a 2-m stainless steel column packed with Shin carbon (50/80) mesh. The operating conditions were as described previously [36]. The soluble metabolite products in the fermentation broth were measured using HPLC according to the method described in [37].

Compositional analysis was conducted following the standard methods of the NREL analytical procedure [34]. First, the sugarcane leaves were treated with 72% sulfuric acid at 30 °C for 1 h, followed by 4% sulfuric acid-treated samples at 121 °C for 1 h, and then the liquid and solid were separated by vacuum filtration. Next, the cellulose and hemicellulose contents were calculated from the corresponding sugar concentrations obtained from the measurement of liquid fractions by HPLC with conversion factors of 0.90 for glucose and 0.88, respectively, for glucose and xylose. Next, the acid-insoluble lignin was gravimetrically determined from the solid fractions using acid-insoluble lignin and ash. In contrast, the acid-soluble lignin content was determined by measuring the absorbance of the liquid fractions at 205 nm using a spectrophotometer.

2.11. Statistical Analysis

Statistical analysis was conducted with the IBM SPSS statistics program version 21. The means were compared using one-way ANOVA analysis with the Duncan test as a post hoc test. In addition, the means difference of the three samples were compared using an independent-samples *t*-test. All statistical tests were performed with a 95% confidence level.

2.12. Calculations

The hydrogen and methane volumes were calculated using the mass balance equation [38] and the, hydrogen and methane yields were expressed in mL H_2 or CH_4/g-VS_{added}. Next, the modified Gompertz equation (Equation (1)) was used to fit the cumulative hydrogen and methane yield curves [39]. Finally, a modified Gompertz equation was used to fit the cumulative hydrogen production curves and obtain the hydrogen production P, the hydrogen production rate R, and λ.

$$H = Pexp\left\{-exp\left[\frac{Rm \times e}{P}\right](\lambda - t) + 1\right\} \quad (1)$$

where H is the cumulative hydrogen or methane production (mL), λ is the lag phase time (h), P is the hydrogen or methane production potential (mL), and Rm is the hydrogen or methane production rate (mL H_2/L d or mL CH_4/L d). The incubation time (t) is reported in hours (h) or days. e is an exponential constant equal to 2.718. The equation was plotted using a nonlinear curve fitting in SigmaPlot 11 (Systat Software Inc., Palo Alto, CA, USA).

3. Results

3.1. Effects of Molar Ratio and Type of Deep Eutectic Solvent (DES) on Sugarcane Leaves

The untreated sugarcane leaves had a lower cellulose, hemicellulose, and lignin percentage than the pretreated ones (Table 1). These results implied that the DES pretreatment efficiently removed the lignin content in the lignocellulosic biomass.

Table 1. Effects of DES types and molar ratio on the pretreatment of sugarcane leaves at a 1:16 substrate to DES solution ratio, 6 h, and 80 °C.

DES	Molar Ratio	pH	Pretreatment			Composition of Residues		
			Cellulose Recovery (%)	Hemicellulose Removal (%)	Lignin Removal (%)	Cellulose (%)	Hemicellulose (%)	Lignin (%)
Untreated	-	-	-	-	-	36.18 ± 0.73 [f]	25.23 ± 0.02 [a]	27.68 ± 0.35 [a]
ChCl/MEA	1:6	14.08	71.86 ± 0.37 [c,d]	69.88 ± 2.93 [b,c]	77.62 ± 0.79 [a]	57.62 ± 0.30 [b,c]	16.84 ± 1.64 [d]	13.73 ± 0.48 [e]
	1:8	14.12	73.49 ± 0.53 [c]	71.07 ± 0.54 [b]	77.50 ± 0.33 [a]	58.92 ± 0.42 [a]	16.21 ± 0.30 [d]	13.80 ± 0.30 [e]
	1:10	14.18	70.96 ± 2.10 [d]	68.21 ± 1.45 [c]	76.41 ± 0.85 [a]	55.33 ± 1.64 [d]	17.28 ± 0.79 [d]	14.07 ± 0.51 [e]
ChCl/G	1:2	7.93	82.37 ± 0.71 [b]	33.60 ± 1.16 [e]	28.98 ± 1.23 [d]	39.40 ± 0.34 [e]	22.15 ± 0.45 [b]	25.99 ± 0.45 [b]
	1:4	6.40	85.08 ± 0.81 [a]	33.57 ± 1.67 [e]	25.11 ± 0.61 [e]	38.72 ± 0.53 [e]	21.09 ± 0.53 [b,c]	26.08 ± 0.21 [b]
	1:6	6.60	82.14 ± 1.49 [b]	37.35 ± 2.39 [d]	29.41 ± 0.80 [d]	39.12 ± 0.71 [e]	20.81 ± 0.79 [c]	25.72 ± 0.29 [b]
ChCl/G/$AlCl_3$	1:2:0.33	0.40	65.65 ± 0.23 [f]	87.00 ± 0.07 [a]	68.57 ± 0.26 [c]	59.28 ± 0.20 [a]	8.19 ± 0.05 [e]	21.72 ± 0.18 [c]
	1:4:0.33	0.38	66.61 ± 0.70 [e,f]	88.54 ± 0.48 [a]	70.27 ± 0.54 [b]	58.74 ± 0.62 [a,b]	7.05 ± 0.29 [f]	20.06 ± 0.37 [d]
	1:6:0.33	0.36	68.11 ± 0.81 [e]	88.21 ± 0.22 [a]	68.39 ± 0.19 [c]	57.37 ± 0.68 [c]	6.92 ± 0.13 [f]	20.37 ± 0.13 [d]

Untreated, untreated sugarcane leaves; ChCl, choline chloride; MEA, monoethanolamine; G, glycerol; $AlCl_3$, aluminum chloride; and DES, deep eutectic solvent. Values marked with the same letters are not significantly different ($p < 0.05$).

The results demonstrated that the molar ratio of ChCl to MEA in sugarcane leaf pretreatment did not affect lignin removal, hemicellulose removal, or cellulose recovery (Table 1). Based on these data, the optimal molar ratio for the pretreatment of sugarcane leaves with ChCl/MEA at ratios of 1:6, 1:8, and 1:10 was determined to be 1:6 because the MEA concentration was the lowest. The highest lignin removal efficiency was observed at molar ratios of 1:2 and 1:6 when sugarcane leaves were pretreated with ChCl/glycerol (G). In contrast, the molar ratio variations of 1:6 and 1:4 resulted in the highest hemicellulose removal and cellulose recovery. Because this study focuses on the ability of DESs to remove lignin from sugarcane leaves, a molar ratio that can remove the majority of lignin was chosen for the pretreatment of sugarcane leaves. Additionally, a low molar ratio leads to a smaller amount of each HBA and HBD being considered. Therefore, the optimum ChCl/G molar ratio was determined to be 1:2. When ChCl/G/$AlCl_3$ was employed as the pretreatment solvent, ChCl/G/$AlCl_3$, at a molar ratio of 1:4:0.33, achieved maximum lignin removal. However, the molar ratio had no significant influence on hemicellulose removal and cellulose recovery. Thus, the optimum molar ratio of ChCl/G/$AlCl_3$ is 1:4:0.33. The molar ratio affects the DES's physical properties, such as freezing point, density, and viscosity.

Lignin removal efficiency ranged from 76.41–77.62% at various molar ratios in DESs with strong bases (pH 14.08–14.18), i.e., ChCl/MEA (Table 1). This is not surprising because lignin is a base-soluble biopolymer [40]; therefore, basic solvents are advantageous for its removal. Furthermore, it was discovered that the more basic the solvent, the more lignin was extracted [40]. Under basic conditions, the breakage of ether links in lignin and ester bonds between lignin and hemicellulose leads to lignin removal [40] and dissolution [41].

At the optimum ChCl/G ratio of 1:2, the lignin removal was only $28.98 \pm 1.23\%$, which is the lowest compared to other DES types (Table 1), indicating that ChCl/G is less effective in removing lignin from sugarcane leaves. This is because ChCl/G has a neutral pH (6.4–7.93), making the ChCl/G solution less efficient at eliminating lignin [42].

DESs with a strong acid i.e., ChCl/G/$AlCl_3$ (pH of 0.8–0.9), removed lignin in the moderate range of 68.39–70.27% but showed the highest hemicellulose removal (Table 1). In general, pretreatment solutions with low pH affect hemicellulose hydrolysis but have an intense effect on lignin dissolution and cellulose hydrolysis [41].

In the pretreatment processes, pH and viscosity were considered the main factors when determining the DES performance. The pH-influenced lignin removal from the lignocellulosic biomass occurs because lignin is a base-soluble polymer. Therefore, a DES with a high pH is considered for use as a pretreatment agent [40]. In contrast, DESs possess a comparatively higher viscosity (>100 cP) at room temperature which significantly limits their extraction applications. Additionally, their high viscosity reduces the mass transfer rate between the sample and extraction phase owing to the formation of extensive H-bond networks between the HBA and HBD components [43].

ChCl: MEA had the highest pH compared to ChCl/G and ChCl/G/$AlCl_3$. Because lignin is a base-soluble polymer, ChCl/MEA basicity makes it more effective for lowering the lignin concentration in the biomass while maintaining high cellulose recovery; hence, it is more advantageous for lignin removal. However, the results showed that ChCl/MEA had an insignificant effect on lignin removal at different molar ratios (Table 1). A higher molar ratio leads to more solvents, which increases the production costs when applied on an industrial scale. Therefore, the selected DESs in subsequent experiments were ChCl/MEA at a 1:6 molar ratio.

3.2. Effects of Pretreatment Time and Pretreatment Temperature

This study was conducted by varying the pretreatment time to 3, 6, and 9 h, while the pretreatment temperatures were varied at 80, 100, and 120 °C. ChCl/MEA at a 1:6 molar ratio was used as the solvent to pretreat the sugarcane leaves. At a pretreatment time of 3 h, the pretreatment temperature of 120 °C shows the lowest lignin content of 8.95 ± 0.15%, while still recovering the highest cellulose (70.17 ± 0.20%) compared to the other pretreatment temperatures (Table 2). Similar results were obtained with pretreatment times of 6 h and 9 h at a pretreatment temperature of 120 °C. At 120 °C, the lowest lignin content (i.e., the highest lignin removal) was attained when the pretreatment time was 9 h. However, the cellulose recovery was not significantly different from the other pretreatment conditions. At 120 °C, with various pretreatment times of 3 h, 6 h, and 9 h, the system energy inputs were 622, 1198, and 1774 kJ/g-biomass, respectively. Therefore, the lowest energy occurred with the 3 h pretreatment time. Thus, the optimum pretreatment time and temperature for the pretreatment of sugarcane leaves with ChCl/MEA were 3 h at 120 °C.

Table 2. Effects of the temperatures and time on the pretreatment of sugarcane leaves at a 1:16 substrate to DES solution ratio using 1:6 molar ratio of ChCls/MEA as DES.

DES	Temp (°C)/ Time (h)	Pretreatment			Composition of Residues		
		Cellulose Recovery (%)	Hemicellulose Removal (%)	Lignin Removal (%)	Cellulose (%)	Hemicellulose (%)	Lignin (%)
Untreated					36.56 ± 0.77 ^{k}	27.31 ± 0.03 ^{a}	27.68 ± 0.35 ^{a}
ChCl/MEA (1:6)	80/3	69.25 ± 0.97 ^{g}	57.76 ± 0.72 ^{h}	73.49 ± 0.86 ^{i}	53.37 ± 0.75 ^{j}	24.32 ± 0.41 ^{b}	15.47 ± 0.50 ^{b}
	80/6	68.69 ± 1.51 ^{g}	66.60 ± 1.90 ^{g}	78.67 ± 0.52 ^{h}	57.61 ± 1.27 ^{h}	20.93 ± 1.19 ^{c}	13.55 ± 0.33 ^{c}
	80/9	68.37 ± 1.13 ^{g}	70.23 ± 1.40 ^{f}	80.81 ± 0.61 ^{g}	60.93 ± 1.01 ^{g}	19.82 ± 0.93 ^{d}	12.95 ± 0.41 ^{d}
	100/3	71.54 ± 0.30 ^{f}	70.29 ± 0.30 ^{f}	83.45 ± 0.02 ^{e,f}	64.25 ± 0.27 ^{f}	19.93 ± 0.20 ^{d}	11.25 ± 0.01 ^{e}
	100/6	72.81 ± 0.42 ^{e}	75.05 ± 0.96 ^{a,b}	85.96 ± 0.28 ^{d}	67.52 ± 0.39 ^{d}	17.28 ± 0.66 ^{f,g}	9.86 ± 0.20 ^{f}
	100/9	72.04 ± 0.40 ^{e,f}	76.57 ± 0.19 ^{a}	87.06 ± 0.01 ^{c}	69.64 ± 0.39 ^{b,c}	16.92 ± 0.14 ^{g,h}	9.47 ± 0.01 ^{f,g}
	120/3	78.75 ± 0.22 ^{a}	70.64 ± 0.51 ^{e,f}	86.73 ± 0.22 ^{c}	70.17 ± 0.20 ^{b}	19.54 ± 0.34 ^{d}	8.95 ± 0.15 ^{h,i}
	120/6	78.32 ± 0.38 ^{a}	72.57 ± 0.37 ^{d}	87.95 ± 0.53 ^{b}	71.91 ± 0.34 ^{a}	18.12 ± 0.24 ^{e,f}	8.07 ± 0.35 ^{j}
	120/9	78.80 ± 0.89 ^{a}	74.54 ± 0.79 ^{b,c}	89.18 ± 0.29 ^{a}	71.33 ± 0.81 ^{a}	17.22 ± 0.54 ^{f,gh}	7.42 ± 0.20 ^{k}

Untreated: untreated sugarcane leaves; ChCl: choline chloride; and MEA: monoethanolamine. Values marked with the same letters are not significantly different ($p < 0.05$).

3.3. Effect of the Substrate to DES Solution Ratio

The effect of the substrate to DES solution ratio was determined at the optimum conditions using ChCl/MEA at a 1:6 molar ratio, 120 °C, and 3 h. The substrate to DES solution ratios were 1:8, 1:12, 1:16, 1:20, and 1:24. The lowest lignin contents of 8.13 ± 0.57% and 8.15 ± 0.33%, respectively, were achieved at the substrate to DES solution ratios of 1:16 and 1:20, with lignin removal of 88.23 ± 0.83% and 88.06 ± 0.48% (Table 3). The energy inputs at 1:16 and 1:20 substrate to DES solution under the optimum conditions of 1:6 molar ratio ChCl/MEA, 120 °C, and 3 h, were 622 and 776 kJ/g-biomass, respectively. Therefore, the ratio of 1:16 substrates to DES solution is the optimal ratio when the lignin content after pretreatment is considered the criterion.

Table 3. Effects of the substrate to DES solution ratio using a 1:6 molar ratio of ChCls/MEA to DES at a temperature of 120 °C and time of 3 h on the pretreatment of sugarcane leaves.

The Substrate to DES Solution Ratio	Pretreatment			Composition of Residues		
	Cellulose Recovery (%)	Hemicellulose Removal (%)	Lignin Removal (%)	Cellulose (%)	Hemicellulose (%)	Lignin (%)
Untreated				36.56 ± 0.77 [d]	27.31 ± 0.03 [a]	27.68 ± 0.35 [a]
1:8	81.84 ± 2.23 [b]	72.40 ± 0.68 [c]	86.86 ± 0.16 [b]	69.81 ± 0.62 [c]	17.59 ± 0.43 [b]	8.49 ± 0.10 [b,c]
1:12	84.13 ± 0.77 [a]	73.98 ± 0.42 [b]	86.37 ± 0.36 [b]	72.54 ± 0.67 [b]	16.76 ± 0.27 [c]	8.90 ± 0.23 [b]
1:16	81.44 ± 1.21 [b]	74.62 ± 0.41 [b]	88.23 ± 0.83 [a]	74.27 ± 1.11 [a]	17.29 ± 0.28 [b]	8.13 ± 0.57 [c]
1:20	81.31 ± 1.11 [b]	75.78 ± 0.45 [a]	88.06 ± 0.48 [a]	73.34 ± 1.00 [a,b]	16.32 ± 0.30 [c]	8.15 ± 0.33 [c]
1:24	84.38 ± 0.15 [a]	76.29 ± 0.10 [a]	86.42 ± 0.09 [b]	74.27 ± 0.13 [a]	15.59 ± 0.07 [d]	9.05 ± 0.06 [b]

Untreated: untreated sugarcane leaves; values marked with the same letters are not significantly different ($p < 0.05$).

At ratios of 1:12 and 1:24 substrate to DES solution, the cellulose recovery was $84.13 \pm 0.77\%$ and $84.38 \pm 0.02\%$, respectively (Table 3). However, at a ratio of 1:24 substrate to DES solution, the results indicate the highest lignin content compared to other variations in the substrate to DES solution ratio. Because a ratio of 1:12 substrate (467 kJ/g-biomass) to DES solution had a lower energy input than at 1:24 (930 kJ/g-biomass), a 1:12 substrate to DES solution ratio was considered the optimum condition for cellulose recovery.

The ratio of 1:8 substrate to DES solution yielded the lowest cellulose concentration of $69.81 \pm 0.62\%$. This could be because the solvent cannot break more ether bonds in lignin-carbohydrate complexes because of the high substrate concentrations used at this ratio, resulting in minimal lignin removal (Table 3). A high ratio of substrates to the DES solution may also lead to the accumulation of solid particles and an increase in system viscosity [44]. Additionally, the viscous system limits the mass transfer between the samples and the extraction phase. Consequently, the 1:8 substrate to DES solution ratio is not optimal because it diminishes lignin removal and cellulose recovery capabilities.

The substrate to DES solution ratio of 1:16 resulted in greater lignin removal. In contrast, at a substrate to DES solution ratio of 1:12, the cellulose recovery was higher with a low hemicellulose content of $16.76 \pm 0.2\%$. At a substrate to DES solution ratio of 1:12, the energy input was lower than that at 1:16, resulting in energy inputs of 467 kJ/g-biomass and 622 kJ/g-biomass, respectively. The results indicated that the optimum substrate to DES solution ratio was 1:12, resulting in low lignin content and high cellulose content.

3.4. Enzymatic Hydrolysis of DES Pretreated Sugarcane Leaves by the Cellulolytic Consortium

Sugarcane leaves were pretreated using DESs at a molar ratio of 1:6 and 120 °C for 3 h, and a substrate to DES solution ratio of 1:12 before being subjected to enzymatic hydrolysis. Untreated sugarcane leaves were used as the controls. The highest reducing sugar concentrations of 1.49 ± 0.28 g/L and 2.36 ± 0.02 g/L were obtained from untreated and DES-pretreated sugarcane leaves on day 7 (Table 4). However, it should be noted that the substrate loading should not be greater than 4 g of volatile solids $(VS)_{added}$ because solid accumulation can occur, limiting the mass transfer between the substrate and enzyme. Additionally, at the high loading of the pretreated biomass, enzymatic digestibility and fermentation efficiency were dramatically reduced as mixing became increasingly difficult as viscosity increased [44]. Therefore, a 4 g-VS_{added} was chosen as the optimum substrate loading for enzymatic hydrolysis by the cellulolytic consortium using DES-pretreated sugarcane leaves as substrates.

Table 4. The effects of substrate loading on the enzymatic hydrolysis of untreated and DES pretreated sugarcane leaves by the cellulolytic consortium.

Substrate Loading (g-VS$_{added}$)	Reducing Sugar Concentration (g/L)							
	Day 0	Day 1	Day 2	Day 3	Day 4	Day 5	Day 6	Day 7
U-1	0.15 ± 0.02 ^{a}	0.40 ± 0.01 ^{e}	0.28 ± 0.01 ^{e}	0.40 ± 0.04 ^{d}	0.36 ± 0.01 ^{e}	0.44 ± 0.01 ^{d}	0.37 ± 0.01 ^{f}	0.26 ± 0.24 ^{f}
U-2	0.16 ± 0.02 ^{a}	0.21 ± 0.01 ^{g}	0.43 ± 0.00 ^{e}	0.50 ± 0.00 ^{d}	0.78 ± 0.05 ^{d}	0.88 ± 0.05 ^{c}	0.96 ± 0.02 ^{d}	0.85 ± 0.04 ^{e}
U-3	0.17 ± 0.02 ^{a}	0.81 ± 0.03 ^{c}	0.50 ± 0.02 ^{d}	0.72 ± 0.02 ^{c}	0.90 ± 0.02 ^{c,d}	0.98 ± 0.00 ^{c}	1.70 ± 0.01 ^{d}	1.13 ± 0.01 ^{d}
U-4	0.17 ± 0.02 ^{a}	0.35 ± 0.02 ^{f}	0.36 ± 0.00 ^{f}	0.74 ± 0.02 ^{c}	1.12 ± 0.20 ^{b}	1.66 ± 0.33 ^{a}	1.64 ± 0.29 ^{c}	1.49 ± 0.28 ^{c}
P-1	0.17 ± 0.02 ^{a}	0.41 ± 0.02 ^{e}	0.36 ± 0.03 ^{f}	0.49 ± 0.02 ^{d}	0.47 ± 0.09 ^{e}	0.60 ± 0.14 ^{d}	0.61 ± 0.11 ^{e}	0.35 ± 0.07 ^{f}
P-2	0.17 ± 0.02 ^{a}	0.70 ± 0.01 ^{d}	0.57 ± 0.02 ^{c}	0.86 ± 0.15 ^{b}	0.95 ± 0.11 ^{c}	1.43 ± 0.02 ^{b}	1.65 ± 0.06 ^{c}	1.85 ± 0.15 ^{b}
P-3	0.16 ± 0.00 ^{a}	1.18 ± 0.03 ^{a}	1.52 ± 0.01 ^{a}	1.56 ± 0.02 ^{a}	1.71 ± 0.02 ^{a}	1.78 ± 0.02 ^{a}	1.97 ± 0.08 ^{b}	1.95 ± 0.07 ^{b}
P-4	0.17 ± 0.02 ^{a}	0.95 ± 0.03 ^{b}	1.10 ± 0.03 ^{b}	0.92 ± 0.03 ^{b}	1.25 ± 0.03 ^{b}	1.88 ± 0.01 ^{a}	2.18 ± 0.02 ^{a}	2.35 ± 0.02 ^{a}

Untreated: untreated sugarcane leaves; P: pretreated sugarcane leaves; values marked with the same letters are not significantly different ($p < 0.05$).

3.5. The Two-Stages Hydrogen and Methane Production by SHF and SSF

3.5.1. Hydrogen Production

The SHF and SSF processes were carried out to determine an efficient fermentation process to produce hydrogen and methane through a two-stage hydrogen and methane production process. Various substrate loadings, 1–4 g-VS, of DES-pretreated sugarcane leaves under the optimum conditions of a 1:6 molar ratio, 120 °C, 3 h, and a substrate to DES solution ratio of 1:12 were used as the substrate. Additionally, untreated sugarcane leaves were used as a control at various substrate loadings of 1–4 g-VS.

The hydrogen production rate and hydrogen production potential of the DES-pretreated sugarcane leaves by SSF of the two-stage hydrogen and methane production processes were higher than those of untreated sugarcane leaves (Table 5). The highest hydrogen production and production rates of 3187 ± 202 mL H_2/L and 317.5 ± 20.5 mL H_2/L.d, respectively, were achieved by SSF using DES-pretreated sugarcane leaves at a substrate loading of 4 g VS (Table 5). Due to the presence of lignin in the untreated sugarcane leaves, their hydrogen production was lower than that of the DES-pretreated sugarcane leaves.

Table 5. Comparison of hydrogen production by SSF and SHF of the two-stage hydrogen production process.

Substrate Loading (g-VS$_{added}$)	SSF-Two Stages Hydrogen Production					SHF-Two Stages Hydrogen Production				
	Hydrogen Production (mL/L)	Hydrogen Production Rate (mL/L d)	Lag Phase (λ) (d)	Hydrogen Yield (mL/g-VS$_{added}$)	R^2	Hydrogen Production (mL/L)	Hydrogen Production Rate (mL/L d)	Lag Phase (λ) (d)	Hydrogen Yield (mL/g-VS$_{added}$)	R^2
U_1	21 ± 4 ^{e}	1.8 ± 0.8 ^{h}	0.0	1.0 ± 0.1 ^{f}	0.9826	17 ± 1 ^{e}	1.5 ± 0.2 ^{h}	0.5	0.8 ± 0.1 ^{f}	0.9844
U_2	66 ± 5 ^{e}	10.0 ± 2.0 ^{g,h}	2.0	1.7 ± 0.1 ^{f}	0.9935	32 ± 15 ^{e}	6.2 ± 6.4 ^{g,h}	4.0	0.8 ± 0.4 ^{f}	0.9814
U_3	133 ± 14 ^{e}	22.0 ± 3.3 ^{f,g}	2.0	2.2 ± 0.2 ^{f}	0.9958	52 ± 10 ^{e}	14.2 ± 5.2 ^{g,h}	5.0	0.9 ± 0.2 ^{f}	0.9892
U_4	229 ± 20 ^{e}	34.7 ± 3.9 ^{e,f}	2.0	2.9 ± 0.2 ^{f}	0.9962	99 ± 81 ^{e}	20.2 ± 15.4 ^{f,g,h}	5.0	1.2 ± 1.0 ^{f}	0.9885
P_1	152 ± 7.0 ^{e}	20.0 ± 2.2 ^{f,g}	0.5	7.6 ± 0.3 ^{e}	0.9947	48 ± 19 ^{e}	7.9 ± 3.6 ^{g,h}	3.0	2.4 ± 1.0 ^{f}	0.9902
P_2	821 ± 48 ^{d}	85.9 ± 5.1 ^{d}	4.0	20.5 ± 1.2 ^{c}	0.9956	682 ± 154 ^{d}	40.9 ± 10.7 ^{e}	4.0	17.0 ± 3.8 ^{d}	0.9916
P_3	1840 ± 201 ^{c}	211.2 ± 14.7 ^{b}	4.0	30.7 ± 3.3 ^{b}	0.9978	1700 ± 136 ^{c}	112.4 ± 7.2 ^{c}	5.0	28.1 ± 1.9 ^{b}	0.9962
P_4	3187 ± 209 ^{a}	317.5 ± 20.5 ^{a}	5.0	39.8 ± 2.6 ^{a}	0.9941	2135 ± 315 ^{b}	197.2 ± 20.5 ^{b}	7.0	26.7 ± 3.9 ^{a}	0.9952

Untreated: untreated sugarcane leaves; P: pretreated sugarcane leaves; values marked with the same letters are not significantly different ($p < 0.05$).

Similar findings on hydrogen production were obtained with SHF, and similar findings on hydrogen production were obtained with the SHF process. DES-pretreated sugarcane leaves had higher hydrogen production, a higher hydrogen production rate, and a shorter lag phase than the untreated (Table 5). The highest hydrogen production and hydrogen production rate were achieved at 4 g VS of DES-pretreated sugarcane leaves, resulting in 2135 ± 315 mL H_2/L and 197.2 ± 20.5 mL H_2/L.d, respectively. The untreated sugarcane leaves at 4 g VS produced hydrogen and hydrogen production rates of 99 ± 81 mL H_2/L and 20.2 ± 15.4 mL H_2/L, respectively.

3.5.2. Methane Production

The hydrogenic effluent from the first stage was used to produce methane during the second stage. Methane production (3583 ± 128 mL CH_4/L) from the SHF process using the optimum substrate loading of 4 g VS of DES-pretreated sugarcane leaves was lower than that from the SSF process (5923 ± 251 mL CH_4/L) (Table 6). Substrate loading influences VFAs production. The maximum methane content under the optimum condition obtained from the SHF and SSF processes were 51% and 50 %, respectively (data not shown). The results revealed that 4 g-VS of DES-pretreated sugarcane leaves gave the highest VFAs production (Figure 1) and hydrogen (Table 5) and methane production (Table 6) in both SSF and SHF processes.

Table 6. Comparison of methane production by SSF and SHF of the two-stage methane production process.

Substrates Loading (g-VS_{added})	SSF Two-Stages Methane Production					SHF Two-Stages Methane Production				
	Methane Production (mL/L)	Methane Production Rate (mL/L d)	Lag Phase (λ) (d)	Methane Yield (mL/ g-VS_{added})	R^2	Methane Production (mL/L)	Methane Production Rate (mL/L d)	Lag Phase (λ) (d)	Methane Yield (mL/ g-VS_{added})	R^2
U_1	1448 ± 354 ^{f}	29.9 ± 10.4 ^{e}	14	115.9 ± 28.3 ^{c,d,e,f}	0.9931	1468 ± 441 ^{f}	33.3 ± 13.9 ^{d,e}	13	117.4 ± 35.3 ^{c,d,e,f}	0.9921
U_2	2633 ± 390 ^{c,d,e}	64.7 ± 9.7 ^{c,d,e}	11	105.3 ± 15.6 ^{d,e,f}	0.9968	1974 ± 545 ^{e,f}	43.1 ± 15.6 ^{d,e}	12	91.1 ± 21.4 ^{e,f}	0.9972
U_3	3009 ± 516 ^{b,c}	63.1 ± 15.6 ^{c,d,e}	12	80.3 ± 13.8 ^{e,f}	0.9949	2847 ± 530 ^{b,c,d}	53.8 ± 3.1 ^{d,e}	10	75.9 ± 14.1 ^{e,f}	0.9950
U_4	3311 ± 806 ^{b,c}	76.8 ± 18.8 ^{b,c,d,e}	19	66.2 ± 16.1 ^{e,f}	0.9954	3097 ± 244 ^{b,c}	70.1 ± 3.6 ^{b,c,d,e}	18	61.9 ± 4.9 ^{f}	0.9926
P_1	2654 ± 277 ^{c,d,e}	101.8 ± 27.7 ^{b,c}	6	212.3 ± 22.2 ^{a}	0.9950	2077 ± 308 ^{d,e,f}	68.4 ± 25.4 ^{b,c,d,e}	6	166.1 ± 24.6 ^{a,b,c,d}	0.9873
P_2	3402 ± 760 ^{b,c}	109.7 ± 68.4 ^{b,c}	3	178.5 ± 72.0 ^{a,b,c}	0.9660	3644 ± 473 ^{b}	77.5 ± 25.3 ^{b,c,d,e}	16	201.2 ± 113.8 ^{a,b}	0.9795
P_3	5179 ± 291 ^{a}	160.1 ± 15.2 ^{a}	1	138.1 ± 7.8 ^{b,c,d,e}	0.9751	3221 ± 417 ^{b,c}	113.8 ± 30.9 ^{b}	1	85.9 ± 11.1 ^{e,f}	0.9797
P_4	5923 ± 251 ^{a}	159.3 ± 19.0 ^{a}	1	118.5 ± 5.0 ^{c,d,e,f}	0.9828	3583 ± 128 ^{b}	79.3 ± 9.8 ^{b,c,d}	2	71.7 ± 2.6 ^{e,f}	0.9778

Untreated: untreated sugarcane leaves; P: pretreated sugarcane leaves; values marked with the same letters are not significantly different ($p < 0.05$).

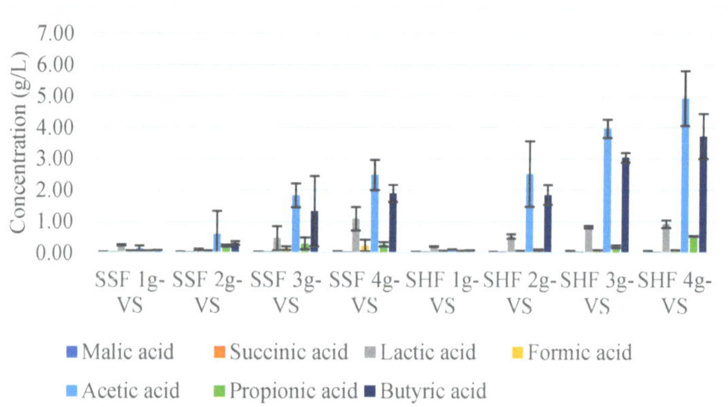

Figure 1. Comparisons of soluble metabolite products from SSF and SHF of two-stage hydrogen and methane production using DES-pretreated sugarcane leaves at different loading substrates.

3.6. One-Stage Methane Production by SHF and SSF

For the SHF process, DES-pretreated sugarcane leaves at 3 g-VS gave the highest methane production of 3723 ± 340 mL CH_4/L (Table 7). Under these conditions, the methane content was 53 % (data not shown). Methane production increased with increasing substrate concentrations. However, methane production was reduced when the substrate concentration reached 4-gVS. This may be because of the accumulation of VFAs during AD. The results in Figure 2 show that at 4 g VS, acetic acid, butyric acid, and propionic acid accumulated in the methanogenic effluent.

Table 7. Effects of substrate loading on enzymatic hydrolysis of untreated and DES pretreated sugarcane leaves by the cellulolytic consortium.

Substrates Loading (g-VS$_{added}$)	SSF One-Stage Methane Production					SHF One-Stage Methane Production				
	Methane Production (mL/L)	Methane Production Rate (mL/L d)	Lag Phase (λ) (d)	Methane Yield (mL/ g-VS$_{added}$)	R^2	Methane Production (mL/L)	Methane Production Rate (ml/L d)	Lag Phase (λ) (d)	Methane Yield (mL/ g-VS$_{added}$)	R^2
U_1	1632 ± 481 ^{f,g}	27.4 ± 14.0 ^{e}	9	130.5 ± 38.5 ^{a,b,c}	0.9848	1597 ± 12 ^{f,g}	21.5 ± 6.3 ^{e}	10	127.8 ± 33.5 ^{a,b,c}	0.9918
U_2	2015 ± 370 ^{e,f,g}	30.0 ± 7.0 ^{d,e}	9	80.6 ± 14.8	0.9816	2179 ± 405 ^{e,f,g}	33.4 ± 5.1 ^{d,e}	18	87.2 ± 16.2 ^{d,e,f,g}	0.9938
U_3	2735 ± 186 ^{c,d,e}	42.6 ± 3.1 ^{c,d,e}	16	76.3 ± 5.5 ^{d,e,f,g}	0.9907	1941 ± 558 ^{e,f,g}	29.2 ± 11.3 ^{e}	16	51.7 ± 14.9 ^{e,f,g}	0.9629
U_4	2988 ± 112 ^{b,c,d}	46.8 ± 4.9 ^{c,d,e}	12	59.8 ± 2.2 ^{e,f,g}	0.9940	2304 ± 304 ^{d,e,f}	35.9 ± 7.5 ^{c,d,e}	15	46.1 ± 6.1 ^{g}	0.9867
P_1	2029 ± 104 ^{e,f,g}	41.8 ± 3.7 ^{c,d,e}	4	162.4 ± 8.3 ^{a}	0.9948	1853 ± 242 ^{f,g}	34.5 ± 7.9 ^{d,e}	9	148.3 ± 19.3 ^{a,b}	0.9948
P_2	2155 ± 449 ^{e,f,g}	59.3 ± 31.2 ^{b,c}	2	113.0 ± 44.9 ^{b,c,d}	0.9905	1478 ± 796 ^{g}	44.6 ± 17.7 ^{c,d,e}	2	74.7 ± 34.3 ^{d,e,f,g}	0.9600
P_3	3474 ± 613 ^{a,b,c}	79.8 ± 15.8 ^{b}	1	92.6 ± 16.3 ^{c,d,e,f}	0.9875	3723 ± 340 ^{a,b}	60.3 ± 13.4 ^{b,c}	12	99.3 ± 9.1 ^{c,d,e}	0.9965
P_4	4067 ± 319 ^{a}	120.5 ± 9.5 ^{a}	4	81.3 ± 6.4 ^{d,e,f,g}	0.9936	3349 ± 415 ^{a,b,c}	54.7 ± 16.6 ^{c,d}	3	67.0 ± 8.3 ^{e,f,g}	0.9793

Untreated: untreated sugarcane leaves; P: pretreated sugarcane leaves; values marked with the same letters are not significantly different ($p < 0.05$).

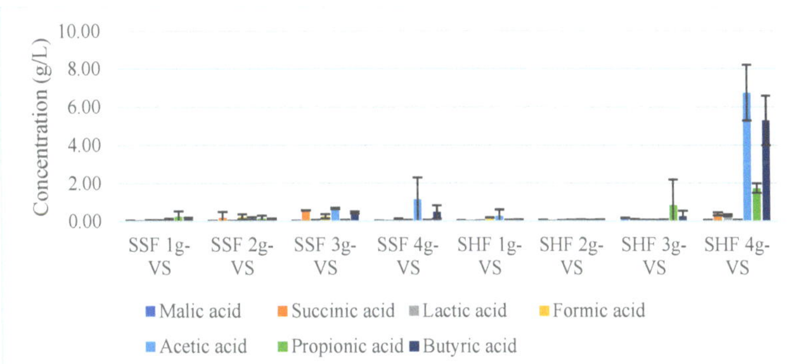

Figure 2. Comparison of soluble metabolite production in SSF and SHF one-stage methane production using DES-pretreated sugarcane leaves with substrate loading variations.

One-stage methane production by SSF showed that the highest methane production of 4067 ± 319 mL CH_4/L was achieved using 4 g-VS DES-pretreated sugarcane leaves as the substrate (Table 7). Under these conditions, the methane content was approximately 56%. This result indicates that SSF supports a higher concentration of the substrate to produce methane than SHF for one-stage methane production. This might be because the soluble metabolite products produced by SSF were directly converted to methane during the methanogenesis stage, as shown in Figure 2. Therefore, there was no accumulation of VFAs which can inhibit and reduce methane production.

4. Discussion

4.1. The Effect of Molar Ratio and Type of Deep Eutectic Solvents (DESs) on Sugarcane Leaves

An efficient pretreatment can be achieved via the breakage of the lignin structure, which is the main biomass protective barrier, thereby enhancing the accessibility of solvents or enzymes into the biomass for sugar hydrolysis [45]. Furthermore, the DES mechanisms in biomass delignification are due to their ability to selectively cleave ether bonds without affecting C–C linkages [45]. Moreover, the ability of DES to donate and accept protons and electrons could reduce cellulose crystallinity owing to the disruption of H-bonds in the lignocellulosic biomass [45].

A strong basicity solvent is expected to remove lignin via saponification of intermolecular ester linkages, crosslinking xylan, hemicelluloses, and other components, such as lignin and other hemicelluloses. With the elimination of crosslinks, the porosity of lignocellulosic materials rises [46]. The basic solvent used in pretreatment operations causes swelling, which leads to an increase in the internal surface area, a decrease in the degree of polymerization, a reduction in crystallinity, separation of structural links between lignin and carbohydrates, and disruption of the lignin structure [46].

The H-bonds present in ChCl/G were identified as Ch^+ and glycerol (cationic H-bond), chloride ions (Cl^-) and glycerol (anionic H-bond), ChCl ion pairs (doubly ionic H-bonds), and glycerol-glycerol (neutral H-bonds) [47]. The close to neutral pH of ChCl/G is due to the Cl^- ion being surrounded by glycerol, resulting in a stronger glycerol-glycerol bond than the Cl^--glycerol bond. Moreover, the H-bond strength of lignin was stronger than that of ChCl/G lignin (lignin-lignin > Cl^-/glycerol > Ch^+/glycerol) [47]. Moreover, the H-bond energy of β-O-4 ether linkages in lignin is greater than that of ChCl/G [47]. Thus, ChCl/G, which has a weak H-bond interaction, cannot break the ether contained in the biomass [47].

The method of using strong acid solvents to disrupt the van der Waals forces, H-bonds, and covalent bonds that hold the cellulose and hemicellulose structures together in the lignocellulosic biomass, results in hemicellulose solubilization and cellulose reduction [48]. Furthermore, the acid catalyst hydronium ions cause long cellulose and hemicellulose chains to break down into sugar monomers [49]. Although using acid in the lignocellulosic pretreatment process removes hemicellulose and increases the lignocellulose pore size, the main disadvantage is the formation of inhibitors such as hydroxymethylfurfural (HMF), furfural, and other by-products, such as phenylic compounds and aliphatic carboxylic acids [41]. Additionally, the downstream process requires pH neutralization and significant biomass size reduction [50].

4.2. The Effect of Pretreatment Time and Pretreatment Temperature

The results of the variations in pretreatment temperature at each pretreatment time indicated that increasing the pretreatment temperature could increase the DES's effectiveness in removing lignin from the lignocellulosic biomass. During this process, the cellulose content increased as the lignin content in the biomass decreased. This is because the application of heat in the DES pretreatment process can reduce the biomass mechanical strength and break the β-O-4-aryl ether bonds in the lignin-carbohydrate structure. Breaking this bond causes lignin to dissolve in the DES, causing a decrease in the lignin content in the biomass [51]. Additionally, an increase in the pretreatment temperature means that more energy is provided in the pretreatment system, which can increase the ability of the DES to break ether bonds [52]. Therefore, it is necessary to consider the energy used during the pretreatment process to determine the optimum pretreatment time and temperature.

4.3. Effects of the Substrate to DES Solution Ratio

The number of substrates used must be considered when determining the optimal substrate to DES solution ratio. The main benefit of using numerous substrates is that it improves the process efficiency because there is more biomass in the reaction system. Using more substrates can lower the amount of energy used in the pretreatment process [53] and utilizing a lower substrate to DES solution ratio can cause the solvent to break more ether bonds of lignin-carbohydrate complexes because of the high substrate concentrations used at this ratio, resulting in minimal lignin removal. A high ratio of substrates to the DES solution may also lead to the accumulation of solid particles and an increase in system viscosity [44]. In addition, the viscous system limits the mass transfer between the samples and the extraction phase.

4.4. The Enzymatic Hydrolysis of DES Pretreated Sugarcane Leaves by the Cellulolytic Consortium

The DES-pretreated sugarcane leaves had a higher reducing sugar content than untreated sugarcane leaves in all substrate loadings used (Table 4). This might be due to the complex structure of untreated sugarcane leaves that inhibits the cellulolytic consortium enzyme from accessing the cellulose contained in the lignocellulosic biomass structure of sugarcane leaves [37]. This causes the enzymatic hydrolysis process to be inefficient, as proven by the low sugar content of untreated sugarcane leaves as substrates for enzymatic hydrolysis. Therefore, pretreatment is essential for increasing the efficiency of the enzymatic hydrolysis process. Moreover, solid accumulation can occur due to the utilization of high substrate loading which limits the mass transfer between the substrates and the enzyme. Additionally, at a high loading of pretreated biomass, enzymatic digestibility and fermentation efficiency were dramatically reduced as mixing became increasingly difficult as viscosity increased [44].

4.5. The Two Stages of Hydrogen and Methane Production by SHF and SSF

4.5.1. Hydrogen Production

The untreated sugarcane leaves showed a lower hydrogen production and a longer lag phase than the DES-pretreated sugarcane leaves, which was caused by lignin in untreated sugarcane leaves. Lignin acts as a physical barrier to prevent enzyme access to hydrolyze cellulose and reduces hydrolysis efficiency [37]. Therefore, the cellulolytic consortium cannot access the cellulose in sugarcane leaves, resulting in less availability of fermentable sugars. Consequently, hydrogen production was low.

A comparison of the two-stage process of hydrogen production by SHF and SSF revealed that at high substrate loading, SSF was more efficient in producing hydrogen than SHF (Table 5). This is because of the accumulation of hydrolysis products in the SHF. In contrast, for SSF, the hydrolysis products are simultaneously produced and consumed. Thus, there is no accumulation of hydrolysis products that may inhibit the process.

4.5.2. Methane Production

Our results showed that SHF produced more acetic and butyric acid than SSF (Figure 1), which can cause a decrease in the pH of the SHF reactor system and result in low methane production (Table 6). Low pH adversely affects the activity of methanogens, resulting in low methane production during the AD process [37]. Methanogenic bacteria have been found to prefer a pH range of 6.5 to 7.5 to generate methane [54]. They have modest growth rates and are particularly sensitive to environmental changes. A reactor with a pH less than 6.0 is frequently related to reduced methane generation [55].

Substrate loading influences VFA production. The results revealed that 4 g VS of DES-pretreated sugarcane leaves gave the highest VFA production (Figure 2) and hydrogen (Table 5) and methane production (Table 6) in both SSF and SHF processes. It should be noted that utilizing a higher substrate loading of over 4 g VS could lead to system failure owing to the fast generation of VFAs [56]. Additionally, high substrate loading could drive the process to incomplete organic matter degradation owing to inhibition by overloading [56]. The results showed that increasing the substrate loading increased VFA production (Figure 2). This might be due to the increased availability of organic compounds, that is, fermentable sugars, at high substrate loading.

A literature search reveals insufficient information on two-stage and one-stage methane production from pretreated sugarcane leaves. Therefore, the methane yield obtained from this study (SSF and SHF) was compared with the literature search using a one-stage process (Table 8). The maximum methane yield of 118 and 81 mL/g-VS$_{added}$ from DES-pretreated sugarcane leaves under optimum conditions for SSF two-stage hydrogen and methane production and one-stage methane production is less than that of sugarcane leaves pretreated with other pretreatment methods (Table 8). The results may have been affected by the disparity between the inoculum source and the initial substrate load. For instance, Luo et al., 2018 [57] utilized a NaOH pretreated sugarcane leaf with an initial concentration of 65 g-TS/L to produce methane, but our study utilized only 4 g-VS/L DES pretreated sugarcane leaf (equal to 4.62 g-TS/L). Pretreatment duration in this study is less than that of the ammonium fiber explosion (AFEX), NaOH, KOH, liquid hot water (LHW), and dilute acid (DA) methods.

Table 8. The comparison between the methane yield obtained in this study with the literature searches using pretreated sugarcane leaves as the substrate.

Substrate	Pretreatment Method	Pretreatment Conditions	Fermentation Mode	Methane Yield	References
Sugarcane leaves	Ammonium fiber explosion (AFEX) pretreatment	80–120 °C, 60 min	N/A-One-stage	336 mL/g-VS_{added}	[58]
	Sodium hydroxide (6% NaOH) pretreatment	25 °C, 3 days	N/A-One-stage	287 mL/g-TS_{added}	[57]
	Potassium hydroxide (KOH) pretreatment	170 °C, 60 min	N/A-One-stage	205 mL/g-TS_{added}	[59]
	Liquid hot water (LHW) pretreatment	190 °C, 60 min	N/A-One-stage	162 mL/g-TS_{added}	[59]
	Dilute acid (DA) pretreatment	170 °C, 15 min	N/A-One-stage	156 mL/g-TS_{added}	[59]
	DES Pretreatment	120 °C, 3 h	SSF-Two-stage	118 mL/g-VS_{added}	This study
	DES Pretreatment	120 °C, 3 h	SSF-One-stage	81 mL/g-VS_{added}	This study

N/A means the data was not applicable.

Moreover, the pretreatment temperature used in this research was lower than the KOH, LHW, and DA pretreatment. Results revealed the advantages of the DES pretreatment in terms of energy efficiency. Furthermore, DES can be recovered and regenerated for subsequent pretreatment. AFEX pretreatment, on the other hand, utilizes more complicated equipment, which is likely to increase initial production costs, as well as high pressure in explosive systems, which can reduce the efficiency of the pretreatment process.

4.6. One-Stage Methane Production by SHF and SSF

The results in Figure 2 show that at 4 g VS, acetic acid, butyric acid, and propionic acid accumulated in the methanogenic effluent. The accumulation of this product is due to a separate enzymatic hydrolysis process that converts the cellulose in DES-pretreated sugarcane leaves to glucose, which will later be converted in the acidogenesis process to produce VFAs. After the enzymatic hydrolysis process, glucose production increases the VFAs in acidogenesis, which causes the accumulation of VFAs, thereby reducing methane production at the end of AD [60]. Moreover, acetic acid accumulation may be due to the cellulolytic consortium producing such VFAs during cellulose degradation. In contrast, for SSF, the VFAs were directly consumed by methanogenic bacteria to produce methane. Therefore, the VFA products in the SSF were lower than those in the SHF, leading to higher methane production in the SSF.

As shown in Figure 2, the results indicate that SSF supports a higher substrate loading concentration to produce methane than SHF for one-stage methane production. This might be because the soluble metabolite products produced by SSF were directly converted to methane during the methanogenesis stage. Therefore, there was no accumulation of VFAs which can inhibit and reduce methane production. The loading substrates in the AD system are critical factors that can affect the efficiency of the AD system. At low substrate loadings, there is a possibility that the biogas produced will make the overall process inefficient. However, at the same time, if the loading substrate is too high, the products produced during the AD process will accumulate, which will cause AD process inhibition [61].

Lignocellulosic biomass biodegradability increases with decreasing lignin content [62]. Therefore, reducing the lignin content can improve the effectiveness of enzymatic hydrolysis by a cellulolytic consortium as the enzymes can easily access the cellulose in pretreated substrates. In this study, the lignin content of sugarcane leaves pretreated with DES was significantly reduced, resulting in a significant improvement in biogas production. Therefore, it can be concluded that sugarcane leaves pretreated with DES at 4-g-VS under optimal conditions and the addition of a cellulose consortium for enzymatic hydrolysis can increase methane production through one-stage methane production by SSF.

5. Conclusions

A DES pretreatment efficiently removes lignin from lignocellulosic biomass. ChCl: MEA is the most effective at eliminating the lignin in the sugarcane leaves, with a lignin removal efficiency of 76.41–77.62% for every molar ratio. The two-stage hydrogen and methane production from DES-pretreated sugarcane leaves hydrolyzed by a cellulolytic consortium demonstrated increased hydrogen production compared to untreated sugarcane leaves in SSF and SHF processes. Methane production in a single stage by SSF and SHF from DES-treated sugarcane leaves and hydrolysis by a cellulolytic consortium was greater than methane production from untreated sugarcane leaves. SSF is an optimal fermentation method for two-stage hydrogen and methane production and one-stage methane production. This study demonstrates the utilization of sugarcane leaves to produce bioenergy (hydrogen and methane) while mitigating the PM 2.5 issues associated with burning sugarcane leaves.

Author Contributions: This study was made possible through the collaboration of all authors. Conceptualization, A.R. and S.S.; methodology, A.R.; software, A.K.M.; validation, S.S. and A.R.; formal analysis, A.K.M.; investigation, A.K.M.; resources, A.R.; data curation, A.R.; writing—original draft preparation, A.K.M., S.S., T.I., A.S. and A.R.; writing—review and editing, A.K.M., S.S., T.I., A.S. and A.R.; visualization, A.K.M.; supervision, A.R.; project administration, A.R.; funding acquisition, A.R. All authors have read and agreed to the published version of the manuscript.

Funding: This research was funded by Thailand Science and Research Innovation (TSRI) Senior Research Scholar (Grant No. RTA6280001) and Research and Graduate Studies, Khon Kaen University, Thailand.

Institutional Review Board Statement: Not applicable.

Informed Consent Statement: Not applicable.

Data Availability Statement: Not applicable.

Acknowledgments: A.K.M. would like to thank the Khon Kaen University Scholarship for ASEAN and GMS's Countries' Personnel of Academic Year 2020, the Department of Graduate and Research Studies, Khon Kaen University, the Faculty of Technology Scholarship of Academic Year 2020, and the international research funding by the Department of Biotechnology for the scholarship.

Conflicts of Interest: The authors declare no conflict of interest.

References

1. Olguin-Maciel, E.; Singh, A.; Chable-Villacis, R.; Tapia-Tussell, R.; Ruiz, H.A. Consolidated Bioprocessing, an Innovative Strategy towards Sustainability for Biofuels Production from Crop Residues: An Overview. *Agronomy* **2020**, *10*, 1834. [CrossRef]
2. Wilkie, A.C. Biomethane from Biomass, Biowaste, and Biofuels. In *Bioenergy*; ASM Press: Washington, DC, USA, 2008; pp. 195–205. [CrossRef]
3. Zhu, X.; Yellezuome, D.; Liu, R.; Wang, Z.; Liu, X. Effects of co-digestion of food waste, corn straw and chicken manure in two-stage anaerobic digestion on trace element bioavailability and microbial community composition. *Bioresour. Technol.* **2022**, *346*, 126625. [CrossRef] [PubMed]
4. Kumar, S.; Ankaram, S. Waste-to-Energy Model/Tool Presentation. In *Current Developments in Biotechnology and Bioengineering*; Elsevier: Amsterdam, The Netherlands, 2019. [CrossRef]
5. Manyi-Loh, C.E.; Mamphweli, S.N.; Meyer, E.L.; Okoh, A.I.; Makaka, G.; Simon, M. Microbial Anaerobic Digestion (Bio-Digesters) as an Approach to the Decontamination of Animal Wastes in Pollution Control and the Generation of Renewable Energy. *Int. J. Environ. Res. Public Health* **2013**, *10*, 4390–4417. [CrossRef] [PubMed]
6. Xu, N.; Liu, S.; Xin, F.; Zhou, J.; Jia, H.; Xu, J.; Jiang, M.; Dong, W. Biomethane Production from Lignocellulose: Biomass Recalcitrance and Its Impacts on Anaerobic Digestion. *Front. Bioeng. Biotechnol.* **2019**, *7*, 191. [CrossRef] [PubMed]
7. Uddin, M.; Wright, M.M. Anaerobic digestion fundamentals, challenges, and technological advances. *Phys. Sci. Rev.* **2022**. [CrossRef]
8. Ambaye, T.G.; Vaccari, M.; Bonilla-Petriciolet, A.; Prasad, S.; van Hullebusch, E.D.; Rtimi, S. Emerging technologies for biofuel production: A critical review on recent progress, challenges and perspectives. *J. Environ. Manag.* **2021**, *290*, 112627. [CrossRef]
9. Dodo, C.M.; Mamphweli, S.; Okoh, O. Bioethanol production from lignocellulosic sugarcane leaves and tops. *J. Energy South. Afr.* **2017**, *28*, 1–11. [CrossRef]

10. Chandel, A.K.; da Silva, S.S.; Carvalho, W.; Singh, O.V. Sugarcane bagasse and leaves: Foreseeable biomass of biofuel and bio-products. *J. Chem. Technol. Biotechnol.* **2012**, *87*, 11–20. [CrossRef]
11. Moodley, P.; Kana, E.G. Comparative study of three optimized acid-based pretreatments for sugar recovery from sugarcane leaf waste: A sustainable feedstock for biohydrogen production. *Eng. Sci. Technol. Int. J.* **2018**, *21*, 107–116. [CrossRef]
12. Moodley, P.; Kana, E.G. Optimization of xylose and glucose production from sugarcane leaves (Saccharum officinarum) using hybrid pretreatment techniques and assessment for hydrogen generation at semi-pilot scale. *Int. J. Hydrogen Energy* **2015**, *40*, 3859–3867. [CrossRef]
13. Oliveira, D.M.; Mota, T.R.; Grandis, A.; de Morais, G.R.; de Lucas, R.C.; Polizeli, M.L.; Marchiosi, R.; Buckeridge, M.S.; Ferrarese-Filho, O.; dos Santos, W.D. Lignin plays a key role in determining biomass recalcitrance in forage grasses. *Renew. Energy* **2020**, *147*, 2206–2217. [CrossRef]
14. Prasad, R.K.; Chatterjee, S.; Mazumder, P.B.; Gupta, S.K.; Sharma, S.; Vairale, M.G.; Datta, S.; Dwivedi, S.K.; Gupta, D.K. Bioethanol production from waste lignocelluloses: A review on microbial degradation potential. *Chemosphere* **2019**, *231*, 588–606. [CrossRef] [PubMed]
15. Saravanan, A.; Kumar, P.S.; Jeevanantham, S.; Karishma, S.; Vo, D.-V.N. Recent advances and sustainable development of biofuels production from lignocellulosic biomass. *Bioresour. Technol.* **2022**, *344*, 126203. [CrossRef]
16. Yoo, C.G.; Meng, X.; Pu, Y.; Ragauskas, A.J. The critical role of lignin in lignocellulosic biomass conversion and recent pretreatment strategies: A comprehensive review. *Bioresour. Technol.* **2020**, *301*, 122784. [CrossRef] [PubMed]
17. Yoo, C.G.; Li, M.; Meng, X.; Pu, Y.; Ragauskas, A.J. Effects of organosolv and ammonia pretreatments on lignin properties and its inhibition for enzymatic hydrolysis. *Green Chem.* **2017**, *19*, 2006–2016. [CrossRef]
18. Kellock, M.; Maaheimo, H.; Marjamaa, K.; Rahikainen, J.; Zhang, H.; Holopainen-Mantila, U.; Ralph, J.; Tamminen, T.; Felby, C.; Kruus, K. Effect of hydrothermal pretreatment severity on lignin inhibition in enzymatic hydrolysis. *Bioresour. Technol.* **2019**, *280*, 303–312. [CrossRef]
19. Yu, Z.; Gwak, K.-S.; Treasure, T.; Jameel, H.; Chang, H.-M.; Park, S. Effect of Lignin Chemistry on the Enzymatic Hydrolysis of Woody Biomass. *ChemSusChem* **2014**, *7*, 1942–1950. [CrossRef]
20. Zoghlami, A.; Paës, G. Lignocellulosic Biomass: Understanding Recalcitrance and Predicting Hydrolysis. *Front. Chem.* **2019**, *7*, 874. [CrossRef]
21. Gunny, A.A.N.; Arbain, D. Ionic Liquids: Green Solvent for Pretreatment of Lingnocellulosic Biomass. *Adv. Mater. Res.* **2013**, *701*, 399–402. [CrossRef]
22. Chen, Y.; Mu, T. Application of deep eutectic solvents in biomass pretreatment and conversion. *Green Energy Environ.* **2019**, *4*, 95–115. [CrossRef]
23. Wang, B.; Qin, L.; Mu, T.; Xue, Z.; Gao, G. Are Ionic Liquids Chemically Stable? *Chem. Rev.* **2017**, *117*, 7113–7131. [CrossRef] [PubMed]
24. Cao, Y.; Chen, Y.; Lu, L.; Xue, Z.; Mu, T. Water Sorption in Functionalized Ionic Liquids: Kinetics and Intermolecular Interactions. *Ind. Eng. Chem. Res.* **2013**, *52*, 2073–2083. [CrossRef]
25. Cao, Y.; Mu, T. Comprehensive Investigation on the Thermal Stability of 66 Ionic Liquids by Thermogravimetric Analysis. *Ind. Eng. Chem. Res.* **2014**, *53*, 8651–8664. [CrossRef]
26. Ishizaki, H.; Hasumi, K. Ethanol Production from Biomass. In *Sustainable Biomass Systems*; Elsevier: Amsterdam, The Netherlands, 2013. [CrossRef]
27. Shamurad, B.; Sallis, P.; Petropoulos, E.; Tabraiz, S.; Ospina, C.; Leary, P.; Dolfing, J.; Gray, N. Stable biogas production from single-stage anaerobic digestion of food waste. *Appl. Energy* **2020**, *263*, 114609. [CrossRef]
28. Van, D.P.; Fujiwara, T.; Tho, B.L.; Toan, P.P.S.; Minh, G.H. A review of anaerobic digestion systems for biodegradable waste: Configurations, operating parameters, and current trends. *Environ. Eng. Res.* **2020**, *25*, 1–17. [CrossRef]
29. Holdeman, L.V.; Moore, W.E.C. *Anaerobe Laboratory Manual*; Virginia Polytechnic Institute and State University, Anaerobe Laboratory: Blacksburg, VA, USA, 1975.
30. Feng, Y.; Yu, Y.; Wang, X.; Qu, Y.; Li, D.; He, W.; Kim, B.H. Degradation of raw corn stover powder (RCSP) by an enriched microbial consortium and its community structure. *Bioresour. Technol.* **2011**, *102*, 742–747. [CrossRef]
31. Owen, W.F.; Stuckey, D.C.; Healy, J.B., Jr.; Young, L.Y.; McCarty, P.L. Bioassay for Monitoring Biochemical Methane Potential And Anaerobic Toxicity. *Water Res.* **1979**, *13*, 485–492. [CrossRef]
32. Ghose, T.K. Measurement of Cellulase Activities. *Pure Appl. Chem.* **1987**, *59*, 257–268. [CrossRef]
33. APHA. *Standard Methods for the Examination of Water and Wastewater*, 23rd ed.; American Public Health Association: Washington, DC, USA, 2017.
34. Sluiter, A.; Hames, B.; Ruiz, R.; Scarlata, C.; Sluiter, J.; Templeton, D.; Crocker, D.L. Determination of Structural Carbohydrates and Lignin in Biomass. *Natl. Renew. Energy Lab.* **2012**, *17*, 1–13.
35. Miller, G.L. Use of Dinitrosalicylic Acid Reagent for Determination of Reducing Sugar. *Anal. Chem.* **1959**, *31*, 426–428. [CrossRef]
36. Sitthikitpanya, S.; Reungsang, A.; Prasertsan, P. Two-stage thermophilic bio-hydrogen and methane production from lime-pretreated oil palm trunk by simultaneous saccharification and fermentation. *Int. J. Hydrogen Energy* **2018**, *43*, 4284–4293. [CrossRef]

37. Jomnonkhaow, U.; Sittijunda, S.; Reungsang, A. Enhanced simultaneous saccharification and fermentation of Napier grass and Napier silage for two stage bio-hydrogen and methane production using organosolv and hydrothermal. *Mater. Chem. Phys.* **2021**, *267*, 124614. [CrossRef]
38. Zheng, X.-J.; Yu, H.-Q. Inhibitory effects of butyrate on biological hydrogen production with mixed anaerobic cultures. *J. Environ. Manag.* **2005**, *74*, 65–70. [CrossRef]
39. Khanal, S. Biological hydrogen production: Effects of pH and intermediate products. *Int. J. Hydrogen Energy* **2004**, *29*, 1123–1131. [CrossRef]
40. Zhao, Z.; Chen, X.; Ali, M.F.; Abdeltawab, A.A.; Yakout, S.M.; Yu, G. Pretreatment of wheat straw using basic ethanolamine-based deep eutectic solvents for improving enzymatic hydrolysis. *Bioresour. Technol.* **2018**, *263*, 325–333. [CrossRef]
41. Galbe, M.; Wallberg, O. Pretreatment for biorefineries: A review of common methods for efficient utilisation of lignocellulosic materials. *Biotechnol. Biofuels* **2019**, *12*, 294. [CrossRef]
42. Kwon, G.-J.; Bandi, R.; Yang, B.-S.; Park, C.-W.; Han, S.-Y.; Park, J.-S.; Lee, E.-A.; Kim, N.-H.; Lee, S.-H. Choline chloride based deep eutectic solvents for the lignocellulose nanofibril production from Mongolian oak (Quercus mongolica). *Cellulose* **2021**, *28*, 9169–9185. [CrossRef]
43. Ling, J.K.U.; Hadinoto, K. Deep Eutectic Solvent as Green Solvent in Extraction of Biological Macromolecules: A Review. *Int. J. Mol. Sci.* **2022**, *23*, 3381. [CrossRef]
44. Triwahyuni, E.; Muryanto; Sudiyani, Y.; Abimanyu, H. The Effect of Substrate Loading on Simultaneous Saccharification and Fermentation Process for Bioethanol Production from Oil Palm Empty Fruit Bunches. *Energy Procedia* **2015**, *68*, 138–146. [CrossRef]
45. Loow, Y.-L.; New, E.K.; Yang, G.H.; Ang, L.Y.; Foo, L.Y.W.; Wu, T.Y. Potential use of deep eutectic solvents to facilitate lignocellulosic biomass utilization and conversion. *Cellulose* **2017**, *24*, 3591–3618. [CrossRef]
46. Sun, Y.; Cheng, J. Hydrolysis of lignocellulosic materials for ethanol production: A review. *Bioresour. Technol.* **2002**, *83*, 1–11. [CrossRef]
47. Xia, Q.; Liu, Y.; Meng, J.; Cheng, W.; Chen, W.; Liu, S.; Liu, Y.; Li, J.; Yu, H. Multiple hydrogen bond coordination in three-constituent deep eutectic solvents enhances lignin fractionation from biomass. *Green Chem.* **2018**, *20*, 2711–2721. [CrossRef]
48. Amin, F.R.; Khalid, H.; Zhang, H.; Rahman, S.U.; Zhang, R.; Liu, G.; Chen, C. Pretreatment methods of lignocellulosic biomass for anaerobic digestion. *AMB Express* **2017**, *7*, 72. [CrossRef]
49. Baruah, J.; Nath, B.K.; Sharma, R.; Kumar, S.; Deka, R.C.; Baruah, D.C.; Kalita, E. Recent Trends in the Pretreatment of Lignocellulosic Biomass for Value-Added Products. *Front. Energy Res.* **2018**, *6*, 141. [CrossRef]
50. Niju, S.; Swathika, M.; Balajii, M. Pretreatment of lignocellulosic sugarcane leaves and tops for bioethanol production. In *Lignocellulosic Biomass to Liquid Biofuels*; Academic Press: Cambridge, MA, USA, 2019. [CrossRef]
51. Nhuchhen, D.; Basu, P.; Acharya, B. A Comprehensive Review on Biomass Torrefaction. *Int. J. Renew. Energy Biofuels* **2014**, *2014*, 506376. [CrossRef]
52. Lismeri, L.; Darni, Y.; Sanjaya, M.D.; Immadudin, M.I. Pengaruh Suhu Dan Waktu Pretreatment Alkali Pada Isolasi Selulosa Limbah Batang Pisang. *J. Chem. Process Eng.* **2019**, *4*, 18–22. [CrossRef]
53. Modenbach, A.A.; Nokes, S.E. The use of high-solids loadings in biomass pretreatment—A review. *Biotechnol. Bioeng.* **2012**, *109*, 1430–1442. [CrossRef]
54. Lay, J.-J.; Li, Y.-Y.; Noike, T. Influences of pH and moisture content on the methane production in high-solids sludge digestion. *Water Res.* **1997**, *31*, 1518–1524. [CrossRef]
55. Taconi, K.A.; Zappi, M.E.; French, W.T.; Brown, L.R. Methanogenesis under acidic pH conditions in a semi-continuous reactor system. *Bioresour. Technol.* **2008**, *99*, 8075–8081. [CrossRef]
56. Magdalena, J.A.; Greses, S.; González Fernández, C. Impact of Organic Loading Rate in Volatile Fatty Acids Production and Population Dynamics Using Microalgae Biomass as Substrate. *Sci. Rep.* **2019**, *9*, 18374. [CrossRef]
57. Luo, J.; Meng, H.; Yao, Z.; Wachemo, A.C.; Yuan, H.; Zhang, L.; Li, X. Anaerobic co-digestion of sodium hydroxide pretreated sugarcane leaves with pig manure and dairy manure. *Int. J. Agric. Biol. Eng.* **2018**, *11*, 197–204. [CrossRef]
58. Mokomele, T.; Da CostaSousa, L.; Balan, V.; van Rensburg, E.; Dale, B.E.; Görgens, J.F. Incorporating anaerobic co-digestion of steam exploded or ammonia fiber expansion pretreated sugarcane residues with manure into a sugarcane-based bioenergy-livestock nexus. *Bioresour. Technol.* **2019**, *272*, 326–336. [CrossRef]
59. Ketsub, N.; Latif, A.; Kent, G.; Doherty, W.O.; O'Hara, I.M.; Zhang, Z.; Kaparaju, P. A systematic evaluation of biomethane production from sugarcane trash pretreated by different methods. *Bioresour. Technol.* **2021**, *319*, 124137. [CrossRef]
60. Ellacuriaga, M.; Cascallana, J.G.; González, R.; Gómez, X. High-Solid Anaerobic Digestion: Reviewing Strategies for Increasing Reactor Performance. *Environments* **2021**, *8*, 80. [CrossRef]
61. Wang, B.; Strömberg, S.; Li, C.; Nges, I.A.; Nistor, M.; Deng, L.; Liu, J. Effects of substrate concentration on methane potential and degradation kinetics in batch anaerobic digestion. *Bioresour. Technol.* **2015**, *194*, 240–246. [CrossRef]
62. Zheng, Y.; Zhao, J.; Xu, F.; Li, Y. Pretreatment of lignocellulosic biomass for enhanced biogas production. *Prog. Energy. Combust. Sci.* **2014**, *42*, 35–53. [CrossRef]

fermentation

MDPI

Review

Current Trends in Biological Valorization of Waste-Derived Biomass: The Critical Role of VFAs to Fuel A Biorefinery

Corine Nzeteu †, Fabiana Coelho †, Emily Davis †, Anna Trego * and Vincent O'Flaherty *

Microbial Ecology Laboratory, Microbiology, School of Natural Sciences, University of Galway, University Road, H91 TK33 Galway, Ireland
* Correspondence: annachristine.trego@universityofgalway.ie (A.T.); vincent.oflaherty@universityofgalway.ie (V.O.); Tel.: +353-(0)91-493734 (A.T. & V.O.)
† These authors contributed equally to this work.

Abstract: The looming climate and energy crises, exacerbated by increased waste generation, are driving research and development of sustainable resource management systems. Research suggests that organic materials, such as food waste, grass, and manure, have potential for biotransformation into a range of products, including: high-value volatile fatty acids (VFAs); various carboxylic acids; bioenergy; and bioplastics. Valorizing these organic residues would additionally reduce the increasing burden on waste management systems. Here, we review the valorization potential of various sustainably sourced feedstocks, particularly food wastes and agricultural and animal residues. Such feedstocks are often micro-organism-rich and well-suited to mixed culture fermentations. Additionally, we touch on the technologies, mainly biological systems including anaerobic digestion, that are being developed for this purpose. In particular, we provide a synthesis of VFA recovery techniques, which remain a significant technological barrier. Furthermore, we highlight a range of challenges and opportunities which will continue to drive research and discovery within the field. Analysis of the literature reveals growing interest in the development of a circular bioeconomy, built upon a biorefinery framework, which utilizes biogenic VFAs for chemical, material, and energy applications.

Keywords: anaerobic digestion; biorefinery; fermentation; VFAs; biomass valorization

Citation: Nzeteu, C.; Coelho, F.; Davis, E.; Trego, A.; O'Flaherty, V. Current Trends in Biological Valorization of Waste-Derived Biomass: The Critical Role of VFAs to Fuel A Biorefinery. *Fermentation* **2022**, *8*, 445. https://doi.org/10.3390/fermentation8090445

Academic Editor: Sanjay Nagarajan

Received: 16 August 2022
Accepted: 4 September 2022
Published: 7 September 2022

1. Introduction

As both climate change and the global energy crisis escalate, it becomes ever more critical to implement sustainable resource management strategies such as biorefineries and resource recovery systems. These systems typically utilize innovative resource recovery technologies and novel renewable materials. The valorization of biomass can play a foundational role within these systems, supporting the generation of energy (biofuel) as well as a wide range of bio-based products through the biorefinery concept [1–4]. Biomass can be broadly classified as either an *energy crop* or *residue*. Energy crops are specifically cultivated for energy generation. These crops are typically cultivated using intensive farming practices, and since they are often edible, using them for energy generation results in less food and less food-crop land. In contrast, biomass residues are non-edible and are generally composed of waste products or agro-industrial side streams.

Among biomass residues, food waste and agricultural waste have demonstrated their promising potential for biorefinery applications [2,3,5–7]. In Europe, these biomasses are valorized using biological, chemical, and thermochemical methods. However, variability in the quantity and composition of biomass limits the technological and economic viability of these valorization methods. Therefore, these highly variable biomass resources are better suited for processes such as anaerobic digestion (AD), which is able to convert a wide range of organics into products such as volatile fatty acids (VFAs), biohydrogen, polyhydroxyalkanoates (PHAs), and bioenergy. AD can serve as a sustainable and economically attractive biological pretreatment for lignocellulosic biomass, facilitating its conversion into bio-based products by exposing lignin and undigested fibers for further valorization.

The ultimate purpose of an AD-based biorefinery system is to optimize resource-use efficiency while minimizing waste; this is typically accomplished by maximizing energy/biogas production. The generation of alternative valuable by-products, in addition to biogas, represents a new opportunity to enhance resource recovery (Figure 1). An important value-added by-product, VFAs, are produced during the initial phases of AD in a process known as acidogenic fermentation. VFAs have a wide range of potential applications in the biorefinery industry where they can be used as feedstock for various bio-based products. For instance, VFAs are considered a potential platform for the production of biodegradable PHA polymers [8]. Currently, synthetically produced VFAs are used in the food and beverage industries, as well as in pharmaceutical and synthetic chemistry. The ratios of the specific volatile fatty acids that are produced via acidogenic fermentation are dependent on the feedstock biomass' composition, the extent of hydrolysis, operational conditions, reactor design, and the structure of the microbial community. Research investigating these parameters is being carried out and promises to greatly improve the efficiency and stability of the acid-forming stage.

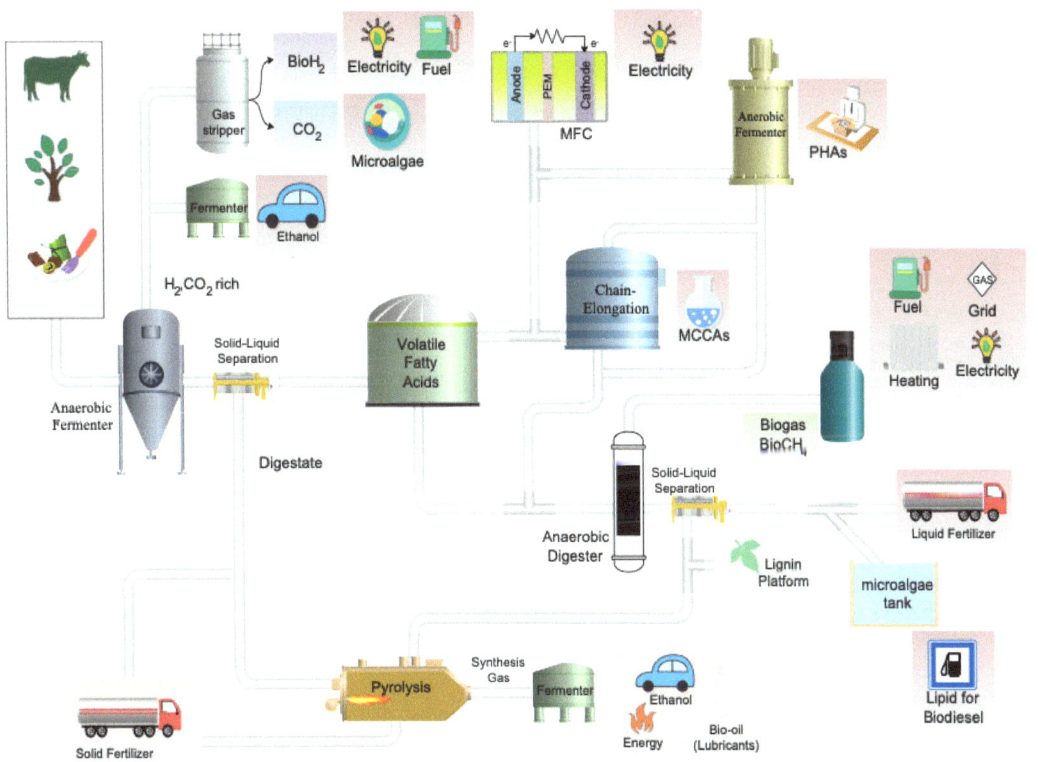

Figure 1. Potential biorefinery process focusing on maximizing VFA production. The process begins with the valorization of residual feedstocks and culminates in the potential production of various high-value end-products (highlighted in pink).

This review elucidates the potential for low-value biomass to be used as feedstock for VFA generation within a biorefinery model. Specifically, we discuss various techniques and system designs which optimize energy use and product yield. Finally, we discuss current trends and challenges for a biorefinery concept, and the outstanding research necessary to support a functioning bioeconomy.

2. Methods of Valorizing Low-Value Feedstocks

Biomass valorization has gained popularity and traction in recent years given its potential to sustainably meet regulatory requirements in terms of energy, chemicals, and materials. Research suggests that within the biorefinery concept, biomass valorization—where all fractions are processed selectively towards a variety of products—is generally achieved using two strategies: thermochemical and/or biological conversion [9]. While this review focuses on biological processes, we will also briefly mention trends in thermochemical techniques.

2.1. Thermochemical Approach

Among the thermochemical conversion processes, gasification and pyrolysis are commonly used to produce heat, biochar, and syngas from lignocellulosic biomass. Conventional gasification technologies include fixed beds, fluidized beds, and entrained flow reactors [10]. However, these technologies still struggle with process inefficiencies related to biomass moisture content and tar production. Recently, efforts have been made to mitigate these factors—using technologies such as pyrolysis and supercritical water gasification [11–16]. Unlike gasification, pyrolysis is a technology that converts biomass into bio-oil, syngas, and biochar in the absence of oxygen [17]. Pyrolysis can be used to valorize different types of recalcitrant biomass, such as agricultural residues and wood wastes. The resulting syngas can then be converted by anaerobic bacteria into biochemicals and biofuels independent of the original biomass composition; a process known as hybrid thermochemical-biochemical [18]. However, the high cost and safety risk of the pyrolysis process make it unviable for large scale applications [17].

Recently, supercritical water gasification has been considered as a potential technique to valorize lignocellulosic biomass and wastes with high moisture content—up to 80% wet weight [16]. Supercritical water gasification is being studied especially for hydrogen production, as the composition of the resulting synthesis gas is higher in hydrogen and lower in carbon monoxide [19]; in addition, the low production of tar and char is an advantage compared to other technologies [20]. However, supercritical water gasification is a technology still unfeasible for large scale applications in biorefineries—its implementation requires improvements in terms of pump energy efficiency [16], and reactor designs which can withstand corrosion [16] and high pressure [21].

2.2. Biological Approach

Biological conversion processes encompass both AD and fermentation and are commonly used to valorize biomass such as food waste, agricultural residues, organic fraction of the municipal solid waste (OFMSW), and energy crops. Unlike the thermochemical conversion method where the primary product is biofuel, the biological conversion of biomass can produce biofuel and chemicals. Due to the high moisture content of most biomass, direct valorization using thermochemical technologies is challenging. Therefore, biological conversion technologies are reported to be more eco-friendly and appropriate for waste biomass with high moisture content [22].

AD is a well-established process for the sustainable management of solid organic feedstock [22]. AD can be used to convert various organic substrates into methane-rich gas destined for energy generation. In this context, organic residues are conveniently used to meet global energy demand while reducing the burden of fuel consumption and waste disposal. In Europe, the success of AD is witnessed by its dynamic ascent with a total of 18,202 biogas installations, producing 11,082 MW, and 63,511 GWh worth of biogas as recorded in 2018 [23,24]. Despite this continued growth, AD technology is still not cost-competitive with natural gas without fiscal incentives. This is due to high costs associated with biogas production, whereas natural gas is available at lower cost worldwide. Therefore, increasing the efficiency of AD processes is critical to improving its economic attractiveness. To this end, feedstock pre-treatment, reactor configuration, and feedstock co-digestion have been studied as potential means of improving resource recovery [25,26].

2.3. *Valorization—Selecting a Method*

The selection of a particular valorization method is highly dependent on the biomass characteristics and composition. For instance, biological approaches are suitable for readily degradable, high-moisture-content biomass such as food waste. Thermochemical methods are more commonly used for recalcitrant feedstocks such as lignocellulosic biomass. While both treatment options entail installation and operation costs, research and application suggest that the biological approach may be more flexible in terms of feedstock and products. Moreover, AD and fermentative processes result in fewer undesirable effects such as tar production.

3. Sustainable Feedstock Types

3.1. *Food Waste*

Food waste makes up a significant portion of anthropogenically derived organic waste and constitutes an environmental burden where landfill disposal is employed. One third of all food produced in the world for human consumption goes to waste [27], with 14% of food waste occurring during production processes alone [28]. While post-consumer waste can be minimized through prevention campaigns, production wastage (peelings, damaged or diseased matter, inedible plant parts) is likely to remain at similar or increasing values. Generally, food waste is composed of fruits, vegetables, and tubers [28]. These materials all have relatively high moisture and energy contents and, therefore, qualify as high value feedstock for AD [29]. Through AD, this waste stream can be converted into a renewable resource while simultaneously reducing waste-related challenges in the long term [30].

Food waste composition varies greatly but is fundamentally a mix of carbohydrates, proteins, and lipids. The ratio of these three biomolecules largely determines the material's energy generation potential. Lipids have higher energy content than carbohydrates and proteins; however, they have been reported to be difficult to breakdown in AD bioreactors, even destabilizing digesters at high concentrations [31–33]. Most food waste is primarily composed of complex carbohydrates, including lignocellulosic and/or hemicellulosic compounds (25–30% of total solids (TS)) [34,35]. These carbohydrates originate from plant matter and are challenging to hydrolyze. Indeed, hydrolysis is frequently reported as the rate limiting step in AD [36]. Efforts to facilitate hydrolysis have been made, primarily the investigation of various pre-treatment methods including alkaline [25,37], thermal [38], acid [25] and enzymatic pre-treatments [39]. However, these treatments all increase operational costs. Whereas biological strategies, such as tailoring operational conditions to promote the growth and persistence of key microbial hydrolysers within AD bioreactors, represent a promising alternative [35].

3.2. *Agricultural Residues*

Agricultural residual biomasses comprise crop and plant residues, vegetable waste, forest residues, grass, and livestock manure [40]. These are largely composed of lignocellulose which can be converted via AD and fermentation to bioenergy and biochemicals. The efficiency of these conversions is determined by specific lignocellulosic characteristics such as lignin content, degree of polymerization, hemicellulose structure, cellulose crystallinity, porosity, and specific area [41,42].

Many studies in the literature review the use of pre-treatments that would decrease the recalcitrance of this biomass by improving the accessibility of cellulose to cellulases. This is achieved either by decreasing the hemicellulose content (e.g., dilute acids and bases) or by applying physical treatment (e.g., high temperature and pressure) to disrupt the lignin matrix [40,42]. Of course, all pre-treatment processes entail a trade-off between the cost of pre-treatment and the desired end-product yield increase [43]. While ionic liquids and deep-eutectic solvents have been recently investigated, full-scale biorefineries generally employ steam explosion, organosolv, or dilute acids [43,44]. The use of these pretreatment technologies may negatively impact the indigenous microbiome of the feedstock, which can be critical to the fermentative process.

The use of lignocellulosic waste as feedstock for biogas production through AD is well established. However, the potential of lignocellulosic waste for VFA production has been garnering increased attention [45]. In a biorefinery context, carboxylic acids are a desirable product with high market value [46–49]. Among lignocellulosic wastes, grass is an abundant, renewable, and cheap feedstock that has been largely employed to produce biogas in AD [50]. Relatedly, silage is grass which has been fermented to facilitate preservation during storage. During fermentation, lactic acid bacteria use soluble carbohydrates present on the surface of grass in the production of lactic acid, causing a decrease in pH, which allows the feedstock to be preserved for animal feed without risk of spoilage [51].

While grass is considered a sustainable feedstock due to its carbon-sequestering capacity, co-digestion with other agricultural residues, such as cattle slurry, may further enhance the sustainability of the process [50,52]. Cattle slurry is an abundant agricultural waste and is cheaper and richer in nutrients than grassland feedstocks. Furthermore, by co-digesting this waste stream with grass, greenhouse gas (GHG) emissions from slurry are reduced [50]. The co-digestion of different grassland forages with grass could also improve AD yields due to an improvement in nutrient availability for the microbial community [53]. Moreover, the combined growth of multi-species grassland mixtures (herbs, legumes, and grass) in intensively managed grassland may enhance yield while mitigating disturbances, such as drought and environmental impact, when compared to monocultures [52].

3.3. *Animal Residues*

Animal manure is a primary contributor to environmental pollution in rural areas. This is usually due to emissions from land-spreading and manure storage facilities, which release harmful substances to the soil, water, and atmosphere. Animal manure/slurry has high concentrations of nutrients (such as nitrogen and phosphorus) and metals (such as copper, zinc, arsenic, and cadmium). Leaching of these metals into the surrounding environment increases phytotoxicity, reduces soil fertility and productivity, and increases toxicity of crops and food products grown on the contaminated soil [54]. Meanwhile, leaching of nutrients contributes to water quality degradation and eutrophication. Moreover, the storage and land-spreading of animal manure/slurry can release GHG, such as methane, nitrous oxide, and ammonia, into the air contributing to climate change [55,56].

To mitigate the environmental burden of the manure/slurry, researchers have engaged in developing techniques for its sustainable treatment. Although composting, incineration, pyrolysis, and gasification have been evaluated, AD is outstanding in its capacity to reduce pollution while generating valuable by-products such as fertilizers and renewable energy [54]. However, there are some factors at play which limit the use of slurry/manure fed AD: (i) slurry bio-methane potential is low due to high moisture and low organic content [55]; (ii) a large volume of feedstock, usually collected from multiple sites, is required for efficiencies of scale [54]; and (iii) slurry has a low C:N ratio, which tends to inhibit methane production.

These issues can be mitigated or avoided entirely by employing co-digestion, specifically, by making a mixed feedstock composed of slurry with other organic wastes/residues/ energy crops that have a high C:N ratio. Several researchers have reported that co-digestions improved biogas production [57–59] or VFA production [60,61]. Although the co-digestion of manure/slurry with other feedstock provides a means to increase economic feasibility, the nutrient and metal-rich liquid digestate remains an issue in an AD-based facility. Therefore, complete valorization of the manure/slurry within an AD-based biorefinery concept could result in a more desirable digestate product.

4. State-of-the-Art System Designs

4.1. Single-Stage System Design and Application

Anaerobic bioreactors may be designed to optimize the processing of a selected biomass and for the production of a specific desired product. Many bioreactor types have the capacity to produce VFAs, hydrogen associated with VFA as a by-product, or biogas. Several reactors, including the continuous stirred tank reactor (CSTR), the packed bed biofilm column reactor, leach bed reactor (LBR), two-stage anaerobic bioreactor, and continuous flow fermentation reactors, have been used to produce VFAs (Table 1). Studies using solid feedstocks generally use CSTRs or LBRs and have generated promising results. The CSTR is perhaps the most widely used single-stage wet AD design [62]. CSTRs are suitable for materials with solids content up to 10% [63] and work by thoroughly mixing feedstock and microbes in the presence of suspended solids [64]. Previous studies have reported successful production of VFAs from food waste and OFMSW using the CSTR configuration (Table 1). However, this reactor design has significant inherent inefficiencies, including (i) a tendency for biomass washout, (ii) the need for size reduction of the substrates, (iii) energy input required for continuous stirring, and (iv) the low solids content (<10%) requirement [65,66]. In an attempt to overcome these limitations, a novel CSTR design consisting of a solid–liquid separator was proposed to retain undigested biomass with the active community in the system [67]. This approach addresses the issue of biomass washout, but not the limitations for feedstock processing (size reduction, low solids) or energy consumption.

LBRs are a promising alternative to the CSTR for VFA production from high-solid waste such as food waste, OFMSW and vegetable waste, and grass (Table 1). Compared to CSTRs, these reactors have been reported to permit higher loads and high VFA production [35,68]. In LBRs, solid material is loaded into the reactor and irrigated with water, which is recycled through the system continuously. Hydrolysis occurs in the solid bed, while fermentation occurs in the liquid phase, thus, decoupling the hydrolysis and fermentation processes. The recirculation mechanism allows for the dilution of inhibitory compounds and increases the moisture in the solid bed which facilitates micro-organism growth and activity, all with a relatively low water requirement [69,70].

Compared to CSTRs, LBRs have several financial advantages—less instrumentation, maintenance, and investment are required, making it an attractively low-cost, high-solids AD reactor [71]. However, since LBRs process solid feedstock which is not stirred, VFA product accumulation can occur. Furthermore, high levels of VFA can inhibit micro-organisms involved in the hydrolysis and fermentation stages [72–74]. In-line VFA separation, which could remove VFA product from the LBR leachate, is currently being investigated [75,76]. However, there is currently no consensus or single outstanding technology being used to recover VFA from fermentation liquor. Therefore, researchers have focused on developing two-stage systems in which VFAs generated in LBRs are removed and valorized through processes such as chain elongation (CE), PHA production, or even biogas production.

Table 1. Chemical characteristics of feedstocks and inoculum sources used for VFA production.

Raw Material	TS	VS	COD/TOC	Lignocellulosic	Remarks	Ref
	(%ww)	(%ww)	($gO_2.kg^{-1}ww$)	(%TS)		
Feedstock:						
Cattle manure	5.0–9.5	7.0–7.3	44–54	n/r	TKN: 1.9–3.6 $gN.kg^{-1}ww$	[3]
Ryegrass silage	35–40	31.3–36.0	312–360	n/r	TKN: 4.7–5.9 $gN.kg^{-1}ww$	[3]
Napier grass	15.12 $g.L^{-1}$	12.65 $g.L^{-1}$	0.92 $g.g^{-1}$	Cel: 36.81%, Hem: 26.16%, Lig: 8.27%		[77]
Ryegrass silage	25.5	24.1	n/a	Cel: 34.3%, Hem: 29.6%, Lig: 8.6%	Ddata based on fresh ryegrass before ensiling	[78]
Food waste	42.46 ± 0.78	38.45 ± 1.87	53.02 ± 2.29% [a]	n/a	TKN: 2.10 ± 0.17%	[79]
Dried farmland grass	83.6 ± 0.6	72.8 ± 1.1	n/a	n/a		[80]
OFMSW	12 ± 1.4	10.7 ± 0.7	102.8 ± 13.0	n/a	Mix of feed, inoculum, and tap water to a TS of 7–8 %ww. TKN: 3.12 ± 0.51 $gN.kg^{-1}ww$	[81]
OFMSW	28.14 ± 4.01	25.98 ± 2.29	312.6 ± 120.8	n/a	A mix of OFMSW and water was used as inoculum after it was acclimatized to 55 °C. TKN: 8.16 ± 1.83 $gN.kg^{-1}ww$	[82]
Food waste	28.19 ± 2.32	25.96 ± 2.08	376.4 ± 51.3 [b], 28.6 ± 2.3 [c]	Cel: 2.82 ± 0.95%, Hem: 32.58 ± 4.48%	Lipids: 27.50 ± 1.45 %ww; Protein: 20.69 ± 1.17 %ww	[35]
Food waste	17.8	17.1	320 $gO_2.L^{-1}$ [b], 95 $gO_2.L^{-1}$ [c]	n/a	Lipids: 0.59 %ww; Protein: 3.79 %ww	[83]
Food waste	16.5 ± 0.2	15.5 ± 0.7	264 ± 27 $gO_2.L^{-1}$ [b]	n/a	Sludge inoculum acclimatized for 5 days at 37 °C. Inoculum was treated with BES to inhibit methanogenesis	[84]
Kitchen waste	128.92 ± 2.33 $g.L^{-1}$	115.91 ± 2.84 $g.L^{-1}$	n/a	n/a		[85]

Table 1. *Cont.*

Raw Material	TS	VS	COD/TOC	Lignocellulosic	Remarks	Ref
	(%ww)	(%ww)	($gO_2.kg^{-1}ww$)	(%TS)		
Inoculum:						
Cow manure	16.81 $g.L^{-1}$	11.78 $g.L^{-1}$	0.19 $g.g^{-1}$	Cel: 18.29%, Hem: 9.07%, Lig: 11.77%		[77]
Liquid digestate from the co-digestion of pig manure and grass silage	2.30%	1.60%	0.58 $gO_2.L^{-1}$ [c]	n/a	Stored at 35 °C until CH4 production was minimal.	[78]
Cow manure	18.22 ± 0.77	16.33 ± 0.76	50.76 ± 2.96% [a]	n/a	TKN: 1.40 ± 0.02%	[79]
Anaerobic digested food waste	2.98 ± 0.00	2.67 ± 0.00	34.80 ± 0.98% [a]	n/a	TKN: 1.99 ± 0.03%	[79]
Anaerobic granular sludge	9.01 ± 0.09	7.85 ± 0.04	n/a	n/a		[35]
Anaerobic digestion sludge	0.4 ± 0.1	0.3 ± 0.1	10.5 ± 1.2 $gO_2.L^{-1}$ [b], 5.1 ± 0.8 $gO_2.L^{-1}$ [c]	n/a		[84]
Anaerobic digestion sludge	31.31 ± 0.49 $g.L^{-1}$	19.67 ± 0.35 $g.L^{-1}$	n/a	n/a	Pre-treated with heat-shock at 70 °C for 30 min	[85]

Acronyms: TS—total solids, vs.—volatile solids, COD—chemical oxygen demand, TOC—total organic carbon, TKN—total kjendahl nitrogen, Cel—cellulose, Hem—hemicellulose, Lig—lignin, BES—2-bromoethane sulfonic, and n/r—not reported. Notes: a—TOC, b—Total COD, and c—Soluble COD.

4.2. Multi-Stage System Design and Application

A multi-stage bioreactor is, broadly, any system with two or more bioreactors. This design facilitates the segregation of different microbial processes into separate reactors, allowing the environment of each reactor to be optimized for a specific functional microbiome. Such systems are capable of efficiently treating organic waste in terms of degradation yield and biogas production [86], and of producing valuable products such as VFA, lactic acid, alcohols, and medium-chain carboxylic acids (MCCAs) [87–89]. In a multi-stage system, hydrolysis and acidification stages occur in one reactor, while CE, PHA production, and methanogenesis occur in a separate reactor. In this way, the inhibition of the methanogens is avoided in the first reactor and different operating conditions can be used in each stage to maximize yields. This approach has been found to be more stable than single-stage systems in treating organic waste with high solid content [90,91]. The observed enhanced performance is reportedly due to the flexibility in process control offered by two-stage systems [8,92].

The number of multi-stage systems throughout Europe was expected to rise due to their ability to handle higher loading rates and improved process stability and flexibility. However, less than 10% of AD capacity in Europe are multi-stage systems [93,94]. This discrepancy is likely due to the complexity and cost of building and operating such systems. Nevertheless, the versatility and potential of multi-stage systems to improve process performance has encouraged ongoing research, especially within the biorefinery context. The viability of the multi-stage bioreactor systems was evaluated in a previous study in which one- and two-stage systems for the enzymatic hydrolysis of a municipal solid waste were compared using a techno-economic assessment (TEA) approach. The authors reported, on average, a 15–22% return on investment (ROI) and a 4–6 year payback period (PP) for two-stage systems, compared to 4–7% and 13–25 years for one-stage systems [95]. Regalado et al. (2022) pointed out that a multi-stage processing system, in which biogas is simultaneously recovered with other value-added products, offers a possible solution for achieving a more robust circular economy [96]. In addition, multi-stage systems allow for the treatment of large quantities of recalcitrant biomass which otherwise could not be treated with one-stage systems. This enhances the carbon-neutral energy output.

5. Process Optimization for Carboxylic Acid Production

5.1. Producing Carboxylic Acids

Carboxylic acids could serve as the foundation for a circular bioeconomy; here we focus especially on lactic acid and VFAs. Lactic acid can be used by the cosmetic, dairy, and pharmaceutical industries. It is also a precursor for the synthesis of bioplastics (polylactic acid) and MCCAs [80,97]. VFAs are aliphatic organic acids with less than six carbons that, likewise, have applications in the pharmaceutical, dairy, food, animal food, textile, and cosmetic industries [97]. However, the commercialization of biogenic, mixed VFAs is still challenging, as a commercially feasible technology to recover and purify individual VFAs is still needed [98]. Alternatively, many studies have proposed the use of mixed VFAs as building blocks in a biorefinery to produce MCCAs, polyesters, PHAs, bioenergy, and electricity [98,99].

The fermentative process for VFA production requires further development. As compared to established chemical routes, the process has lower productivity and yield with higher production costs [100]. Of course, the chemical synthesis and petrochemical route produces more GHG emissions and consumes unrenewable energy sources [98]. Our review of the literature indicates that VFA production costs could be minimized by valorizing low-cost residual biomass, such as food waste, grass, and manure. Moreover, productivity yields and concentration can be improved by optimizing operational parameters such as inoculum, feedstock, temperature, pH, organic loading rate, and leachate dilution (Tables 1 and 2). Therefore, it seems that from a sustainability standpoint, the biological production of VFA is the most promising option.

Table 2. Summary of the operational conditions for VFA production using the raw materials described in Table 1.

Reactor Operation	Leachate Dilution	Recirculation (L h^{-1})	pH	Biomass Degradation (%VS)	VFA Total (gCOD.L^{-1})	VFA Profile (%)					Ref
						Ace	*Prop*	*But*	*Val*	*Cap*	
AF [a], loading 18 kgCOD m^{-3} day^{-1} for 120 days at 35 °C	n/r	n/r	5.5	25.9–37.0	6.4	77.6	11.3	6.3	1.9	0.5	[3]
					10.3	71.1	12.0	7.3	2.0	1.2	[3]
LBR for 28 days at 28 ± 3 °C	No dilution	4.0	6.6 ± 1.2	~55	22.8 g	35.6	15.5	32.1	16.8	n/r	[3]
	2x, 3-day interval	4.0	6.2 ± 0.8	~65	46.9 g	54.2	14.0	20.4	11.5	n/r	[77]
	2x, 3-day interval	4.0	6.0 ± 0.5	~35	25.9 g	36.3	28.7	20.2	14.9	n/r	[77]
LBR, loading 0.5–1.0 kg m^{-3} day^{-1} for 24–32 days at 28 ± 3 °C	2x, 6-day interval	0.2	6.5 [b]	62.1–66.3	0.3–0.4 gCOD $gsCOD^{-1}$	38–41	27–31	25–28	3-5	n/r	[78]
LBR, loading 12.8 kg m^{-3} day^{-1} for 17 days at 35 °C	2x every sampling	n/r	6.0 [c]	68.05 ± 2.14	5.8	22.6	11.3	66.0	n/r	n/r	[79]
CE [a], loading 25 mL for 30 days at 32 °C	n/r	n/r	5.5-6.3 [d]	n/r	12 g L^{-1}	33.4	2.1	7.5	20.9	34.1	[80]
CSTR, loading 14–15 kgVS m^{-3} day^{-1} for 180 days at 37 °C	n/r	n/r	6.6 ± 0.2	42	24.4 ± 0.2	14.4	28.6	15.4	27.2	n/r	[81]
CSTR, loading 17 kgVS m^{-3} day^{-1} for 100 days at 55 °C	n/r	n/r	5.3 ± 0.1	83	13.9 ± 0.5	n/r	n/r	n/r	n/r	n/r	[82]
CSTR, loading 80 gVS.L^{-1} for 7 days at 37 °C	15x [e]	0.05	7.0 [f]	68-76	73.5 ± 1.9	28.4	12.1	24.8	0.0	29.7	[35]
CSTR, loading 40 gVS.L^{-1} for 12 days at 37 °C	n/r	n/r	6 [g]	55	48 g	~37	~13	~17	~9	~21	[83]
LBR, loading 21.7 gVS.L^{-1} for 14 days at 22 °C	n/r	4.4	6	81	24 ± 0.2 [h]	29.2	4.2	66.7	n/r	n/r	[84]
			7	84	28 ± 0.6 [h]	39.3	7.1	50.0	n/r	n/r	[84]
			8	87	27 ± 0.2 [h]	51.9	18.5	29.6	n/r	n/r	[84]
CSTR, loading 5 gVS L^{-1} day^{-1} for 30 days at 37 °C	n/r	n/r	7.0 ± 0.3	n/r	19.6-24.8 g L^{-1}	35–48	12–21	24–30	3–13	2–17	[85]

Acronyms: AF—acidogenic fermentation, CE—chain elongation, COD—chemical oxygen demand, CE—chain elongation, LBR—leach-bed reactor, CSTR—continuous stirred tank reactors, n/r—not reported. Notes: a—reactor configuration was not disclosed, b—adjusted every 6 days, c—adjusted every sampling, d—not adjusted, e—at the beginning of each cycle, f—at the beginning, g—controlled every 2 days, and h—VFA production reported in terms of acetic, propionic, and butyric acids.

5.2. Inoculum—Providing an Appropriate Microbial Community

Fermentative processes are carried out by microbial communities, and their composition and activities are directly related to reactor operational conditions. VFAs can be produced through fermentation employing either pure cultures or mixed cultures as inoculum. The use of pure culture fermentation, however, requires synthetic media, pure substrates, and media sterilization, increasing production costs [101]. Alternatively, the use of residual biomass feedstocks during mixed-culture AD does not require energy for sterilization or pure substrate supplementation, thus, providing greater cost-efficiency [99].

AD is catalyzed by a microbial consortium composed of different hydrolytic and fermentative bacteria, and methanogenic archaea [102]. VFAs are produced in the second step (acidogenesis) of the AD process, alongside lactic acid, CO_2, and H_2. However, complete mineralization involves the consumption of VFAs by methanogenic archaea producing acetic acid, CO_2, and H_2 and, ultimately biogas [50]. Therefore, to optimize VFA production, methanogenic archaea must be inhibited using chemical (e.g., BES) [83] or physical (e.g., heat-shock) [85] pre-treatments, or by manipulating operational conditions (e.g., pH) [103]. Balancing the trade-off between VFA productivity and the cost of these pre-treatments is necessary to ensure process feasibility.

The use and choice of inoculum is another crucial consideration in VFA production. The absence of inoculum in grass fermentation led to very low biomass degradation and VFA production (Table 2) [77,78]. The use of cow manure as inoculum increased VFA production as well as the degradation of grass. The quantity of inoculum was also shown to be an important consideration; adding more than 20% of cow manure to the solid fraction proved to negatively impact the process due to the high solid content inside the reactor [77]. The digestate from an anaerobic co-digestion of pig manure and silage was responsible for increasing VFA production from grass silage, enhancing biomass degradation [78]. Additionally, rumen bacteria from cow manure caused a shift in the profile of VFAs obtained from food waste, leading to increased propionic, butyric acid, and ethanol concentrations [79].

5.3. Two-Stage Design for Optimized Carboxylic Acid Production

As discussed previously, two-stage systems are preferable for the accumulation of VFAs in AD since the optimal operational conditions of acidogenesis and methanogenesis are drastically different. Moreover, a second stage reactor can also be used to further convert VFAs into MCCAs, PHAs, and other valuable products. A two-stage strategy for the AD of grass was found to increase hydrolysis of grass silage and biogas productivity [50]. Additionally, a two-stage AD of OFMSW using a mesophilic CSTR produced 24.4 g COD.L^{-1} of VFA while maximizing acidification [81]. While this study was designed to optimize biogas production in the second stage, the observed accumulation of propionic and valeric acids in the first stage highlight the potential for PHA production. Another two-stage study recorded grass fermentation and subsequent microbial CE of the lactic acid produced [80]. The native micro-organisms on the surface of the grass were responsible for the lactic acid production (9.36 g L^{-1}) at low pH, while caproic acid, acetic acid, and butyric were obtained in a CE reactor. Caproic acid has very low solubility in water and is immiscible at concentrations above 11 g.L^{-1} (20 °C). Therefore, optimizing the operational conditions to obtain caproic acid above this concentration would be beneficial for the process, not only due to its high market value but to decrease costs associated with product recovery.

5.4. Leachate Dilution in LBRs Affects VFA Production

A recent study demonstrated that diluting LBR leachate resulted in increased VFA production and grass solubilization [77]. Reactors with undiluted leachate had lower VFA production; the degradation of grass was also limited, represented by the low soluble COD produced (51.5 g). Meanwhile, leachate dilution led to a higher production of VFAs—observed as an accumulation of acetic acid (54.17%). The accumulation of higher chain VFAs (e.g., butyric) can have inhibitory effects on micro-organisms, negatively impacting overall acid production and feedstock degradation [77,78]. As an alternative to leachate dilution, in-line selective VFA extraction may remove higher-chain VFAs with a higher market value to produce PHAs and MCCAs, while separating and concentrating the acetic acid for biomethane production in a high-rate reactor [3,97,98,104,105].

5.5. pH Directly Affects Biomass Degradability and VFA Profile

In fermentative processes, it has been found that the pH directly affects the microbial community, as well as biomass degradability [79]. Further, pH is directly correlated with VFA and H_2 production in the fermentation of food waste [82]. Low pH (below 6.5) effectively inhibits methanogenesis but may also inhibit hydrolysis when lower than 4 [78,79,103]. Recently, high VFA production during fermentation of OFMSW was attributed to an operational pH of 6.6, which may have been high enough to avoid the inhibitory effects of acidic environments [81]. Moreover, studies showed that pH values ranging from 6 to 7 improved food waste hydrolysis and maize silage solubilization while increasing VFA accumulation [106,107]. In the fermentation of grass silage, slightly lower pH levels were responsible for higher VFA yield with stable operational conditions and suppressed methane generation [108]. In the fermentation of grass pellets, controlling the pH levels at 5.50 using 6 M sodium hydroxide led to an efficient degradation of grass to produce 4.5 g/L^{-1} of VFAs [105]. In two-stage systems, the recycling of anaerobic leachate from a second reactor can eliminate the need for external buffering agents to control the pH in a first stage LBR—resulting in overall reductions in operational costs, downstream processing, and environmental impacts [50].

5.6. Temperature Implications for VFA Accumulation

Temperature is also an important factor when considering the digestion of high-solid biomass for the production of VFAs. Although it does not affect the VFA profile as significantly as pH, temperature can affect the microbial community dynamics [101]. Lower temperatures, in particular, are reported to reduce the hydrolysis of grass but may have a positive effect on VFA accumulation [47]. Mesophilic temperatures (37 °C) during the fermentation of OFMSW prevented sudden drops in pH, which consequently prevented inhibition of VFA production [81]. Conversely, thermophilic temperatures (55 °C) during the fermentation of food waste in a similar CSTR reactor with no inoculum addition led to a 43% reduction in VFA production [82,107]. A final consideration when deciding upon an operational temperature is the cost of heating the system—additional heating can significantly add to the operational costs of the treatment. In light of these considerations, low-temperature operational conditions may be ideal for VFA generation.

5.7. Organic Loading Rate and Hydraulic Retention Time

Organic loading rate (OLR) and hydraulic retention time (HRT) are also important parameters in the production of VFAs. At higher HRTs, micro-organisms are retained in the bioreactor for longer periods, leading to more thorough conversion of the biomass. Although, a high HRT will also increase operational costs [101]. OLR is an important parameter in VFA production [109,110] — A higher OLR translates into increased overall substrate availability and may also cause a decrease in pH, inhibiting methanogenesis [103]. Indeed, increasing the OLR has been shown to successfully inhibit the methane production without the aid of additional methanogen inhibitors [83]. This inhibition and nutrient availability leads to a increased accumulation of VFAs, e.g., during fermentation of OFMSW in plug-flows, doubling the OLR resulted in a 150% increase in VFA production and 30% decrease in specific biogas production [110]. On the other hand, the pH decrease that comes with a high OLR may also function to hinder hydrolysis [78]. It was also observed that increasing the OLR and pH levels lower than 5 led to a predominance of ethanol production instead of VFAs in the fermentation of food waste in leach-bed reactors [109]. Therefore, an optimal pH range associated with both a balanced and appropriate OLR is necessary to support optimal production.

5.8. Feedstock Choice Influences VFA Profile

As discussed in Section 3, feedstock has a direct impact on VFA production (Table 1) similar to OLR. Characteristics, such as ammonia, COD, pH, and micronutrient availability, could affect the microbial conversion of the biomass to the desired acids [101]. Notably, the hydrolysis of biomasses that have higher total COD than soluble COD is directly impaired due to the degradation of particulate compounds. Feedstock type is also known to affect the profile of VFAs produced due to its characteristics and/or indigenous microbial community [80,101,107]. For example, propionic and valeric acids were the main VFAs obtained in the fermentation of food waste at 5-day HRT, 14–15 kgVS.m^{-3}.day^{-1} [81], while butyric and acetic acids were the main VFAs produced from fermentation of grass and grass silage (Table 2) [77].

5.9. The Challenge: VFA Recovery and Concentration

The choice of which operational conditions to employ in VFA generation must be informed by the intended purpose of those VFAs. For example, when designing a second stage to produce biomethane or electricity using microbial fuel cells, conditions should maximize the accumulation of acetic acid [101]. However, if the primary function is to produce PHAs in a second reactor, conditions should support the accumulation of butyric acid or propionic and valeric acids. In this way, optimizing the operational conditions to produce VFAs is extremely important. Although focus has been placed on VFA production and optimization, the separation of individual VFAs remains a substantial challenge [97].

The separation of individual VFAs is inherently challenging due to (i) the similar physical properties of VFAs, (ii) their potential to form azeotropes with water, and (iii) their oftentimes low concentrations in fermentation media [99]. In one recent study attempting to isolate caproic acid, a VFA-rich stream was treated via electrochemical extraction. It successfully concentrated caproic acid above its solubility concentration in water, which resulted in the formation of a hydrophobic layer making the separation feasible with 70%wt purity [80]. However, scaled application would require external electricity supplementation. In a recent study, two techniques were investigated: a combination of ultrafiltration and reverse osmosis membrane and liquid-liquid extraction with diethyl ether and methyl-isobutyl-ketone [97]. Another combination of techniques using solid screening, microfiltration, pervaporation, and electrodialysis was successful in recovering 4.5 g.L^{-1} of VFAs from a 80 L reactor fermenting grass while separating the solids [105]. However, to date, the selective separation of individual acids is still unavailable, which directly impacts the commercial feasibility of isolating specific VFAs to sell as individual products [98,99].

Moreover, VFA separation from complex media, such as fermentation liquor, is primarily limited to lab-scale investigations which implement a wide variety of technologies. Analyzing trends in this work may facilitate consensus building and eventual pilot-scale and real-world applications. Current trends suggest that separation methods often consist of multiple steps and technologies, making up process cascades (Figure 2). Most require one or more media preparation steps, such as pH correction or solids removal. These steps each require equipment and operational input and greatly influence the overall efficiency of the pipeline. For instance, approximately 75% of these processes used a solid removal method (Figure 2). Most used centrifugation or filtration as opposed to more efficient, passive solids removal technologies such as tank separation. Additionally, while pH correction was implemented in most fermentation systems, it was only considered a step in the VFA separation process if (i) it was corrected to a very high or very low pH [111,112], (ii) the pH was specifically referenced as facilitating separation [113], or (iii) if the fermentation broth was pH-corrected after collection from reactor [114].

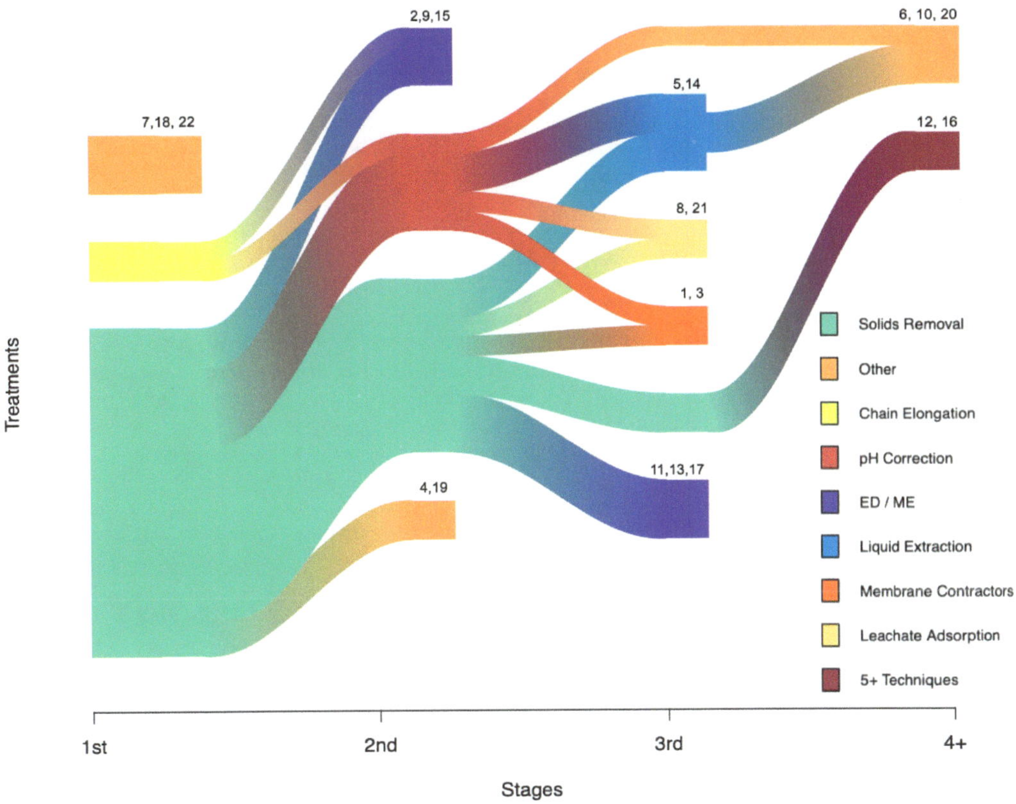

Figure 2. Processes used to separate VFAs from fermentation liquors. Sankey plot showing the variety of VFA harvesting processes as a flow of stages (1st, 2nd, 3rd, and 4+). Technologies using similar principles and materials were grouped together; for example, electrodialysis and membrane electrolysis were grouped together since they both use membranes and current driven separation of solutes. Technologies were classified as "other" if they were used by only a single study at any given stage. References: 1—[115]; 2—[75], 3—[114], 4—[116], 5—[117], 6—[46]; 7—[113]; 8—[118]; 9—[119]; 10—[97]; 11—[120]; 12—[105]; 13—[121]; 14—[88]; 15—[80]; 16—[122]; 17—[123]; 18—[124]; 19—[125]; 20—[112]; 21—[111]; and 22—[126].

CE was used in only 10% of these treatments. Generally, CE alone is not sufficient to accomplish VFA separation. Also referred to as a secondary fermentation, it is a microbially mediated process that lengthens the carbon chains of fatty acids, making them more hydrophobic and easier to separate from an aqueous solution. Often CE requires electron donor supplementation (e.g., ethanol). However, one study instead relied on donors produced during primary fermentation [80]. Because fermentation liquors have significant levels of electron donors (e.g., lactic acid) they may be well-suited for use as feedstock for CE processes.

In VFA separation processes, research suggests that the last step is the most intensive and effective. Roughly 30% of treatments terminated with the use of electrodialysis or membrane electrolysis, 20% with membrane contractors/membrane based solvent extraction/membrane-based reactant extraction, 14% used leachate absorption, and 14% used liquid extraction (Figure 2). Only 14% of the terminal treatments were classified as "other" technologies which were used two or fewer times (e.g., distillation, pervaporation). Each separation technology generally entails a trade-off between cost and productivity [97]. The cost of using membranes with electricity is dependent upon the price of the membrane material. Meanwhile, the use of organic solvents is controversial in terms of sustainability since most of these solvents are fossil-fuel derived chemicals. The use of distillation to recover these solvents and VFAs could be feasible provided a low-cost energy source is available. As mentioned previously, some studies have avoided separation altogether by converting mixed VFAs into the desired products directly from the fermentation liquor [99,101,127–129]. However, more work needs to be done in this area to maximize productivity and improve commercial feasibility.

6. Innovative VFA Applications

6.1. VFAs for Bioplastic Production

PHAs are biodegradable thermoplastic polymers synthesized by micro-organisms from VFA (Figure 1) and, therefore, are considered an environmentally friendly substitution to fossil fuel-derived plastics [101,128,130]. The characteristics of the final bioplastic are directly related to the polymer chain-length and the monomers and co-monomers used in its formation [128,130,131]. Although PHAs are already commercially produced, high operational costs still hinder large-scale production of these bioplastics [130]. Most commercial productions are performed by pure or genetically modified cultures, consequently resulting in high operational costs due to downstream processing (separation, filtration, and centrifugation), energy input (media and reactor sterilization), substrate formulation (pure VFAs), and equipment cost [101,130]. However, many studies have shifted focus to the development of processes using mixed cultures and low-cost biomass, thereby improving economic feasibility [101,128,129,131,132]. The profitability of PHA production seems to be associated with selecting (i) low-cost feedstocks and (ii) specialized mixed cultures (as opposed to pure cultures) to decrease operational costs. Optimizing the conditions for PHA production would also decrease the costs associated with product recovery [101].

6.2. VFAs for Chain Elongation Chemicals

MCCAs are aliphatic and straight carboxylic acids composed of 6–12 atoms of carbon [127]. MCCAs are produced from the biological CE of ethanol and/or short-chain carboxylic acids (SCCAs) through the reverse β-oxidation pathway by anaerobic bacteria [133]. The MCCAs have longer carbon chains and are hydrophobic as compared to SCCAs. Thus, the recovery of MCCAs from liquid media is easier, translating into lower downstream separation costs for higher-market-value compounds [127]. As with PHA production, the use of mixed-species cultures rather than pure cultures would increase commercial feasibility. Therefore, many studies on optimizing the production of MCCAs from low-cost feedstock and VFA-rich streams have been conducted [35,80,115,127,134–138]. Challenges remain in terms of optimizing operational conditions to selectively produce the MCCAs of interest and concentrate them to a solubility level that would facilitate separation from the liquid media. Moreover, the development of effective and cheap separation processes would also increase the competitiveness of MCCAs.

6.3. VFAs for Bioenergy and Biofuel

Biogenic VFAs can also be converted into bioenergy and biofuels such as biogas, biomethane, biohydrogen, and electricity (Figure 1). Although not as profitable an application as the synthesis of PHAs and MCCAs, bioenergy is essential to the function of a biorefinery. In order to support chemical platforms of the biorefinery, VFAs can be converted to energy and used to sustainably maintain the biorefinery processes. Surplus energy could be sold to the grid, while biogas can be used as a source of heat, for combined heat and power (CHP) plants [139], or upgraded to biomethane [140]. Biohydrogen has also attracted attention as a fossil fuel substitute for transport due to its clean combustion (generating water) and high energetic value [141]. VFAs can be used to generate biohydrogen through photo fermentation and microbial electrolysis cell [101,142,143]. Alternatively, VFAs can also be used to produce electricity from microbial fuel cells [101]. These processes, however, have not yet achieved a commercial state and are highly dependent upon an acetic acid rich VFA stream to maximize efficiency.

7. Fermentative Microbial Communities

7.1. Microbial Communities—Why Bother?

Microbiology provides a validation that communities develop and adapt to conditions inside engineered systems [35,144–147]. Understanding their responses and ongoing development can give us confidence around applications of new technologies. Understanding that there are degradative processes that can be linked to biological processes and further linked to design [148,149] is critical for the future of the field. Furthermore, understanding how microbial communities develop under these conditions and underpin efficient conversions, supports the application of biotechnologies under conditions which were previously considered unsuitable. Gaining a deeper understanding of the interactions, ongoing development, and functions of built-ecosystem microbiomes will move us one step further toward harnessing their transformative metabolisms at full capacity—resulting in more efficient systems and a wider range of bioproducts at higher yields.

7.2. Key Fermentative Groups

The structure and function of a bioreactor's microbial consortium directly depend upon the applied operating conditions; thus, community profiles vary from study to study. For example, operational choices, such as pre-treatment [150], temperature [151], inoculum [152], and even digester design [148] all induce shifts in microbial community structure and function. However, while fermentative systems support a diverse range of community profiles, several common trends and notable findings reoccur. Namely, *Clostridia* are consistently cited for efficient production of VFA across a wide range of studies with differing operational conditions [35,48,150,153]. Other important players in terms of efficient VFA production are *Sporanaerobacter*, *Tissierella*, *Bacillus*, and *Firmicutes* [152,153]. Interestingly, however, one study noted that *Chloroflexi* were negatively associated with increased VFA yields [152].

8. Future Perspectives: Biorefinery Concept, Application, and Challenges

Most existing AD plants generate biofuel and/or biochemicals in single production chains, generating low-value products or residues which are treated as waste or land spread. In contrast, AD plants can function within a biorefinery as part of a zero-waste strategy, resulting in the complete conversion of wastes into valuable products. The biorefinery concept is parallel to the refinery process in the oil industry, where crude oil is taken in and separated into a myriad of petrochemical products including fuels, lubricants, waxes, asphalt, and polymer production chemicals. Similarly, a biorefinery is a facility that takes in biogenic feedstock to produce biofuels, power, and biochemicals [154].

However, unlike an oil refinery, biorefineries must cope with an extensive variability of feedstock in terms of carbohydrate composition, recalcitrance, ash content, etc. In order to optimally produce zero-waste end-products the biorefinery must integrate physical, chemical, biological, and thermochemical processes to convert each fraction into product [2,7]. Via process integration, these systems convert heterogenous biogenic and waste streams into a multitude of value-added products. AD has great potential as a valuable core technology within the biorefinery concept (Figure 1) due to its diverse functionality. It can carry out waste remediation, bioenergy production, bio-based product synthesis, and biological pretreatment of lignocellulosic biomass [2,3,7].

AD is a proven process that generally produces energy-rich biogas as the main attractive product. However, biogas production is usually not stable in digester systems dealing with heterogeneous feedstock (e.g., food waste, grass, and slurry). Therefore, the yield of biogas is reduced, further reducing its already low economical added value. An alternative approach is to bioengineer the AD process for the production of carboxylic acids alongside biogas and other products, thereby converting biogas plants into biorefineries.

Carboxylic acids, such as lactic, succinic, and VFA, have been successfully generated from the initial anaerobic fermentation of food waste, agricultural waste (silage and cattle liquid manure), and OFMSW (Table 1). These carboxylic acids are valuable products when separated from the fermentation broth. However, due to their high solubility in water, the recovery has proven difficult and economically unattractive [125,155]. These carboxylic acids may be further processed into biogas or converted through biological and chemical process into alcohol-based fuels (e.g., ethanol and butanol) or other value-added products (e.g., PHAs and MCCAs), or they can be used directly to generate electricity in microbial fuel cells [3,7,156]. In addition to organic acids, gaseous molecular hydrogen and carbon dioxide are normally produced during the anaerobic fermentation of organic substrates. These can be biologically converted to methane [157] or chemically processed into methanol [158].

The solid residue obtained in the anaerobic treatment of biomass, known as digestate, has been viewed as a low-value product, conventionally managed as a fertilizer or animal bedding. However, recent studies have proposed innovative concepts and techniques for its valorization to biogas and bio-based products [2,3,159,160]. The digestate, together with a low VFA-liquor and biogas, can be processed to a methane-rich biogas which, in turn, can be used to supply heat and electricity for facility operation. Furthermore, studies have demonstrated that AD can act as a biological pre-treatment for lignocellulosic feedstock as it degrades hemicellulose faster than cellulose, thereby facilitating the subsequent enzymatic hydrolysis of cellulose in downstream processes [161,162]. Following the cellulose extraction, the lignin-rich residue can be thermochemically processed to biofuel or other valuable products.

Interestingly, a significant opportunity exists to valorize the nutrient-rich liquor from the digester for macro- and micro-algae production. This approach not only allows for the production of algal biomass, which can be further processed into biofuels and bio-based products, but also accomplishes nutrient-removal from AD effluents which can then be recycled back as process-water into the AD plant [7,163]. Despite the potential value offered by biomass biorefineries technological, spatial, and logistical barriers impact its economic viability, thus hindering widespread application [164]. Notably, the technological barrier seems to be the most pressing challenge as it directly affects production yield. To maximize product yield from biomass, various pre-treatment methods and enzymatic hydrolysis techniques have been used within the AD-based biorefinery. However, these techniques present many limitations (e.g., low efficiency and high cost). Apart from the technological barriers, challenges such as the recovery of products from effluent, the transportation of the fuel and feedstock also impact the success of AD-based biorefinery (Figure 3).

Addressing these challenges by developing novel, sustainable, and economically viable technologies will contribute towards the development of an economically attractive biomass biorefinery. For instance, some studies have pointed out potential nanotechnology applications in pre-treatment methods [165]. The enzymatic hydrolysis pre-treatment method could be improved by using magnetic nanoparticles to immobilize hydrolytic enzymes, thus allowing them to be re-used in multiple cycles of hydrolysis [166]. Novel nanotechnological solutions should be further investigated, with an aim to improve product yield and process efficiency.

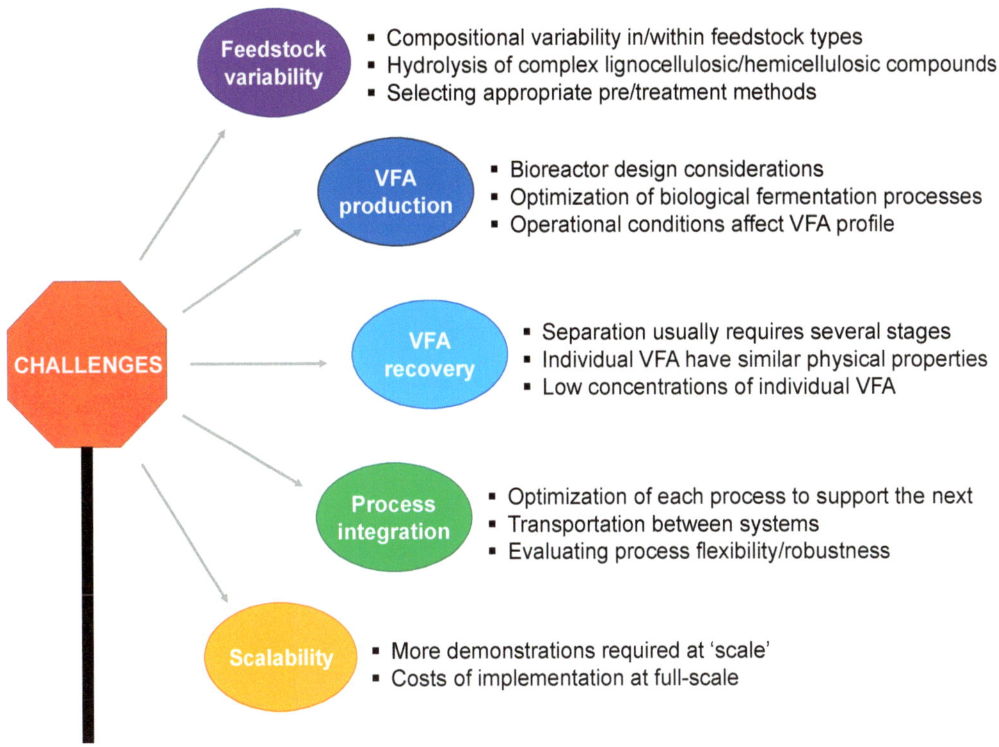

Figure 3. Challenges and opportunities for future research. The promise of a biorefinery-based bioeconomy will rely on innovative and interdisciplinary solutions. The literature suggests that key challenges with respect to feedstock variability, VFA production and recovery, process integration, and scalability all need to be tackled.

9. Conclusions

Although there is clearly room for technological advancements and for closing research gaps, literature suggests that the use of residual biomass within the biorefinery framework is paving the way for a closed-loop bioeconomy. Within such a framework, VFA production could serve as the core platform to produce both energy and a range of bio-based products. This would propel us towards a functioning circular economy that minimizes waste and maximizes production.

Author Contributions: All authors designed the review. C.N., F.C., E.D. and A.T. reviewed the literature and drafted the manuscript and V.O. revised the document. All authors approve the paper and agree for accountability of the work therein. All authors have read and agreed to the published version of the manuscript.

Funding: This work was financially supported by grants from the Higher Education Authority (HEA) of Ireland through: the Programme for Research at Third Level Institutions, Cycle 5 (PRTLI-5), co-funded by the European Regional Development Fund (ERDF); the Enterprise Ireland Technology Centres Programme (TC/2014/0016) and Science Foundation Ireland (14/IA/2371 and 16/RC/3889).

Institutional Review Board Statement: Not applicable.

Informed Consent Statement: Not applicable.

Data Availability Statement: Not applicable.

Acknowledgments: The authors would like to thank the funding bodies which supported this work.

Conflicts of Interest: The authors declare no conflict of interest.

References

1. Badgujar, K.C.; Bhanage, B.M. Dedicated and Waste Feedstocks for Biorefinery: An Approach to Develop a Sustainable Society. In *Waste Biorefinery*; Elsevier: Amsterdam, The Netherlands, 2018; pp. 3–38.
2. Coma, M.; Martinez-Hernandez, E.; Abeln, F.; Raikova, S.; Donnelly, J.; Arnot, T.C.; Allen, M.J.; Hong, D.D.; Chuck, C.J. Organic Waste as a Sustainable Feedstock for Platform Chemicals. *Faraday Discuss.* **2017**, *202*, 175–195. [CrossRef] [PubMed]
3. Righetti, E.; Nortilli, S.; Fatone, F.; Frison, N.; Bolzonella, D. A Multiproduct Biorefinery Approach for the Production of Hydrogen, Methane and Volatile Fatty Acids from Agricultural Waste. *Waste Biomass Valorization* **2020**, *11*, 5239–5246. [CrossRef]
4. Wainaina, S.; Lukitawesa; Kumar Awasthi, M.; Taherzadeh, M.J. Bioengineering of Anaerobic Digestion for Volatile Fatty Acids, Hydrogen or Methane Production: A Critical Review. *Bioengineered* **2019**, *10*, 437–458. [CrossRef] [PubMed]
5. Lytras, G.; Lytras, C.; Mathioudakis, D.; Papadopoulou, K.; Lyberatos, G. Food Waste Valorization Based on Anaerobic Digestion. *Waste Biomass Valorization* **2021**, *12*, 1677–1697. [CrossRef]
6. Michalopoulos, I.; Lytras, G.M.; Mathioudakis, D.; Lytras, C.; Goumenos, A.; Zacharopoulos, I.; Papadopoulou, K.; Lyberatos, G. Hydrogen and Methane Production from Food Residue Biomass Product (FORBI). *Waste Biomass Valorization* **2020**, *11*, 1647–1655. [CrossRef]
7. Surendra, K.C.; Sawatdeenarunat, C.; Shrestha, S.; Sung, S.; Khanal, S.K. Anaerobic Digestion-Based Biorefinery for Bioenergy and Biobased Products. *Ind. Biotechnol.* **2015**, *11*, 103–112. [CrossRef]
8. Demirer, G.N.; Chen, S. Two-Phase Anaerobic Digestion of Unscreened Dairy Manure. *Process Biochem.* **2005**, *40*, 3542–3549. [CrossRef]
9. Ning, P.; Yang, G.; Hu, L.; Sun, J.; Shi, L.; Zhou, Y.; Wang, Z.; Yang, J. Recent Advances in the Valorization of Plant Biomass. *Biotechnol. Biofuels* **2021**, *14*, 102. [CrossRef]
10. Farzad, S.; Mandegari, M.A.; Görgens, J.F. A Critical Review on Biomass Gasification, Co-Gasification, and Their Environmental Assessments. *Biofuel Res. J.* **2016**, *3*, 483–495. [CrossRef]
11. Bonilla, J.; Gordillo, G.; Cantor, C. Experimental Gasification of Coffee Husk Using Pure Oxygen-Steam Blends. *Front. Energy Res.* **2019**, *7*, 127. [CrossRef]
12. Guo, Y.; Wang, S.; Huelsman, C.M.; Savage, P.E. Products, Pathways, and Kinetics for Reactions of Indole under Supercritical Water Gasification Conditions. *J. Supercrit. Fluids* **2013**, *73*, 161–170. [CrossRef]
13. Heidenreich, S.; Foscolo, P.U. New Concepts in Biomass Gasification. *Prog. Energy Combust. Sci.* **2015**, *46*, 72–95. [CrossRef]
14. Kalinci, Y.; Hepbasli, A.; Dincer, I. Exergoeconomic Analysis and Performance Assessment of Hydrogen and Power Production Using Different Gasification Systems. *Fuel* **2012**, *102*, 187–198. [CrossRef]
15. Sikarwar, V.S.; Zhao, M.; Clough, P.; Yao, J.; Zhong, X.; Memon, M.Z.; Shah, N.; Anthony, E.J.; Fennell, P.S. An Overview of Advances in Biomass Gasification. *Energy Environ. Sci.* **2016**, *9*, 2939–2977. [CrossRef]
16. Lee, C.S.; Conradie, A.V.; Lester, E. Review of Supercritical Water Gasification with Lignocellulosic Real Biomass as the Feedstocks: Process Parameters, Biomass Composition, Catalyst Development, Reactor Design and Its Challenges. *Chem. Eng. J.* **2021**, *415*, 128837. [CrossRef]
17. Hu, X.; Gholizadeh, M. Biomass Pyrolysis: A Review of the Process Development and Challenges from Initial Researches up to the Commercialisation Stage. *J. Energy Chem.* **2019**, *39*, 109–143. [CrossRef]
18. Latif, H.; Zeidan, A.A.; Nielsen, A.T.; Zengler, K. Trash to Treasure: Production of Biofuels and Commodity Chemicals via Syngas Fermenting Microorganisms. *Curr. Opin. Biotechnol.* **2014**, *27*, 79–87. [CrossRef]
19. Gemechu, E.D.; Kumar, A. *Chapter 12—The Environmental Performance of Hydrogen Production Pathways Based on Renewable Sources*; Ren, J.B.T.-R.-E.-D.F., Ed.; Academic Press: Cambridge, MA, USA, 2021; pp. 375–406. ISBN 978-0-12-820539-6.
20. Casademont, P.; García-Jarana, M.B.; Sánchez-Oneto, J.; Portela, J.R.; Martínez de la Ossa, E.J. Supercritical Water Gasification: A Patents Review. *Rev. Chem. Eng.* **2017**, *33*, 237–261. [CrossRef]
21. De Blasio, C.; Järvinen, M. *Supercritical Water Gasification of Biomass*; Elsevier: Oxford, UK, 2017; pp. 171–195, ISBN 978-0-12-804792-7.
22. Sevillano, C.A.; Pesantes, A.A.; Peña Carpio, E.; Martínez, E.J.; Gómez, X. Anaerobic Digestion for Producing Renewable Energy—The Evolution of This Technology in a New Uncertain Scenario. *Entropy* **2021**, *23*, 145. [CrossRef]
23. Cesaro, A. The Valorization of the Anaerobic Digestate from the Organic Fractions of Municipal Solid Waste: Challenges and Perspectives. *J. Environ. Manag.* **2021**, *280*, 111742. [CrossRef]
24. EBA European Biogas Association. *EBA European Biogas Association Annual Report*; EBA European Biogas Association: Brussels, Belgium, 2020; p. 38.
25. Cheah, Y.-K.; Vidal-Antich, C.; Dosta, J.; Mata-Álvarez, J. Volatile Fatty Acid Production from Mesophilic Acidogenic Fermentation of Organic Fraction of Municipal Solid Waste and Food Waste under Acidic and Alkaline PH. *Environ. Sci. Pollut. Res.* **2019**, *26*, 35509–35522. [CrossRef]
26. Menzel, T.; Neubauer, P.; Junne, S. Role of Microbial Hydrolysis in Anaerobic Digestion. *Energies* **2020**, *13*, 5555. [CrossRef]
27. FAO. *Food Loss and Food Waste: Causes and Solutions*; Food and Agriculture Organization: Rome, Italy, 2011; ISBN 9781788975391.

28. FAO. *The State of Food and Agriculture 2019: Moving forward on Food Loss and Waste Reduction*; Food and Agriculture Organization: Rome, Italy, 2019; ISBN 9781315764788.
29. Matharu, A.S.; de Melo, E.M.; Houghton, J.A. Opportunity for High Value-Added Chemicals from Food Supply Chain Wastes. *Bioresour. Technol.* **2016**, *215*, 123–130. [CrossRef] [PubMed]
30. Morales-Polo, C.; Cledera-Castro, M.D.M.; Moratilla Soria, B.Y. Reviewing the Anaerobic Digestion of Food Waste: From Waste Generation and Anaerobic Process to Its Perspectives. *Appl. Sci.* **2018**, *8*, 1804. [CrossRef]
31. Fernández, A.; Sanchez, A.; Font, X. Anaerobic Co-Digestion of a Simulated Organic Fraction of Municipal Solid Wastes and Fats of Animal and Vegetable Origin. *Biochem. Eng. J.* **2005**, *26*, 22–28. [CrossRef]
32. Girault, R.; Bridoux, G.; Nauleau, F.; Poullain, C.; Buffet, J.; Peu, P.; Sadowski, A.G.; Béline, F. Anaerobic Co-Digestion of Waste Activated Sludge and Greasy Sludge from Flotation Process: Batch versus CSTR Experiments to Investigate Optimal Design. *Bioresour. Technol.* **2012**, *105*, 1–8. [CrossRef]
33. Neves, L.; Goncalo, E.; Oliveira, R.; Alves, M.M. Influence of Composition on the Biomethanation Potential of Restaurant Waste at Mesophilic Temperatures. *Waste Manag.* **2008**, *28*, 965–972. [CrossRef] [PubMed]
34. Fisgativa, H.; Tremier, A.; Dabert, P. Characterizing the Variability of Food Waste Quality: A Need for Efficient Valorisation through Anaerobic Digestion. *Waste Manag.* **2016**, *50*, 264–274. [CrossRef] [PubMed]
35. Nzeteu, C.O.; Trego, A.C.; Abram, F.; O'Flaherty, V. Reproducible, High-Yielding, Biological Caproate Production from Food Waste Using a Single-Phase Anaerobic Reactor System. *Biotechnol. Biofuels* **2018**, *11*, 108. [CrossRef] [PubMed]
36. Ma, J.; Frear, C.; Wang, Z.; Yu, L.; Zhao, Q.; Li, X.; Chen, S. A Simple Methodology for Rate-Limiting Step Determination for Anaerobic Digestion of Complex Substrates and Effect of Microbial Community Ratio. *Bioresour. Technol.* **2013**, *134*, 391–395. [CrossRef] [PubMed]
37. Fang, Q.; Ji, S.; Huang, D.; Huang, Z.; Huang, Z.; Zeng, Y.; Liu, Y. Impact of Alkaline Pretreatment to Enhance Volatile Fatty Acids (VFAs) Production from Rice Husk. *Biochem. Res. Int.* **2019**, *2019*, 8489747. [CrossRef] [PubMed]
38. Zhang, D.; Jiang, H.; Chang, J.; Sun, J.; Tu, W.; Wang, H. Effect of Thermal Hydrolysis Pretreatment on Volatile Fatty Acids Production in Sludge Acidification and Subsequent Polyhydroxyalkanoates Production. *Bioresour. Technol.* **2019**, *279*, 92–100. [CrossRef]
39. Kim, J.K.; Oh, B.R.; Chun, Y.N.; Kim, S.W. Effects of Temperature and Hydraulic Retention Time on Anaerobic Digestion of Food Waste. *J. Biosci. Bioeng.* **2006**, *102*, 328–332. [CrossRef] [PubMed]
40. Paudel, S.R.; Banjara, S.P.; Choi, O.K.; Park, K.Y.; Kim, Y.M.; Lee, J.W. Pretreatment of Agricultural Biomass for Anaerobic Digestion: Current State and Challenges. *Bioresour. Technol.* **2017**, *245*, 1194–1205. [CrossRef] [PubMed]
41. Bhatia, S.K.; Jagtap, S.S.; Bedekar, A.A.; Bhatia, R.K.; Rajendran, K.; Pugazhendhi, A.; Rao, C.V.; Atabani, A.E.; Kumar, G.; Yang, Y.-H. Renewable Biohydrogen Production from Lignocellulosic Biomass Using Fermentation and Integration of Systems with Other Energy Generation Technologies. *Sci. Total Environ.* **2020**, *765*, 144429. [CrossRef] [PubMed]
42. Zoghlami, A.; Paës, G. Lignocellulosic Biomass: Understanding Recalcitrance and Predicting Hydrolysis. *Front. Chem.* **2019**, *7*, 874. [CrossRef] [PubMed]
43. Galbe, M.; Wallberg, O. Pretreatment for Biorefineries: A Review of Common Methods for Efficient Utilisation of Lignocellulosic Materials. *Biotechnol. Biofuels* **2019**, *12*, 294. [CrossRef] [PubMed]
44. Baruah, J.; Nath, B.K.; Sharma, R.; Kumar, S.; Deka, R.C.; Baruah, D.C.; Kalita, E. Recent Trends in the Pretreatment of Lignocellulosic Biomass for Value-Added Products. *Front. Energy Res.* **2018**, *6*, 141. [CrossRef]
45. Sun, J.; Zhang, L.; Loh, K.-C. Review and Perspectives of Enhanced Volatile Fatty Acids Production from Acidogenic Fermentation of Lignocellulosic Biomass Wastes. *Bioresour. Bioprocess.* **2021**, *8*, 68. [CrossRef]
46. Cerrone, F.; Choudhari, S.K.; Davis, R.; Cysneiros, D.; O'Flaherty, V.; Duane, G.; Casey, E.; Guzik, M.W.; Kenny, S.T.; Babu, R.P.; et al. Medium Chain Length Polyhydroxyalkanoate (Mcl-PHA) Production from Volatile Fatty Acids Derived from the Anaerobic Digestion of Grass. *Appl. Microbiol. Biotechnol.* **2014**, *98*, 611–620. [CrossRef] [PubMed]
47. Cysneiros, D.; Thuillier, A.; Villemont, R.; Littlestone, A.; Mahony, T.; O'Flaherty, V. Temperature Effects on the Trophic Stages of Perennial Rye Grass Anaerobic Digestion. *Water Sci. Technol.* **2011**, *64*, 70–76. [CrossRef] [PubMed]
48. Joyce, A.; Ijaz, U.Z.; Nzeteu, C.; Vaughan, A.; Shirran, S.L.; Botting, C.H.; Quince, C.; O'Flaherty, V.; Abram, F. Linking Microbial Community Structure and Function during the Acidified Anaerobic Digestion of Grass. *Front. Microbiol.* **2018**, *9*, 540. [CrossRef]
49. Tilman, D.; Hill, J.; Lehman, C. Carbon-Negative Biofuels from Low-Input High-Diversity Grassland Biomass. *Science* **2006**, *314*, 1598–1600. [CrossRef]
50. Murphy, J.; Korres, N.; Singh, A.; Smyth, B.; Nizami, A.S.; Thamsiriroj, T. *The Potential for Grass Biomethane as a Biofuel Compressed Biomethane Generated from Grass, Utilised as A Transport Biofuel CCRP Report*; Environmental Protection Agency: Wexford, Ireland, 2013; ISBN 9781840954272.
51. Kung, L.; Shaver, R.D.; Grant, R.J.; Schmidt, R.J. *Silage Review: Interpretation of Chemical, Microbial, and Organoleptic Components of Silages*; Elsevier: Amsterdam, The Netherlands, 2018; Volume 101.
52. Grange, G.; Finn, J.A.; Brophy, C. Plant Diversity Enhanced Yield and Mitigated Drought Impacts in Intensively Managed Grassland Communities. *J. Appl. Ecol.* **2021**, *58*, 1864–1875. [CrossRef]
53. Cong, W.-F.; Moset, V.; Feng, L.; Møller, H.B.; Eriksen, J. Anaerobic Co-Digestion of Grass and Forbs—Influence of Cattle Manure or Grass Based Inoculum. *Biomass Bioenergy* **2018**, *119*, 90–96. [CrossRef]

54. Díaz-Vázquez, D.; Alvarado-Cummings, S.C.; Meza-Rodríguez, D.; Senés-Guerrero, C.; de Anda, J.; Gradilla-Hernández, M.S. Evaluation of Biogas Potential from Livestock Manures and Multicriteria Site Selection for Centralized Anaerobic Digester Systems: The Case of Jalisco, México. *Sustainability* **2020**, *12*, 3527. [CrossRef]
55. Nolan, S.; Thorn, C.E.; Ashekuzzaman, S.M.; Kavanagh, I.; Nag, R.; Bolton, D.; Cummins, E.; O'Flaherty, V.; Abram, F.; Richards, K.; et al. Landspreading with Co-Digested Cattle Slurry, with or without Pasteurisation, as a Mitigation Strategy against Pathogen, Nutrient and Metal Contamination Associated with Untreated Slurry. *Sci. Total Environ.* **2020**, *744*, 140841. [CrossRef]
56. Wolter, M.; Prayitno, S.; Schuchardt, F. Greenhouse Gas Emission during Storage of Pig Manure on a Pilot Scale. *Bioresour. Technol.* **2004**, *95*, 235–244. [CrossRef]
57. Clemens, J.; Trimborn, M.; Weiland, P.; Amon, B. Mitigation of Greenhouse Gas Emissions by Anaerobic Digestion of Cattle Slurry. *Agric. Ecosyst. Environ.* **2006**, *112*, 171–177. [CrossRef]
58. Kougias, P.G.; Kotsopoulos, T.A.; Martzopoulos, G.G. Effect of Feedstock Composition and Organic Loading Rate during the Mesophilic Co-Digestion of Olive Mill Wastewater and Swine Manure. *Renew. Energy* **2014**, *69*, 202–207. [CrossRef]
59. Orive, M.; Cebrián, M.; Zufía, J. Techno-Economic Anaerobic Co-Digestion Feasibility Study for Two-Phase Olive Oil Mill Pomace and Pig Slurry. *Renew. Energy* **2016**, *97*, 532–540. [CrossRef]
60. Tampio, E.A.; Blasco, L.; Vainio, M.M.; Kahala, M.M.; Rasi, S.E. Volatile Fatty Acids (VFAs) and Methane from Food Waste and Cow Slurry: Comparison of Biogas and VFA Fermentation Processes. *GCB Bioenergy* **2019**, *11*, 72–84. [CrossRef]
61. Yin, D.; Mahboubi, A.; Wainaina, S.; Qiao, W.; Taherzadeh, M.J. The Effect of Mono-and Multiple Fermentation Parameters on Volatile Fatty Acids (VFAs) Production from Chicken Manure via Anaerobic Digestion. *Bioresour. Technol.* **2021**, *330*, 124992. [CrossRef]
62. Moran, S. *An Applied Guide to Process and Plant Design*; Elsevier: Amsterdam, The Netherlands, 2019; ISBN 0128148616.
63. Van, D.P.; Fujiwara, T.; Leu Tho, B.; Song Toan, P.P.; Hoang Minh, G. A Review of Anaerobic Digestion Systems for Biodegradable Waste: Configurations, Operating Parameters, and Current Trends. *Environ. Eng. Res.* **2020**, *25*, 1–17. [CrossRef]
64. Pastor-Poquet, V.; Papirio, S.; Steyer, J.-P.; Trably, E.; Escudié, R.; Esposito, G. High-Solids Anaerobic Digestion Model for Homogenized Reactors. *Water Res.* **2018**, *142*, 501–511. [CrossRef]
65. Kariyama, I.D.; Zhai, X.; Wu, B. Influence of Mixing on Anaerobic Digestion Efficiency in Stirred Tank Digesters: A Review. *Water Res.* **2018**, *143*, 503–517. [CrossRef] [PubMed]
66. Lindmark, J.; Thorin, E.; Fdhila, R.B.; Dahlquist, E. Effects of Mixing on the Result of Anaerobic Digestion. *Renew. Sustain. Energy Rev.* **2014**, *40*, 1030–1047. [CrossRef]
67. Karthikeyan, O.P.; Selvam, A.; Wong, J.W.C. Hydrolysis–Acidogenesis of Food Waste in Solid–Liquid-Separating Continuous Stirred Tank Reactor (SLS-CSTR) for Volatile Organic Acid Production. *Bioresour. Technol.* **2016**, *200*, 366–373. [CrossRef] [PubMed]
68. Doğan, E.; Demirer, G.N. Volatile Fatty Acid Production from Organic Fraction of Municipal Solid Waste through Anaerobic Acidogenic Digestion. *Environ. Eng. Sci.* **2009**, *26*, 1443–1450. [CrossRef]
69. Pommier, S.; Chenu, D.; Quintard, M.; Lefebvre, X. A Logistic Model for the Prediction of the Influence of Water on the Solid Waste Methanization in Landfills. *Biotechnol. Bioeng.* **2007**, *97*, 473–482. [CrossRef] [PubMed]
70. Sanphoti, N.; Towprayoon, S.; Chaiprasert, P.; Nopharatana, A. The Effects of Leachate Recirculation with Supplemental Water Addition on Methane Production and Waste Decomposition in a Simulated Tropical Landfill. *J. Environ. Manag.* **2006**, *81*, 27–35. [CrossRef] [PubMed]
71. Degueurce, A.; Trémier, A.; Peu, P. Dynamic Effect of Leachate Recirculation on Batch Mode Solid State Anaerobic Digestion: Influence of Recirculated Volume, Leachate to Substrate Ratio and Recirculation Periodicity. *Bioresour. Technol.* **2016**, *216*, 553–561. [CrossRef] [PubMed]
72. Cadavid-Rodríguez, L.S.; Horan, N.J. Production of Volatile Fatty Acids from Wastewater Screenings Using a Leach-Bed Reactor. *Water Res.* **2014**, *60*, 242–249. [CrossRef] [PubMed]
73. Dahiya, S.; Sarkar, O.; Swamy, Y.V.; Mohan, S.V. Acidogenic Fermentation of Food Waste for Volatile Fatty Acid Production with Co-Generation of Biohydrogen. *Bioresour. Technol.* **2015**, *182*, 103–113. [CrossRef]
74. Hussain, A.; Filiatrault, M.; Guiot, S.R. Acidogenic Digestion of Food Waste in a Thermophilic Leach Bed Reactor: Effect of PH and Leachate Recirculation Rate on Hydrolysis and Volatile Fatty Acid Production. *Bioresour. Technol.* **2017**, *245*, 1–9. [CrossRef]
75. Arslan, D.; Zhang, Y.; Steinbusch, K.J.J.; Diels, L.; Hamelers, H.V.M.; Buisman, C.J.N.; De Wever, H. In-Situ Carboxylate Recovery and Simultaneous PH Control with Tailor-Configured Bipolar Membrane Electrodialysis during Continuous Mixed Culture Fermentation. *Sep. Purif. Technol.* **2017**, *175*, 27–35. [CrossRef]
76. Roume, H.; Arends, J.; Ameril, C.P.; Patil, S.A.; Rabaey, K. Enhanced Product Recovery from Glycerol Fermentation into 3-Carbon Compounds in a Bioelectrochemical System Combined with in Situ Extraction. *Front. Bioeng. Biotechnol.* **2016**, *4*, 73. [CrossRef] [PubMed]
77. Kullavanijaya, P.; Chavalparit, O. The Production of Volatile Fatty Acids from Napier Grass via an Anaerobic Leach Bed Process: The Influence of Leachate Dilution, Inoculum, Recirculation, and Buffering Agent Addition. *J. Environ. Chem. Eng.* **2019**, *7*, 103458. [CrossRef]
78. Xie, S.; Lawlor, P.G.; Frost, J.P.; Wu, G.; Zhan, X. Hydrolysis and Acidification of Grass Silage in Leaching Bed Reactors. *Bioresour. Technol.* **2012**, *114*, 406–413. [CrossRef] [PubMed]
79. Yan, B.H.; Selvam, A.; Wong, J.W.C. Application of Rumen Microbes to Enhance Food Waste Hydrolysis in Acidogenic Leach-Bed Reactors. *Bioresour. Technol.* **2014**, *168*, 64–71. [CrossRef]

80. Khor, W.C.; Andersen, S.; Vervaeren, H.; Rabaey, K. Electricity-Assisted Production of Caproic Acid from Grass. *Biotechnol. Biofuels* **2017**, *10*, 180. [CrossRef]
81. Valentino, F.; Munarin, G.; Biasiolo, M.; Cavinato, C.; Bolzonella, D.; Pavan, P. Enhancing Volatile Fatty Acids (VFA) Production from Food Waste in a Two-Phases Pilot-Scale Anaerobic Digestion Process. *J. Environ. Chem. Eng.* **2021**, *9*, 106062. [CrossRef]
82. Gottardo, M.; Micolucci, F.; Bolzonella, D.; Uellendahl, H.; Pavan, P. Pilot Scale Fermentation Coupled with Anaerobic Digestion of Food Waste—Effect of Dynamic Digestate Recirculation. *Renew. Energy* **2017**, *114*, 455–463. [CrossRef]
83. Lukitawesa; Patinvoh, R.J.; Millati, R.; Sárvári-Horváth, I.; Taherzadeh, M.J. Factors Influencing Volatile Fatty Acids Production from Food Wastes via Anaerobic Digestion. *Bioengineered* **2020**, *11*, 39–52. [CrossRef]
84. Xiong, Z.; Hussain, A.; Lee, J.; Lee, H.-S. Food Waste Fermentation in a Leach Bed Reactor: Reactor Performance, and Microbial Ecology and Dynamics. *Bioresour. Technol.* **2019**, *274*, 153–161. [CrossRef] [PubMed]
85. Swiatkiewicz, J.; Slezak, R.; Krzystek, L.; Ledakowicz, S. Production of Volatile Fatty Acids in a Semi-Continuous Dark Fermentation of Kitchen Waste: Impact of Organic Loading Rate and Hydraulic Retention Time. *Energies* **2021**, *14*, 2993. [CrossRef]
86. Pohland, F.G.; Ghosh, S. Developments in Anaerobic Stabilization of Organic Wastes-the Two-Phase Concept. *Environ. Lett.* **1971**, *1*, 255–266. [CrossRef] [PubMed]
87. Grootscholten, T.I.M.; Strik, D.; Steinbusch, K.J.J.; Buisman, C.J.N.; Hamelers, H.V.M. Two-Stage Medium Chain Fatty Acid (MCFA) Production from Municipal Solid Waste and Ethanol. *Appl. Energy* **2014**, *116*, 223–229. [CrossRef]
88. Kannengiesser, J.; Sakaguchi-Söder, K.; Mrukwia, T.; Jager, J.; Schebek, L. Extraction of Medium Chain Fatty Acids from Organic Municipal Waste and Subsequent Production of Bio-Based Fuels. *Waste Manag.* **2016**, *47*, 78–83. [CrossRef] [PubMed]
89. Xu, J.; Hao, J.; Guzman, J.J.L.; Spirito, C.M.; Harroff, L.A.; Angenent, L.T. Temperature-Phased Conversion of Acid Whey Waste into Medium-Chain Carboxylic Acids Via Lactic Acid: No External e-Donor. *Joule* **2019**, *3*, 885–888. [CrossRef]
90. Fezzani, B.; Cheikh, R. Ben Two-Phase Anaerobic Co-Digestion of Olive Mill Wastes in Semi-Continuous Digesters at Mesophilic Temperature. *Bioresour. Technol.* **2010**, *101*, 1628–1634. [CrossRef]
91. Khalid, A.; Arshad, M.; Anjum, M.; Mahmood, T.; Dawson, L. The Anaerobic Digestion of Solid Organic Waste. *Waste Manag.* **2011**, *31*, 1737–1744. [CrossRef] [PubMed]
92. Cho, J.K.; Park, S.C.; Chang, H.N. Biochemical Methane Potential and Solid State Anaerobic Digestion of Korean Food Wastes. *Bioresour. Technol.* **1995**, *52*, 245–253. [CrossRef]
93. De Baere, L.; Mattheeuws, B. Anaerobic Digestion of the Organic Fraction of Municipal Solid Waste in Europe-Status, Experience and Prospects. In Proceedings of the ISTANBUL3WCONGRESS 2013, Istanbul, Turkey, 22–24 May 2013; Volume 38.
94. Rapport, J.; Zhang, R.; Jenkins, B.M.; Williams, R.B. *Current Anaerobic Digestion Technologies Used for Treatment of Municipal Organic Solid Waste*; California Integrated Waste Management Board: Sacramento, CA, USA, 2008; Volume 236.
95. Climent Barba, F.; Grasham, O.; Puri, D.J.; Blacker, A.J. A Simple Techno-Economic Assessment for Scaling-Up the Enzymatic Hydrolysis of MSW Pulp. *Front. Energy Res.* **2022**, *10*. [CrossRef]
96. Hernández Regalado, R.E.; Häner, J.; Brügging, E.; Tränckner, J. Techno-Economic Assessment of Solid–Liquid Biogas Treatment Plants for the Agro-Industrial Sector. *Energies* **2022**, *15*, 4413. [CrossRef]
97. Jänisch, T.; Reinhardt, S.; Pohsner, U.; Böringer, S.; Bolduan, R.; Steinbrenner, J.; Oechsner, H. Separation of Volatile Fatty Acids from Biogas Plant Hydrolysates. *Sep. Purif. Technol.* **2019**, *223*, 264–273. [CrossRef]
98. Atasoy, M.; Owusu-Agyeman, I.; Plaza, E.; Cetecioglu, Z. Bio-Based Volatile Fatty Acid Production and Recovery from Waste Streams: Current Status and Future Challenges. *Bioresour. Technol.* **2018**, *268*, 773–786. [CrossRef]
99. Ramos-Suarez, M.; Zhang, Y.; Outram, V. Current Perspectives on Acidogenic Fermentation to Produce Volatile Fatty Acids from Waste. *Rev. Environ. Sci. Bio/Technol.* **2021**, *20*, 439–478. [CrossRef]
100. Luo, H.; Yang, R.; Zhao, Y.; Wang, Z.; Liu, Z.; Huang, M.; Zeng, Q. Recent Advances and Strategies in Process and Strain Engineering for the Production of Butyric Acid by Microbial Fermentation. *Bioresour. Technol.* **2018**, *253*, 343–354. [CrossRef]
101. Lee, W.S.; Chua, A.S.M.; Yeoh, H.K.; Ngoh, G.C. A Review of the Production and Applications of Waste-Derived Volatile Fatty Acids. *Chem. Eng. J.* **2014**, *235*, 83–99. [CrossRef]
102. Biosantech, T.A.S.; Rutz, D.; Janssen, R.; Drosg, B. 2—Biomass Resources for Biogas Production. In *Woodhead Publishing Series in Energy*; Wellinger, A., Murphy, J., Baxter, D.B.T.-T.B.H., Eds.; Woodhead Publishing: Cambridge, UK, 2013; pp. 19–51, ISBN 978-0-85709-498-8.
103. Magdalena, J.A.; Greses, S.; González-Fernández, C. Impact of Organic Loading Rate in Volatile Fatty Acids Production and Population Dynamics Using Microalgae Biomass as Substrate. *Sci. Rep.* **2019**, *9*, 18374. [CrossRef] [PubMed]
104. Vartoukian, S.R.; Palmer, R.M.; Wade, W.G. The Division Synergistes. *Anaerobe* **2007**, *13*, 99–106. [CrossRef]
105. Jones, R.J.; Massanet-Nicolau, J.; Fernandez–Feito, R.; Dinsdale, R.M.; Guwy, A.J. Fermentative Volatile Fatty Acid Production and Recovery from Grass Using a Novel Combination of Solids Separation, Pervaporation, and Electrodialysis Technologies. *Bioresour. Technol.* **2021**, *342*, 125926. [CrossRef] [PubMed]
106. Cysneiros, D.; Banks, C.J.; Heaven, S.; Karatzas, K.-A.G. The Effect of PH Control and 'Hydraulic Flush' on Hydrolysis and Volatile Fatty Acids (VFA) Production and Profile in Anaerobic Leach Bed Reactors Digesting a High Solids Content Substrate. *Bioresour. Technol.* **2012**, *123*, 263–271. [CrossRef]
107. Moza, A.; Ram, N.R.; Srivastava, N.K.; Nikhil, G.N. Bioprocessing of Low-Value Food Waste to High Value Volatile Fatty Acids for Applications in Energy and Materials: A Review on Process-Flow. *Bioresour. Technol. Rep.* **2022**, *19*, 101123. [CrossRef]

108. Steinbrenner, J.; Oskina, A.; Müller, J.; Oechsner, H. PH-Depended Flushing in an Automatized Batch Leach Bed Reactor System for Volatile Fatty Acid Production. *Bioresour. Technol.* **2022**, *360*, 127611. [CrossRef]
109. Campuzano, R.; Trejo-Aguilar, G.M.; Cuetero-Martínez, Y.; Ramírez-Vives, F.; Monroy, O. Acidogenesis of Food Wastes at Variable Inlet and Operational Conditions. *Environ. Technol. Innov.* **2022**, *25*, 102162. [CrossRef]
110. Rossi, E.; Pecorini, I.; Paoli, P.; Iannelli, R. Plug-Flow Reactor for Volatile Fatty Acid Production from the Organic Fraction of Municipal Solid Waste: Influence of Organic Loading Rate. *J. Environ. Chem. Eng.* **2022**, *10*, 106963. [CrossRef]
111. Yousuf, A.; Bonk, F.; Bastidas-Oyanedel, J.-R.; Schmidt, J.E. Recovery of Carboxylic Acids Produced during Dark Fermentation of Food Waste by Adsorption on Amberlite IRA-67 and Activated Carbon. *Bioresour. Technol.* **2016**, *217*, 137–140. [CrossRef]
112. Yin, J.; Wang, K.; Yang, Y.; Shen, D.; Wang, M.; Mo, H. Improving Production of Volatile Fatty Acids from Food Waste Fermentation by Hydrothermal Pretreatment. *Bioresour. Technol.* **2014**, *171*, 323–329. [CrossRef]
113. Dessì, P.; Asunis, F.; Ravishankar, H.; Cocco, F.G.; De Gioannis, G.; Muntoni, A.; Lens, P.N.L. Fermentative Hydrogen Production from Cheese Whey with In-Line, Concentration Gradient-Driven Butyric Acid Extraction. *Int. J. Hydrogen Energy* **2020**, *45*, 24453–24466. [CrossRef]
114. Aydin, S.; Yesil, H.; Tugtas, A.E. Recovery of Mixed Volatile Fatty Acids from Anaerobically Fermented Organic Wastes by Vapor Permeation Membrane Contactors. *Bioresour. Technol.* **2018**, *250*, 548–555. [CrossRef]
115. Agler, M.T.; Spirito, C.M.; Usack, J.G.; Werner, J.J.; Angenent, L.T. Chain Elongation with Reactor Microbiomes: Upgrading Dilute Ethanol to Medium-Chain Carboxylates. *Energy Environ. Sci.* **2012**, *5*, 8189–8192. [CrossRef]
116. Bak, C.; Yun, Y.-M.; Kim, J.-H.; Kang, S. Electrodialytic Separation of Volatile Fatty Acids from Hydrogen Fermented Food Wastes. *Int. J. Hydrogen Energy* **2019**, *44*, 3356–3362. [CrossRef]
117. Braune, M.; Yuan, B.; Sträuber, H.; McDowall, S.C.; Nitzsche, R.; Gröngröft, A. A Downstream Processing Cascade for Separation of Caproic and Caprylic Acid from Maize Silage-Based Fermentation Broth. *Front. Bioeng. Biotechnol.* **2021**, *9*, 725578. [CrossRef] [PubMed]
118. Fernando-Foncillas, C.; Cabrera-Rodríguez, C.I.; Caparrós-Salvador, F.; Varrone, C.; Straathof, A.J.J. Highly Selective Recovery of Medium Chain Carboxylates from Co-Fermented Organic Wastes Using Anion Exchange with Carbon Dioxide Expanded Methanol Desorption. *Bioresour. Technol.* **2021**, *319*, 124178. [CrossRef]
119. Hassan, G.K.; Jones, R.J.; Massanet-Nicolau, J.; Dinsdale, R.; Abo-Aly, M.M.; El-Gohary, F.A.; Guwy, A. Increasing 2-Bio-(H2 and CH4) Production from Food Waste by Combining Two-Stage Anaerobic Digestion and Electrodialysis for Continuous Volatile Fatty Acids Removal. *Waste Manag.* **2021**, *129*, 20–25. [CrossRef]
120. Jones, R.J.; Fernández-Feito, R.; Massanet-Nicolau, J.; Dinsdale, R.; Guwy, A. Continuous Recovery and Enhanced Yields of Volatile Fatty Acids from a Continually-Fed 100 L Food Waste Bioreactor by Filtration and Electrodialysis. *Waste Manag.* **2021**, *122*, 81–88. [CrossRef]
121. Jones, R.J.; Massanet-Nicolau, J.; Fernandez-Feito, R.; Dinsdale, R.M.; Guwy, A.J. Recovery and Enhanced Yields of Volatile Fatty Acids from a Grass Fermentation via In-Situ Solids Separation and Electrodialysis. *J. Clean. Prod.* **2021**, *296*, 126430. [CrossRef]
122. Kucek, L.A.; Xu, J.; Nguyen, M.; Angenent, L.T. Waste Conversion into N-Caprylate and n-Caproate: Resource Recovery from Wine Lees Using Anaerobic Reactor Microbiomes and in-Line Extraction. *Front. Microbiol.* **2016**, *7*, 1892. [CrossRef]
123. Tao, B.; Passanha, P.; Kumi, P.; Wilson, V.; Jones, D.; Esteves, S. Recovery and Concentration of Thermally Hydrolysed Waste Activated Sludge Derived Volatile Fatty Acids and Nutrients by Microfiltration, Electrodialysis and Struvite Precipitation for Polyhydroxyalkanoates Production. *Chem. Eng. J.* **2016**, *295*, 11–19. [CrossRef]
124. Yesil, H.; Calli, B.; Tugtas, A.E. A Hybrid Dry-Fermentation and Membrane Contactor System: Enhanced Volatile Fatty Acid (VFA) Production and Recovery from Organic Solid Wastes. *Water Res.* **2021**, *192*, 116831. [CrossRef]
125. Yesil, H.; Tugtas, A.E.; Bayrakdar, A.; Calli, B. Anaerobic Fermentation of Organic Solid Wastes: Volatile Fatty Acid Production and Separation. *Water Sci. Technol.* **2014**, *69*, 2132–2138. [CrossRef] [PubMed]
126. Zhang, Y.; Angelidaki, I. Bioelectrochemical Recovery of Waste-Derived Volatile Fatty Acids and Production of Hydrogen and Alkali. *Water Res.* **2015**, *81*, 188–195. [CrossRef] [PubMed]
127. De Groof, V.; Coma, M.; Arnot, T.; Leak, D.J.; Lanham, A.B. Medium Chain Carboxylic Acids from Complex Organic Feedstocks by Mixed Culture Fermentation. *Molecules* **2019**, *24*, 398. [CrossRef] [PubMed]
128. Mannina, G.; Presti, D.; Montiel-Jarillo, G.; Carrera, J.; Suárez-Ojeda, M.E. Recovery of Polyhydroxyalkanoates (PHAs) from Wastewater: A Review. *Bioresour. Technol.* **2020**, *297*, 122478. [CrossRef]
129. Pagliano, G.; Galletti, P.; Samorì, C.; Zaghini, A.; Torri, C. Recovery of Polyhydroxyalkanoates From Single and Mixed Microbial Cultures: A Review. *Front. Bioeng. Biotechnol.* **2021**, *9*, 54. [CrossRef] [PubMed]
130. Sabapathy, P.C.; Devaraj, S.; Meixner, K.; Anburajan, P.; Kathirvel, P.; Ravikumar, Y.; Zabed, H.M.; Qi, X. Recent Developments in Polyhydroxyalkanoates (PHAs) Production—A Review. *Bioresour. Technol.* **2020**, *306*, 123132. [CrossRef] [PubMed]
131. Zhang, W.; Alvarez-Gaitan, J.P.; Dastyar, W.; Saint, C.P.; Zhao, M.; Short, M.D. Value-Added Products Derived from Waste Activated Sludge: A Biorefinery Perspective. *Water* **2018**, *10*, 545. [CrossRef]
132. Costa, S.S.; Miranda, A.L.; de Morais, M.G.; Costa, J.A.V.; Druzian, J.I. Microalgae as Source of Polyhydroxyalkanoates (PHAs)—A Review. *Int. J. Biol. Macromol.* **2019**, *131*, 536–547. [CrossRef]
133. Angenent, L.T.; Richter, H.; Buckel, W.; Spirito, C.M.; Steinbusch, K.J.J.; Plugge, C.M.; Strik, D.P.B.T.B.; Grootscholten, T.I.M.; Buisman, C.J.N.; Hamelers, H.V.M. Chain Elongation with Reactor Microbiomes: Open-Culture Biotechnology To Produce Biochemicals. *Environ. Sci. Technol.* **2016**, *50*, 2796–2810. [CrossRef] [PubMed]

134. Reddy, M.V.; Mohan, S.V.; Chang, Y.-C. Medium-Chain Fatty Acids (MCFA) Production Through Anaerobic Fermentation Using Clostridium Kluyveri: Effect of Ethanol and Acetate. *Appl. Biochem. Biotechnol.* **2018**, *185*, 594–605. [CrossRef]
135. Roghair, M.; Hoogstad, T.; Strik, D.P.B.T.B.; Plugge, C.M.; Timmers, P.H.A.; Weusthuis, R.A.; Bruins, M.E.; Buisman, C.J.N. Controlling Ethanol Use in Chain Elongation by CO2 Loading Rate. *Environ. Sci. Technol.* **2018**, *52*, 1496–1505. [CrossRef] [PubMed]
136. Roghair, M.; Liu, Y.; Adiatma, J.C.; Weusthuis, R.A.; Bruins, M.E.; Buisman, C.J.N.; Strik, D.P.B.T.B. Effect of N-Caproate Concentration on Chain Elongation and Competing Processes. *ACS Sustain. Chem. Eng.* **2018**, *6*, 7499–7506. [CrossRef] [PubMed]
137. Steinbusch, K.J.J.; Hamelers, H.V.M.; Plugge, C.M.; Buisman, C.J.N. Biological Formation of Caproate and Caprylate from Acetate: Fuel and Chemical Production from Low Grade Biomass. *Energy Environ. Sci.* **2011**, *4*, 216–224. [CrossRef]
138. Nzeteu, C.O.; Coelho, F.; Trego, A.C.; Abram, F.; Ramiro-Garcia, J.; Paulo, L.; O'Flaherty, V. Development of an Enhanced Chain Elongation Process for Caproic Acid Production from Waste-Derived Lactic Acid and Butyric Acid. *J. Clean. Prod.* **2022**, *338*, 130655. [CrossRef]
139. O'Flaherty, V.; Collins, G.; Mahony, T. Anaerobic Digestion of Agricultural Residues. *Environ. Microbiol.* **2010**, *11*, 259–279.
140. Angelidaki, I.; Treu, L.; Tsapekos, P.; Luo, G.; Campanaro, S.; Wenzel, H.; Kougias, P.G. Biogas Upgrading and Utilization: Current Status and Perspectives. *Biotechnol. Adv.* **2018**, *36*, 452–466. [CrossRef]
141. Toledo-Alarcón, J.; Capson-Tojo, G.; Marone, A.; Paillet, F.; Júnior, A.D.N.F.; Chatellard, L.; Bernet, N.; Trably, E. Basics of Bio-Hydrogen Production by Dark Fermentation. In *Bioreactors for Microbial Biomass and Energy Conversion*; Springer: Berlin/Heidelberg, Germany, 2018; pp. 199–220.
142. Kadier, A.; Kalil, M.S.; Abdeshahian, P.; Chandrasekhar, K.; Mohamed, A.; Azman, N.F.; Logroño, W.; Simayi, Y.; Hamid, A.A. Recent Advances and Emerging Challenges in Microbial Electrolysis Cells (MECs) for Microbial Production of Hydrogen and Value-Added Chemicals. *Renew. Sustain. Energy Rev.* **2016**, *61*, 501–525. [CrossRef]
143. Reungsang, A.; Zhong, N.; Yang, Y.; Sittijunda, S.; Xia, A.; Liao, Q. Hydrogen from Photo Fermentation. In *Bioreactors for Microbial Biomass and Energy Conversion*; Springer: Berlin/Heidelberg, Germany, 2018; pp. 221–317.
144. Keating, C.; Hughes, D.; Mahony, T.; Cysneiros, D.; Ijaz, U.Z.; Smith, C.J.; O'Flaherty, V. Cold Adaptation and Replicable Microbial Community Development during Long-Term Low-Temperature Anaerobic Digestion Treatment of Synthetic Sewage. *FEMS Microbiol. Ecol.* **2018**, *94*, fiy095. [CrossRef]
145. Treu, L.; Kougias, P.G.; Campanaro, S.; Bassani, I.; Angelidaki, I. Deeper Insight into the Structure of the Anaerobic Digestion Microbial Community; the Biogas Microbiome Database Is Expanded with 157 New Genomes. *Bioresour. Technol.* **2016**, *216*, 260–266. [CrossRef] [PubMed]
146. Trego, A.C.; McAteer, P.G.; Nzeteu, C.; Mahony, T.; Abram, F.; Ijaz, U.Z.; O'Flaherty, V. Combined Stochastic and Deterministic Processes Drive Community Assembly of Anaerobic Microbiomes During Granule Flotation. *Front. Microbiol.* **2021**, *12*, 1165. [CrossRef] [PubMed]
147. Trego, A.C.; Galvin, E.; Sweeney, C.; Dunning, S.; Murphy, C.; Mills, S.; Nzeteu, C.; Quince, C.; Connelly, S.; Ijaz, U.Z.; et al. Growth and Break-Up of Methanogenic Granules Suggests Mechanisms for Biofilm and Community Development. *Front. Microbiol.* **2020**, *11*, 1126. [CrossRef] [PubMed]
148. McAteer, P.G.; Trego, A.C.; Thorn, C.; Mahony, T.; Abram, F.; O'Flaherty, V. Reactor Configuration Influences Microbial Community Structure during High-Rate, Low-Temperature Anaerobic Treatment of Dairy Wastewater. *Bioresour. Technol.* **2020**, *307*, 123221. [CrossRef] [PubMed]
149. Trego, A.C.; Holohan, B.C.; Keating, C.; Graham, A.; O'Connor, S.; Gerardo, M.; Hughes, D.; Zeeshan Ijaz, U.; O'Flaherty, V. First Proof of Concept for Full-Scale, Direct, Low-Temperature Anaerobic Treatment of Municipal Wastewater. *Bioresour. Technol.* **2021**, *341*, 125786. [CrossRef]
150. Yang, G.; Wang, J. Kinetics and Microbial Community Analysis for Hydrogen Production Using Raw Grass Inoculated with Different Pretreated Mixed Culture. *Bioresour. Technol.* **2018**, *247*, 954–962. [CrossRef] [PubMed]
151. Yuan, Y.; Liu, Y.; Li, B.; Wang, B.; Wang, S.; Peng, Y. Short-Chain Fatty Acids Production and Microbial Community in Sludge Alkaline Fermentation: Long-Term Effect of Temperature. *Bioresour. Technol.* **2016**, *211*, 685–690. [CrossRef]
152. Atasoy, M.; Eyice, O.; Schnürer, A.; Cetecioglu, Z. Volatile Fatty Acids Production via Mixed Culture Fermentation: Revealing the Link between PH, Inoculum Type and Bacterial Composition. *Bioresour. Technol.* **2019**, *292*, 121889. [CrossRef]
153. Wu, Q.-L.; Guo, W.-Q.; Zheng, H.-S.; Luo, H.-C.; Feng, X.-C.; Yin, R.-L.; Ren, N.-Q. Enhancement of Volatile Fatty Acid Production by Co-Fermentation of Food Waste and Excess Sludge without PH Control: The Mechanism and Microbial Community Analyses. *Bioresour. Technol.* **2016**, *216*, 653–660. [CrossRef]
154. Cherubini, F. The Biorefinery Concept: Using Biomass Instead of Oil for Producing Energy and Chemicals. *Energy Convers. Manag.* **2010**, *51*, 1412–1421. [CrossRef]
155. Li, Q.; Wang, D.; Wu, Y.; Li, W.; Zhang, Y.; Xing, J.; Su, Z. One Step Recovery of Succinic Acid from Fermentation Broths by Crystallization. *Sep. Purif. Technol.* **2010**, *72*, 294–300. [CrossRef]
156. Chen, H.; Meng, H.; Nie, Z.; Zhang, M. Polyhydroxyalkanoate Production from Fermented Volatile Fatty Acids: Effect of PH and Feeding Regimes. *Bioresour. Technol.* **2013**, *128*, 533–538. [CrossRef]
157. Leonzio, G. Upgrading of Biogas to Bio-Methane with Chemical Absorption Process: Simulation and Environmental Impact. *J. Clean. Prod.* **2016**, *131*, 364–375. [CrossRef]

158. Yao, Y.; Sempuga, B.C.; Liu, X.; Hildebrandt, D. Production of Fuels and Chemicals from a CO_2/H_2 Mixture. *Reactions* **2020**, *1*, 130–146. [CrossRef]
159. Budzianowski, W.M. High-Value Low-Volume Bioproducts Coupled to Bioenergies with Potential to Enhance Business Development of Sustainable Biorefineries. *Renew. Sustain. Energy Rev.* **2017**, *70*, 793–804. [CrossRef]
160. Khoshnevisan, B.; Tsapekos, P.; Zhang, Y.; Valverde-Pérez, B.; Angelidaki, I. Urban Biowaste Valorization by Coupling Anaerobic Digestion and Single Cell Protein Production. *Bioresour. Technol.* **2019**, *290*, 121743. [CrossRef] [PubMed]
161. MacLellan, J.; Chen, R.; Kraemer, R.; Zhong, Y.; Liu, Y.; Liao, W. Anaerobic Treatment of Lignocellulosic Material to Co-Produce Methane and Digested Fiber for Ethanol Biorefining. *Bioresour. Technol.* **2013**, *130*, 418–423. [CrossRef]
162. Yue, Z.; Teater, C.; MacLellan, J.; Liu, Y.; Liao, W. Development of a New Bioethanol Feedstock—Anaerobically Digested Fiber from Confined Dairy Operations Using Different Digestion Configurations. *Biomass Bioenergy* **2011**, *35*, 1946–1953. [CrossRef]
163. Rajagopal, R.; Mousavi, S.E.; Goyette, B.; Adhikary, S. Coupling of Microalgae Cultivation with Anaerobic Digestion of Poultry Wastes: Toward Sustainable Value Added Bioproducts. *Bioengineering* **2021**, *8*, 57. [CrossRef]
164. Philippini, R.R.; Martiniano, S.E.; Ingle, A.P.; Franco Marcelino, P.R.; Silva, G.M.; Barbosa, F.G.; dos Santos, J.C.; da Silva, S.S. Agroindustrial Byproducts for the Generation of Biobased Products: Alternatives for Sustainable Biorefineries. *Front. Energy Res.* **2020**, *8*, 152. [CrossRef]
165. Ingle, A.P.; Philippini, R.R.; da Silva, S.S. Pretreatment of Sugarcane Bagasse Using Two Different Acid-Functionalized Magnetic Nanoparticles: A Novel Approach for High Sugar Recovery. *Renew. Energy* **2020**, *150*, 957–964. [CrossRef]
166. Rai, M.; Ingle, A.P.; Pandit, R.; Paralikar, P.; Biswas, J.K.; da Silva, S.S. Emerging Role of Nanobiocatalysts in Hydrolysis of Lignocellulosic Biomass Leading to Sustainable Bioethanol Production. *Catal. Rev.* **2019**, *61*, 1–26. [CrossRef]

fermentation

Article

Co-Fermenting Pyrolysis Aqueous Condensate and Pyrolysis Syngas with Anaerobic Microbial Communities Enables L-Malate Production in a Secondary Fermentative Stage

Alberto Robazza [1], Claudia Welter [1], Christin Kubisch [1], Flávio César Freire Baleeiro [1,2], Katrin Ochsenreither [1] and Anke Neumann [1,*]

[1] Institute of Process Engineering in Life Sciences 2: Technical Biology, Karlsruhe Institute of Technology—KIT, 76131 Karlsruhe, Germany
[2] Department of Environmental Microbiology, Helmholtz Centre for Environmental Research—UFZ, 04318 Leipzig, Germany
* Correspondence: anke.neumann@kit.edu

Abstract: The pyrolytic conversion of lignocellulosic biomass into fuels and chemicals is a promising option for the valorization of agricultural and forestry residues. However, technological developments are still needed to maximize product recovery and carbon fixation of the pyrolysis process. The pyrolysis aqueous condensate (PAC), a pyrolysis by-product, has a high water content and is highly toxic, hampering its use. The anaerobic digestion of PAC from different biomasses has been proven a viable technology for PAC valorization and detoxification, but its toxicity limits the methanogenic potential. Alternatively, methanation or VFA production from syngas by anaerobic mixed cultures are technologies of scientific interest. This study investigates the potential of a two-stage process to convert the carbon and energy in syngas and PAC into L-malate. PAC and syngas were co-fermented by two mixed cultures at 37 and 55 °C, identifying kinetic inhibitions and the effects of increasing PAC concentrations on the product pool. The media from selected mixed culture fermentations were then inoculated with *Aspergillus oryzae* for L-malate production. The results show that mixed cultures can perform simultaneous syngas fermentation and PAC detoxification. While PAC concentrations above 2% completely inhibited methanogenesis, CO consumption was inhibited at PAC concentrations above 5%, regardless of the temperature. In fermentations where PAC inhibited methanation, the mixed cultures channelled the carbon and electrons from syngas and PAC to volatile fatty acids or acetate/H_2 production, depending on the incubation temperature. Substantial detoxification of PAC was observed under PAC concentrations up to 10% independently of the rates of syngas metabolism. PAC detoxification enabled the further valorization of the acetate produced via syngas and PAC fermentations into L-malate, achieving yields up to 0.17 mM/mM. These results are promising for the development of an integrated process that simultaneously detoxifies and recovers value from gaseous and aqueous waste streams originating from pyrolysis.

Keywords: open culture; carbon monoxide; gasification; biomass conversion; bioremediation; biomethanation; chain elongation; volatile fatty acids

Citation: Robazza, A.; Welter, C.; Kubisch, C.; Baleeiro, F.C.F.; Ochsenreither, K.; Neumann, A. Co-Fermenting Pyrolysis Aqueous Condensate and Pyrolysis Syngas with Anaerobic Microbial Communities Enables L-Malate Production in a Secondary Fermentative Stage. *Fermentation* **2022**, *8*, 512. https://doi.org/10.3390/fermentation8100512

Academic Editor: Sanjay Nagarajan

Received: 29 August 2022
Accepted: 29 September 2022
Published: 4 October 2022

1. Introduction

Growing concerns for the impact of anthropogenic activities on the environment are shifting the socio-economic interests from a fossil-based economy towards a more sustainable and circular one. According to the Intergovernmental Panel on Climate Change and the International Energy Agency, it is estimated that the total share of biofuels will double in the next decades [1]. The source of all the biomass required to meet the need of an increased bio industry is still an open debate [2]. The energy potential of biomass is enormous considering that the Earth's net biomass production amounts approximately to 2000 EJ/y [3]. However, the diverting part of the energetic reservoir built up by plants towards uses defined by anthropocentric needs could cause undesirable impacts on the environment and on its natural distribution of resources [3]. Similarly, many biofuel crops are competing with food production, and the increasing demands for biofuel could exceed agricultural capacity [2]. The development of new technologies to maximize the energy recovery from wastes and residues of human activities is considered a key step towards carbon-neutrality [4].

The pyrolysis of lignocellulosic waste from municipal and agricultural activities could represent a great opportunity, contributing to meet the needs of a developing bio-based economy [5]. During pyrolysis, the biomass is thermochemically deconstructed at temperatures ranging between 350 and 600 °C in the absence of oxygen [6]. The products of pyrolysis are pyrolysis syngas (PS) (15–20 wt%), a viscous energy-rich pyrolysis organic fraction (POF) (20–30 wt%), an aqueous condensate (PAC) (20–30 wt%) and bio-char (10–30 wt%) [6,7]. Biochar and bio-oil can be either fed back into the pyrolysis reactor or used as fuels. On the other hand, the PAC's use is limited by the high concentrations of various toxic compounds and the high water content [8]. Similarly, the release of PS into the atmosphere should be avoided, due to its high concentrations of greenhouse gases (GHGs). In general, PAC and PS represent about 45 wt% of the total biomass fed into the pyrolysis reactor [7] and up to 41% of the carbon balance [6]. Thus, it might be worth investing into bioprocessing technologies able to convert PAC and syngas into industrially relevant biochemicals.

Several works have already focused on the development of biological processes to valorize the constituents of the PAC. The ability of microorganisms in single culture fermentations to grow on PAC is species-specific due to their varying resistance to toxins contained in PAC [8]. Basaglia et al. [9] studied the toxicity of PAC from fir wood to a wide range of different microbial groups. Out of the 42 strains tested, only 4 fungal strains showed tolerance to pure PAC, whereas several PAC dilutions are required for many bacterial and yeast isolates [9]. However, it appears that PAC must undergo one or more pre-treatment steps to reduce the toxicity, before enabling its bioprocessing in pure culture fermentations [10–17].

Anaerobic digestion is an established technology for the treatment of agricultural residues and industrial wastewaters [18]. The degradation of the organic matter into CH_4 follows four primarily metabolic steps (hydrolysis, acidogenesis, acetogenesis, and methanogenesis) and depends upon mutual and syntrophic interactions between various microorganisms and trophic groups [19]. The wide and diverse genetic spectrum and functional redundancy of thousands of microbial species in anaerobic digesters offer what pure cultures currently cannot achieve: a higher tolerance to environmental stresses and toxicity. Multiple parallel biochemical routes provide greater functional stability because of the potential distribution of the substrate to several populations [20], resulting in a higher community resilience to perturbations [21].

Many studies successfully established anaerobic digestion with pre-treated and raw PACs for biomethane production [22–26], proving how anaerobic mixed culture fermentation is a viable alternative to intricate physiochemical pre-treatments for PAC detoxification and valorization. For example, Zhou et al. [25] studied the tolerance of anaerobic digestion towards increasing concentrations of raw and overlimed PAC as the sole carbon source for biomethane production in batch processes and direct evolution studies, respectively. The batch tests showed that loadings of 3% raw PAC were inhibiting methanogenesis. Extensive studies have been conducted towards a complete integration of pyrolysis and anaerobic digestion for methane production where all the by-products of pyrolysis (PAC and PS included) are fed into an anaerobic digester [24,27,28]. During the anaerobic digestion of PAC derived from corn stalk pellets, volatile fatty acid (VFA) production was observed even though PAC severely inhibited methanogenesis [24]. Giwa et al. [29] evaluated the effects of a real PS generated from a two-stage pyrolysis process treating food waste on methanation rates. The process, designed to minimize POF and PAC, generated syngas with a high H_2-to-CO ratio (60:20%). Methanation rates were enhanced, producing almost

100% more CH_4 than the synthetic syngas control fermentations. The topic has been evaluated also from a techno-economical perspective [30–32]: pairing anaerobic digestion with pyrolysis allows for relevant energy savings in handling pyrolysis by-products and strongly reduces GHGs emissions [30]. Salman et al. [33] estimated a higher annual revenue for the integrated process compared to the sole incineration of green waste.

In anaerobic communities, syngas is commonly metabolized by methanogenic archaea, hydrogenogenic bacteria, acetogenic bacteria, and sulfate-reducing bacteria [34]. By manipulating the fermentation environmental conditions, it is possible to control the syngas conversion towards different catabolic routes [35–38]. Methane is often the primary metabolite having the lowest free energy content per electron, regardless of the temperature range [37]. On the other hand, when methanogenesis is inhibited and in mesophilic environments with high concentrations of reduced compounds such as ethanol and/or lactate, the mixed culture can elongate C1 compounds from syngas into medium-chain carboxylates (MCCs). Such a wide array of metabolic products at mesophilic temperatures is the result of an intricate metabolic network ultimately limited by thermodynamics [39]. Syngas-converting microbial communities at thermophilic temperatures show higher water–gas shift reaction (WGSR) kinetics than mesophilic ones. The high diversity of carboxydotrophic hydrogenogenic bacteria and the thermodynamics of H_2-producing reactions in thermophilic environments favor higher CO conversion rates to produce primarily H_2 and short-chain carboxylates [40]. Hydrogen or MCC production via mixed culture anaerobic fermentation are gaining more scientific and industrial interest [41,42]. However, the success of these technologies is linked to the identification of cheap and recoverable methane inhibitors [43,44].

A. oryzae belongs to the Ascomycetes group and its industrial application spans from food processing to commodity chemical production [45,46]. Several studies have evaluated the potential of producing biochemicals, biofuels or cell biomass (single-cell proteins) with *A. oryzae* from VFA rich waste streams or from acetate [47–49]. Moreover, the fungus was reported to tolerate small concentrations of pyrolysis oils and various PAC components [50] and to be able to grow on the acetate contained in pre-treated PAC from wheat straw [15].

To extend the knowledge about the integration of thermochemical and biochemical processes treating lignocellulose waste, this work evaluates a two-stage process where the products from the co-fermentation of PAC and syngas by anaerobic mixed cultures are fed to an aerobic fermentation to produce L-malate by *A. oryzae*. Several anaerobic mixed culture bottle fermentations were performed at 37 and 55 °C at increasing PAC concentrations in order to understand the effects of PAC on the metabolism of gaseous and liquid compounds. After the syngas fermentation stage, the media from selected mixed culture fermentations were inoculated with *A. oryzae*, focusing on the conversion of acetate from syngas and PAC metabolism into L-malate. Fungal growth, together with the quantification of the removal of selected PAC components, was used to prove the occurrence and the extent of PAC detoxification.

2. Materials and Methods

2.1. Growth Medium

All reagent-grade chemicals were purchased from Sigma-Aldrich (Schnelldorf, Germany) or Carl-Roth (Karlsruhe, Germany). The fermentation medium used in all serum-bottle and flasks experiments was a modified basal anaerobic medium (BA) composed of the following stock solutions: mineral salts solution (NH_4Cl, 161.2 g/L; $MgCl_2 \times 6H_2O$, 5.4 g/L; $CaCl_2 \times 2H_2O$ 6.5 g/L); phosphate buffer solution (KH_2PO_4, 136 g/L); vitamins solution (Biotin, 0.002 g/L; Folic Acid, 0.002 g/L; Pyridoxin, 0.01 g/L; Thiamin, 0.005 g/L; Riboflavin, 0.005 g/L; Nicotinic Acid, 0.005 g/L; Ca-Panthothenate, 0.005 g/L; Vitamin B12, 0.005 g/L; Aminobenzoic Acid, 0.005 g/L; Liponic Acid, 0.005 g/L); trace elements solution ($FeCl_2 \times 4H_2O$, 1.5 g/L; $MnCl_2$, 0.1 g/L; $CoCl_2 \times 6H_2O$, 0.19 g/L; $ZnCl_2$, 0.07 g/L; $CuCl_2 \times 2H_2O$, 0.002 g/L; $NiCl_2 \times 6H_2O$, 0.024 g/L; $Na_2MoO_4 \times 2H_2O$, 0.036 g/L; H_3BO_3, 0.006 g/L; $Na_2SeO_3 \times 5H_2O$, 0.003 g/L; $Na_2WO_4 \times 2H_2O$, 0.02 g/L); reducing agent solution (L-Cysteine, 100 g/L); resazurin solution (Resazurin sodium salt, 1 g/L). For each liter of medium added: 100 mL of mineral salt solution, 800 mL of phosphate buffer solution, 10 mL of vitamins solution, 10 mL of trace elements solution, 5 mL of resazurin solution and 3 mL of reducing agent solution. Once all the solutions were mixed, the pH was adjusted to 6 with 4M NaOH solution as pH-adjusting agent and sodium source. The remaining volume was filled with deionized water to 1 L.

2.2. Inocula and PAC

The anaerobic sludge was collected from an anaerobic digester treating cow manure (Alois & Simon Frey Biogas GbR, Bräunlingen, Germany). Due to the high content of straw residues, right after collection, the sludge was sieved down to 0.5 mm discarding the straw and the retained solids. The sludge was then poured into an anaerobic container and stored in a fridge at 4 °C until needed. The pH, the total suspended solids (TSS) and volatile suspended solids (VSS) concentration of the sieved sludge corresponded to 8.46, 41.36 $\pm$ 2.25 g/L and 12.27 $\pm$ 0.13 g/L, respectively. The TSS and VSS analytics were performed in triplicate and determined as described in [51].

Aspergillus oryzae DSM 1863 was obtained from the DSMZ strain collection (Deutsche Sammlung von Mikroorganismen und Zellkulturen GmbH, Braunschweig, Germany). The cryo-stock of fungal conidia was prepared and stored as described by [49].

The PAC used in this experiment was produced during the fast pyrolysis of miscanthus at BioLiq plant (Karlsruhe Institute of Technology, Karlsruhe, Germany). The chemical oxygen demand (COD) and total organic carbon (TOC) were 253.25 $\pm$ 10.25 g/L and 118.58 $\pm$ 0.11 g/L, respectively. The total nitrogen (TN) was 140.25 $\pm$ 4.24 mg/L. The pH of raw PAC was 2.8, while acetate, propionate and *n*-butyrate concentrations were about 34, 5.07 and 0.5 g/L, respectively. The fast pyrolysis at the BioLiq plant is run as described in [7] and [52]: the flue gases (composed primarily of 20% CO, 25% CO_2, 1.5% H_2, alkanes and N_2) coming from the combustion chamber pass through a hot cyclone to separate the biochar from the product gas stream. Then, the gaseous phase is sent through a series of two quench condensers at ~85–90 °C and at ~30 °C separated by an electrostatic precipitator. The PAC used in this study is the product of the second condensation step.

2.3. Bottle Preparation and Fermentation

Mesophilic and thermophilic experiments with (M-CTRL and T-CTRL) and without methanation (M-BES and T-BES) were run as controls to evaluate the metabolism and the performances of the inoculum grown on synthetic pyrolysis syngas. To inhibit methanogenesis, 50 mM of Sodium 2-bromoethanesulfonate (BES) were dissolved into the BA medium. All experiments not containing PAC were performed in triplicate. To test PAC inhibition, exponentially increasing concentrations ranging from 0.5 to 30% v/v were added in the M-PAC and T-PAC fermentations. As control, abiotic experiments with equal concentrations of PAC (M-PAC-AB and T-PAC-AB) were also prepared and run simultaneously to the corresponding experiments. The mixed culture fermentations and abiotic PAC incubations were performed in 250 mL serum bottles with 50 mL of active volume. Figure 1 and Table 1 summarize the experimental design.

The liquid phase was composed of 5 mL of BA medium, increasing PAC concentrations depending on the experimental design and 4M NaOH as needed to re-adjust the pH of the medium back to 6 after PAC addition. The remaining volume was filled with deionized water up to 45 mL. The serum bottles were stored into an anaerobic tent (5% H_2 in N_2) to anaerobize overnight at room temperature. The bottles were then inoculated with 10% v/v anaerobic sludge and sealed with butyl rubber stoppers and aluminum rings. After sealing the flasks, the bottles were initially flushed and then pressurized with a synthetic pyrolysis gas mixture consisting of 6 kPa H_2, 21 kPa CO, 26 kPa CO_2 and N_2 to a final pressure of 210 kPa_{abs}. Bottle pressurization was performed using a precision pressure indicator GMH 3100 Series (Greisinger, Mainz, Germany) at room temperature.

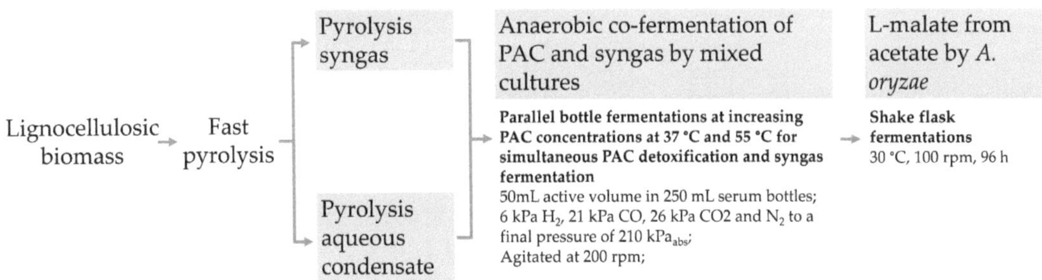

Figure 1. Schematic representation of the experimental design. Pyrolysis aqueous condensate and pyrolysis syngas were co-fermented by two mixed cultures at mesophilic and thermophilic temperatures. The media from selected mixed culture fermentations were centrifuged and inoculated with *A. oryzae* to convert acetate into L-malate.

Table 1. Overview of experiments. MC is mixed culture; AB is abiotic; Asp is *Aspergillus oryzae*.

	T (°C)	Medium	BES (50 mM)	Raw PAC (0.5–30%)	Inoculum	Syngas
			Control Syngas Fermentations			
M-CTRL	37	BA	–	–	MC	+
M-BES	37	BA	+	–	MC	+
T-CTRL	55	BA	–	–	MC	+
T-BES	55	BA	+	–	MC	+
			Mesophilic and Thermophilic PAC Fermentations			
M-PAC	37	BA	–	+	MC	+
T-PAC	55	BA	–	+	MC	+
			Mesophilic and Thermophilic Abiotic Control			
M-PAC-AB	37	BA	–	+	–	+
T-PAC-AB	55	BA	–	+	–	+
			Aspergillus oryzae Fermentations			
M-PAC-Asp	30	from M-PAC	–	detoxified PAC	*A. oryzae*	–
M-PAC-AB-Asp	30	from M-PAC-AB	–	–	*A. oryzae*	–
T-PAC-Asp	30	from T-PAC	–	detoxified PAC	*A. oryzae*	–
T-PAC-AB-Asp	30	from T-PAC-AB	–	–	*A. oryzae*	–

A total of 3 mL of gas phase was sampled daily or depending on the rates of CO or H_2 consumption. The ambient temperature and pressure and the gauge pressure of the bottles were recorded at each sampling, right after taking the bottle from the incubator. When the CO and/or H_2 molar concentrations or the absolute pressure of the serum bottles were about zero or below 190 kPa_{abs}, respectively, then the headspace of the bottle was re-pressurized with the synthetic pyrolysis gas mixture. The maximum possible theoretical uptake rate for CO was about 1.650 mmol/d, while for exogenous H_2 it was 0.570 mmol/d. A total of 1 mL of liquid samples was withdrawn twice a week. The pH of the sample was measured, and the samples were then centrifuged at 17,000× *g* and ambient temperature for 15 min. The resulting supernatant was filtered with 0.2 µm cellulose acetate syringe filters (Restek GmbH, Bad Homburg vor der Höhe, Germany) and stored in a freezer at −20 °C for later analytics. All bottles were incubated in the dark in shaker incubators (multitron incubator shaker, Infors, Bottmingen, Switzerland) at temperatures of 37 or 55 °C. The agitation was set to 200 rpm. All mixed culture fermentations and abiotic controls lasted 39 days of elapsed fermentation time (EFT).

The medium from selected mesophilic and thermophilic fermentations M-PAC and T-PAC (2.5%, 5%, 7.5%, 10%, 20%) and from the corresponding abiotic controls was centrifuged at 4700 × *g* for 8 h. The supernatant was collected, 9 mL of which together with 1 mL fresh BA medium were then poured into 100 mL baffled Erlenmeyer shake flasks. The shake flasks were inoculated with 0.1 mL of the A. oryzae conidia cryo-stock, with spore concentration of 3×10^7 spores/mL. The pH of the medium was not adjusted. All the shake flasks were incubated at 30 °C and 100 rpm. In total, 0.2 mL of liquid samples were taken every 24 h from inoculation for 5 consecutive days. The pH of the sample was measured, and the samples were then stored in a freezer at −20 °C for later analytics. All fermentations with *A. oryzae* were done in triplicate.

2.4. Analytical Methods and Data Processing

The concentration in the fermentation medium of linear and branched monocarboxylates C1-C8 (lactate, acetate, propionate, iso- and *n*-butyrate, iso- and *n*-valerate, iso- and *n*-caproate), of the normal alcohols (ethanol, propanol, butanol and pentanol) and of some selected PAC compounds (2-cyclopenten-1-one, furfural, phenol, guaiacol and *o*-,*m*-,*p*-cresol) were measured by a high-performance liquid chromatography (HPLC) device (Agilent 1100 Series, Agilent, Waldbronn, Germany) operated with an oven set at 55 °C equipped with a Rezex ROA organic acid H + (8%) column (300 by 7.8 mm, 8 μm; Phenomenex, Aschaffenburg, Germany) and a Rezex ROA organic acid H + (8%) guard column (50 by 7.8 mm). The mobile phase was 5 mM H_2SO_4 with a flow of 0.6 mL/min. Short- and medium-chain carboxylates and PAC compound detection was performed with a UV detector at 220 nm at 55 °C, while normal alcohols were detected with an RID detector at 50 °C.

The gas phase samples were analyzed with an Inficon 3000 Micro GC System with a Thermal Conductivity Detector (TCD) equipped with a CP-Molsieve 5 Å column and a PoraPLOT Q column at 80 °C using argon and helium as carrier gases, respectively. The molar composition of the headspace gas of the bottles was computed assuming the ideal gas law after subtracting any air contamination caused by sampling. The accumulation or consumption of each gas was first corrected by a factor accounting for the pressure lost by sampling withdrawal and then cumulated.

The yields and recoveries (in terms of carbon (C-mol) and electron (e-mol) equivalents) for control experiments were calculated using only CO/CO_2 and CO/H_2 as substrates, respectively, as described by Grimalt-Alemany et al. [53]. For M-PAC and T-PAC experiments, CO was accounted as the sole carbon source while CO and H_2 were assumed as electron donors. The multitude of compounds present in PAC interfered with the identification of other metabolites beyond acetate, propionate, and *n*-butyrate. Therefore, only these three acids as well as CO_2, CH_4 and H_2 were accounted as products. The IC50 value was adopted from Zhou et al. [25], indicating the toxicant concentration that causes 50% reduction in cumulative CO consumption or CH_4 production over a fixed period of exposure time. Acetate selectivity is the ratio between acetate and metabolites with carbon atom number greater than 2.

3. Results and Discussion

3.1. Mesophilic and Thermophilic Anaerobic Mixed Microbial Cultures Grown on Pyrolysis Synthetic Syngas

The first set of experiments aimed to understand whether the synthetic pyrolysis syngas used in this study is a suitable carbon and electron source for production of methane, short- and medium-chain carboxylates as well as solvents with mixed microbial cultures. M-CTRL and T-CTRL are bottle fermentations incubated at 37 and 55 °C, respectively, performing syngas methanation. M-BES and T-BES are bottle fermentations at 37 and 55 °C with the addition of 50 mM BES as methanogenesis inhibitor. The metabolism of the communities under M-CTRL, T-CTRL, M-BES and T-BES conditions were characterized and later used as a reference for comparison with the fermentations in the presence of PAC. The initial pH of all control bottles after inoculation was 6.7 ± 0.2. Figure 2 shows C-mol recovery and e-mol recovery from all control experiments.

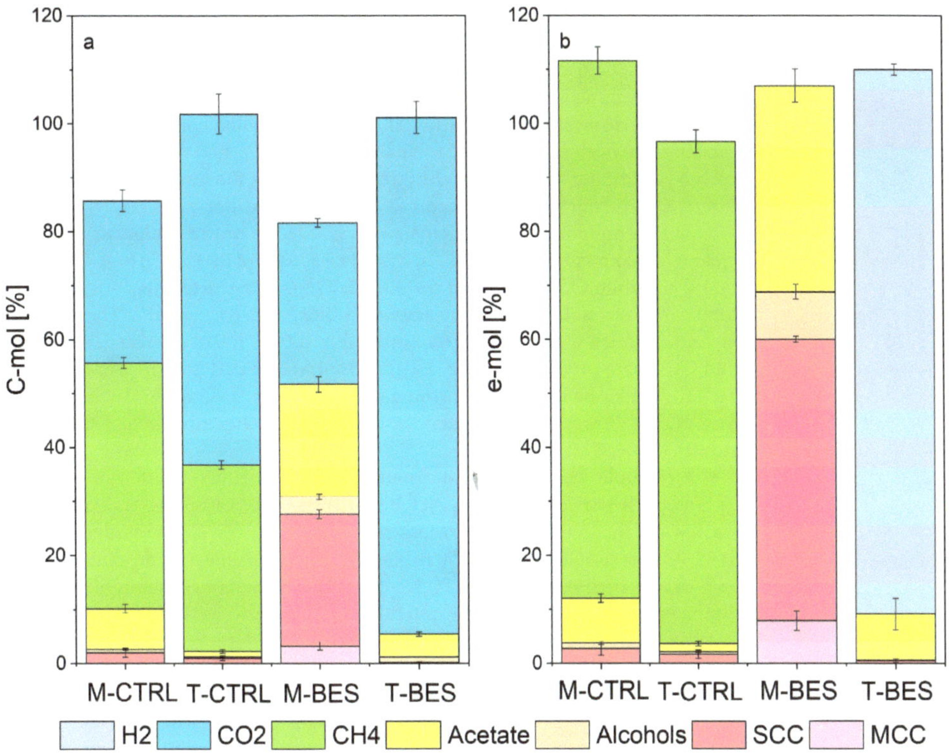

Figure 2. C-mol (**a**) and e-mol (**b**) balances for experiments M-CTRL, T-CTRL, M-BES and T-BES. Conversion factors for electron balances are available in the supplementary materials (Table S1). CO, CO_2 and H_2 were considered as the sole carbon and/or electron donors for all experiments but for T-BES, where CO was the only carbon and electron donor. Alcohols are ethanol, propanol and butanol. Short-chain carboxylates C3- C5 (SCCs) are lactate, iso- and *n*-butyrate, propionate and iso- and *n*-valerate. Medium-chain carboxylates (MCCs) are iso- and *n*-caproate. The productivities of alcohols, some SCCs and MCCs are available in the supplementary materials, Table S2.

During syngas methanation at mesophilic range (M-CTRL), the mixed culture produced primarily CH_4 (45.5 ± 1%) and CO_2 (30.1 ± 2.1%), while 7.6 ± 0.1% of the total carbon metabolized was fixed into acetate. Acetate accounted for 83.7 ± 1.7% of the total C2–C6 metabolites detected in the liquid phase. The carbon stored in carboxylates other than acetate was about 2.6%. The average CO and H_2 uptake rates were 0.34 ± 0.02 mmol/d and 0.28 ± 0.02 mmol/d, while CH_4 was produced at a rate of 0.15 ± 0.01 mmol/d.

From about 20 days EFT, methanogenic rates increased concomitantly to homoacetogenic/hydrogenotrophic activity from exogenous CO_2 and H_2 consumption (Supplementary materials, Figures S1–S4). Simultaneously, decreasing acetate concentrations in the bottles might indicate acetoclastic methanation. However, acetoclastic methanogenesis appears to have barely contributed to the methanation yield. At 37 °C, pH 5.5, and 100 mM acetate hydrogenotrophic methanogenesis has more favorable thermodynamics than acetoclastic methanogenesis [38]. Considering that CO and H_2/CO_2 metabolisms have been reported to have similar kinetics [54], changes in the rates of gases uptake or production might be attributed to shifts within the composition of the microbial population. With the progression of M-CTRL experiments, CO uptake rates lowered the CO partial pressures favoring acetogenic/methanogenic hydrogenotrophism. High CO partial pressures are known to be inhibiting cellular hydrogenase and H_2 uptake [55,56] and might have contributed to the delayed start of H_2/CO_2 metabolism. Liu et al. [57] detected a two-phased process characterized by an initial CO consumption followed by the onset of H_2/CO_2 metabolism to acetate attributed to homoacetogenic microorganisms while performing CO biomethanation with anaerobic granular sludge. In M-CTRL bottles, carboxidotrophic methanation, if any, had a limited contribution towards methane production. Carboxydotrophic methanogens are expected to be easily outcompeted by carboxydotrophic acetogens and hydrogenogens, as the few species that are capable of directly converting CO into CH_4 do so at very low reaction rates [58,59].

The thermophilic syngas methanation (T-CTRL) occurred at higher kinetics but lower yield when compared to M-CTRL. In total, 34.6 ± 0.8% of the carbon from CO was converted into CH_4 while CO_2 accounted for 65 ± 3.7%. Acetate accounted for about 1% for the total carbon from CO and the acetate selectivity was 63 ± 3.51%. The average CO and H_2 uptake rates were 1.48 ± 0.05 mmol/d and 0.56 ± 0.01 mmol/d, respectively. CO_2 and CH_4 were produced at 0.98 ± 0.05 mmol/d and 0.515 ± 0.01 mmol/d, respectively. T-CTRL bottles have been performing primarily carboxidotrophic hydrogenogenesis via the WGSR followed by hydrogenotrophic methane generation, as also described by other studies [34,36,39].

Mesophilic and thermophilic metabolic rates calculated in this study correspond to those reported by Sipma et al. [60], who tested several mesophilic anaerobic sludges from wastewater treatment reactors to convert CO at 30 and 55 °C. The sludges were incubated at 30 °C in serum bottles with 50 mL initial active volume and produced primarily CH_4 and/or acetate. Incubation at 55 °C resulted in the formation of mainly CH_4 and/or H_2 [60]. Sipma et al. detected CO conversion rates ranging between 0.14 and 0.62 mmol/d for the cultures incubated at 30 °C, while thermophilic CO depletion rates varied between 0.73 and 1.32 mmol/d.

The BES addition inhibited all methanogenic pathways in both control mesophilic syngas (M-BES) and control thermophilic syngas (T-BES) fermentations. M-BES fermentations consumed CO at a rate of 0.36 ± 0.03 mmol/d, a similar value to what was calculated for M-CTRL. H_2 uptake rate was 0.03 ± 0.01 mmol/d and CO_2 production rate was 0.11 ± 0.01 mmol/d. HPLC analytics showed that M-BES cultures have been chain elongating CO to *n*-caproate with a net exogenous H_2 consumption to a final caproate concentration of 2.18 ± 0.47 mM. About 60% of the e-mol recovery was accounted for metabolites with a carbon atom number higher than two. CO_2 (29.9 ± 0.8%) and acetate (20.8 ± 1.5%) were the two major carbon sinks.

T-BES experiments showed greater CO consumption kinetics then M-BES. The mixed culture performed almost solely WGSR, generating 1.04 ± 0.33 mmol/d of CO_2 and 1.05 ± 0.31 mmol/d H_2, while the average CO uptake rate was 1.18 ± 0.09 mmol/d. CO_2 accounted for more than 95% of the total carbon fed while acetate was only about 5%. Acetate was the primary metabolite produced by the consortium with selectivities higher than 80%. More than 99% of the e-mol recovery was molecular H_2. These results are corroborated by the work carried out by other research groups. Grimalt-Alemany et al. [39] characterized the conversion of CO by a thermophilic enriched consortium in the presence of BES, resulting in the production of H_2 and acetate as primary metabolites. Slepova et al. [61] traced ^{14}CO to study the metabolism of mixed cultures collected from three pH-neutral hot springs of Uzon Caldera (Kamchatka) under temperatures from 60 to 90 °C. A major part of ^{14}CO was oxidized to $^{14}CO_2$. Samples from the spring with a temperature of 60 °C converted less than 5% of the CO into carboxylates and only 1% in springs with higher temperatures [61]. High acetate selectivities were also reported by Wang et al. [62], showing a 99% acetate selectivity at the end of their thermophilic (55 °C) enrichment process with H_2 and CO_2 as substrates. Shen et al. [63] achieved final acetate selectivity of 96.7% and 96.3% in two hollow fiber membrane bioreactors after 60 days EFT starting from an inoculum from an anaerobic digester. Alves et al. [35] tested different enrichment strategies in bottle experiments at 55 °C and obtained syngas-converting communities able to fix approximately 97% of product recovery into acetate from CO_2 and H_2.

3.2. Co-Fermentation of Syngas and PAC

The effects of increasing PAC concentrations were evaluated on two mixed microbial cultures growing on pyrolysis gas at 37 and 55 °C. The aim was to identify kinetic inhibition and changes in metabolites production patterns of syngas metabolism caused by PAC. Additional interest was to test the PAC detoxification potential of the microbial cultures.

3.2.1. Impact of PAC on the Syngas Metabolism of the Anaerobic Mixed Culture at 37 °C and 55 °C

Figure 3 reports the rates of syngas metabolism at increasing PAC concentration both at mesophilic (37 °C) and thermophilic (55 °C) temperatures. Similar to the control experiments, the initial pH of all M-PAC and T-PAC experiments was 6.7 ± 0.2 after inoculation.

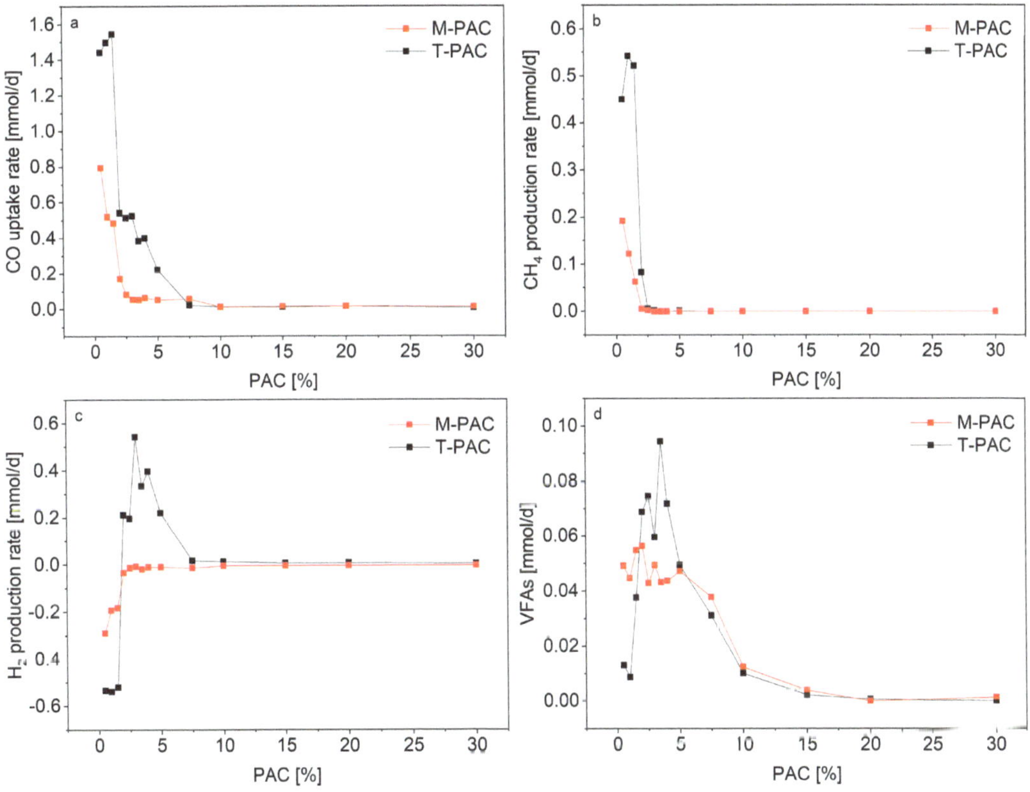

Figure 3. Rates of consumption and/or production of CO (**a**), CH_4 (**b**) and H_2 (**c**) at increasing PAC loadings at mesophilic (37 °C) and thermophilic (55 °C) temperatures. Negative production rates for H_2 indicate consumption. Volatile fatty acids (VFAs) (**d**) are acetate, propionate and *n*-butyrate. Productivities for all experiments are available in the Supplementary materials Tables S3 and S4.

The CO consumption rates for mesophilic fermentations M-PAC at PAC concentrations of 0.5, 1 and 1.5 were all above 0.4 mmol/d. For PAC concentrations higher than 5%, the rates of CO consumption rapidly decreased towards zero. Exogenous H_2 consumption was detected in all M-PAC bottles. Additionally, CO_2 production rates were 60% lower than the stoichiometry of the WGSR, suggesting that the mesophilic mixed culture co-fermented CO and H_2/CO_2. While the methane production rates quickly dropped to zero for concentrations above 1.5% PAC, the VFA daily production decreased only from PAC concentrations above 7.5%.

At thermophilic range, PAC concentrations below 1.5% did not significantly affect CO consumption (Figure 3a). The average CO consumption rates at 55 °C with PAC concentrations from 0.5 to 1.5% were all above 1.4 mmol/d, similar to what was achieved in the control experiments T-CTRL. Above 5% PAC, the kinetics of CO consumption rapidly decreased towards zero. At thermophilic range, methanogenesis was detected for PAC concentrations from 0.5 to 2.5% PAC. The highest CH_4 production rate was 0.54 mmol/d for bottles containing 1% PAC. In T-PAC fermentations with 1.5, 2, 2.5% PAC, the methane production showed a delayed start of about 6 days when compared to T-CTRL (Supplementary materials Figures S5–S8). In Figure 3c, H_2 was consumed to generate methane via hydrogenothropic methanogenesis under conditions with up to 1.5% PAC. At higher PAC loadings, net H_2 production occurred concomitantly to the inhibition of the methanogenic activity. The highest H_2 production rate was detected at 3% PAC with values of 0.54 mmol/d, but it decreased at rates equivalent to CO consumption for higher PAC percentages. Similar to mesophilic bottles, the VFA production rates were low under low PAC loadings and peaked at 3.5% PAC when no methane production was detected.

The kinetics of syngas metabolism for thermophilic PAC fermentations were consistently higher than at mesophilic range, a result consistent with the kinetics of the control experiments. However, Figure 4a shows that, when normalizing M-PAC and T-PAC CO uptake rates to the correponding rates of M-CTRL and T-CTRL, the overall effects of PAC toxicity did not differ between mesophilic and thermophilic experiments. Thus, thermodynamic limitations and different gas solubilities at different temperatures were likely the dominant factors affecting the kinetics of syngas metabolism.

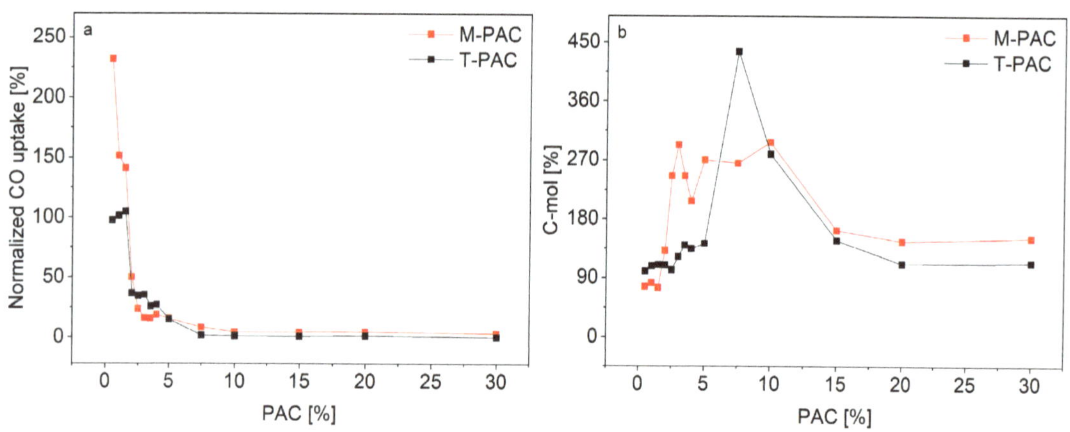

Figure 4. (**a**) M-PAC and T-PAC CO uptake rates normalized to control experiments M-CTRL and T-CTRL, respectively. (**b**) C-mol balances for M-PAC and T-PAC experiments.

Additionally, Figure 4a shows that M-PAC bottles with low PAC concentrations (0.5 to 1.5% PAC) had at least 40% higher CO consumption rates compared to the respective M-CTRL values, peaking at 231% at 0.5% PAC. For bottles with 0.5 to 1.5% M-PAC, from about 20 days EFT, CO oxidation rates higher than 0.36 mmol/d (average CO uptake for M-BES) were detected, matching those of T-CTRL experiments rather than M-CTRL or M-BES (supplementary materials, Figures S1–S4). Factors such as CO and PAC toxicity probably contributed to hinder acetogenic and methanogenic activity at early fermentation stages and the high CO uptake rates might be the result of changes in microbial population consequently to PAC detoxification. However, contrarily to M-CTRL fermentations, the higher kinetics of the WGSR provided enough endogenous CO_2 to all metabolic routes resulting in a net CO_2 production (supplementary materials, Figures S1–S3).

3.2.2. Different PAC Tolerance of Different Trophic Groups

Methane production was inhibited by lower PAC concentrations than CO consumption in both M-PAC and T-PAC cultures. The IC50 values for CO uptake rates at mesophilic range correspond to 2% PAC. Methane production, on the other hand, is halved at PAC concentrations between 1 and 1.5%. At thermophilic range, the IC50 values for CO uptake rates fell within the 2 to 3% PAC range. Regarding methane, the IC50 was found to be between 1.5 and 2% PAC. Zhou et al. [25] reported that the IC50 of mesophilic biomethane potential tests of overlimed PAC was 4.8% PAC. Even though Zhou et al. [25] did not report the IC50 for raw PAC, it could be assumed that the higher tolerance of methanogens towards PAC achieved in their study was the result of the synchrony of the pre-treatment and a lower specific PAC availability, as both factors are known to affect methanation rates [64]. Here, raw PAC loading rates that severely inhibited methanogenesis were 0.41 gCOD/gVSS (2% PAC) at both the mesophilic and thermophilic range, respectively (supplementary materials, Table S4).

When comparing methanogenic versus carboxydotrophic/homoacetogenic activity under PAC influence, homoacetogenesis had a higher tolerance to PAC than methanogenesis. Compounds present in PAC such as furfural, phenol and phenolic compounds can be produced also from the hydrolysis of lignocellulosic matter [8,65,66]. Acetogens are involved in synthropic interactions with other microorganisms during the anaerobic degradation of compounds deriving from the degradation of lignin. Synthetic co-cultures with *Pelobacter acidigallici*, *Acetobacterium woodii*, and *Methanosarcina barkeri* have been reported to convert phenylmethylethers to CH_4 and CO_2 [67]. *A. woodii* metabolizes phenylmethylethers to yield acetate and phenols [68]. Phenols can be degraded to acetate by *P. acidigallici* [69]. In another work studying the degradation of lignin-derived monoaromatic compounds, the initial step was catalyzed by *Sporomusa* spp. to generate acetate via O-demethylation of the methoxylated aromatics. The demethoxylated aromatics were then metabolized into acetate, H_2 and CO_2 by *Firmicutes*. Finally, methane was generated from acetate and H_2/CO_2 by acetoclastic and hydrogenotrophic methanogens, respectively [70]. The latter examples represent interactions between microorganisms that might have occurred in the inoculum in the presence of PAC. Methanogens work at the end of the chain of syntrophic interactions resulting in the production of CH_4 as the primary end-product of the fermentative process. Thus, methanogenic activity is highly influenced by the degradation of those compounds that would otherwise be inhibitory. Low concentrations of lignin derivatives with aldehyde groups or apolar substituents are known to be highly toxic to methanogens [71]. Aromatic carboxylates, on the other hand, were reported to be only mildly toxic. Phenols and their derivatives are known for being methanogenic inhibitors [64,72,73], however, phenolic compounds have been already proven to be degraded to CH_4 [74,75].

Hübner et al. [22] reported longer lag phases at increasing initial PAC concentrations in anaerobic digestion experiments. PAC extended the lag phase of methanogenesis from a few days to some weeks, indicating temporary inhibition [22]. Inhibition of anaerobic digestion by PAC from corn stalk was also observed by Torri and Fabbri [27]. Longer lag-phases at increasing PAC loadings were also detected in this work at mesophilic and thermophilic range for both carboxidotrophism and methanogenesis (Supplementary materials Figures S4 and S8). In general, an extended lag-phase could be related to a lack of acclimatization of the inoculum to an inhibiting organic compounds hard to degrade, therefore requiring enrichment of the microbial community [73]. Alternatively, the inoculation/bioaugmentation of fermentations with cultures collected from particular ecosystems could be a strategy to increase the performances of biological processes [76–79].

3.2.3. PAC Detoxification

The C-mol and e-mol recoveries for bottles with 2.5 to 10% PAC at both temperatures showed balances much higher than 100% (Figure 4b). Most of the VFAs (primariliy acetate) produced in those bottles were not the result of syngas metabolism but from the degradation of aromatc compounds, as proven in other works [67,70,80]. This is also supported by the detoxification efficacy of selected PAC compounds (Figure 5) where high degradation efficacies were recorded at low PAC concentrations.

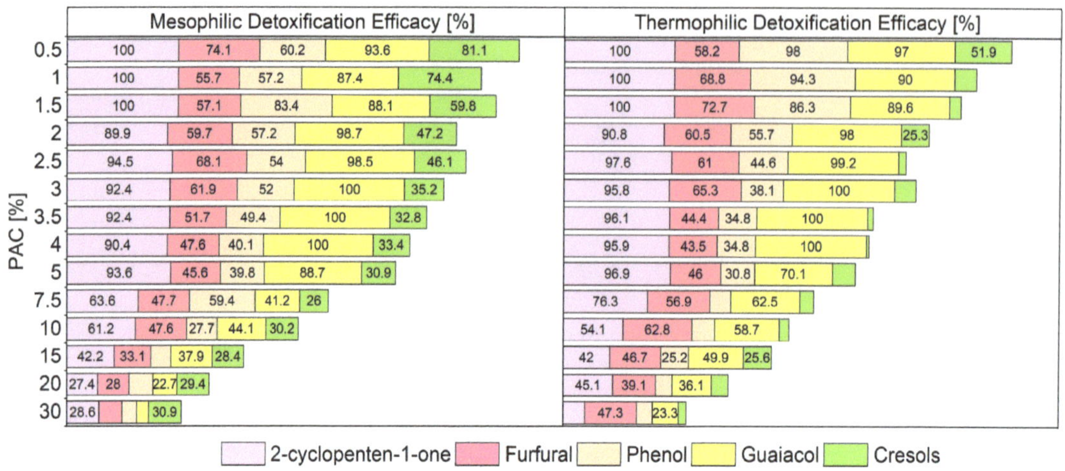

Figure 5. Removal efficacies for some selected PAC compounds after syngas mixed culture fermentations performed at 37 and 55 °C. Numeric values of the removal efficacies below 25% are now shown.

For PAC concentrations above 5%, the efficacy of degradation decreased both at the thermophilic and mesophilic range. A work performed by Fedorak and Hrudey [81] reporting high removal of phenol and *m*- and *p*-cresol from a wastewater of a coal liquefaction plant during anaerobic batch culture experiments supports what was detected here. Hübner and Mumme [22] suggested that low cresols degradation efficacies might be accounting for cresols production via phenol degradation, as cresols and guaiacol are phenol derivates.

Considering that bottles with low CO consumption rates showed high PAC detoxifications efficacies, it can be assumed that PAC detoxification was independent from syngas metabolism and it occurred at concentrations inhibiting carboxidotrophism and homoacetogenesis. On the other hand, the longer lag phases at increasing PAC concentration might suggest that syngas metabolism was dependent on the detoxification of toxins in PAC and it recovered once the concentration of some PAC components fell below toxic levels.

3.3. *A. oryzae Cultivation on Acetate Derived from Syngas Fermentation and PAC Detoxification*

To further test the degree of PAC detoxification and to valorize the carboxylates from the M-PAC and T-PAC experiments, the media from some selected bottles were centrifuged and the resulting supernatant inoculated with *A. oryzae*.

No fungal growth was detected in the media containing the broth from syngas abiotic control experiments with syngas, M-PAC-AB-Asp and T-PAC-AB-Asp (Figure 6). Thus, abiotic incubation over an extensive amount of time did not lower the toxicity levels of PAC towards *A. oryzae*. On the contrary, *A. oryzae* growth was detected in all fermentations up to M-PAC-Asp 10% and T-PAC-Asp 10%. Inhibitory effects of pyrolysis products of wheat straw on *A. oryzae* growth were previously elucidated by Dörsam et al. [50], who studied the toxicity of some selected PAC components. Phenolic compounds such as phenol, *o*-, *m*-, *p*-cresol and guaiacol resulted in a strong inhibition of *A. oryzae* growth even at low concentrations. Although it is known that *A. oryzae* has genes encoding for enzymes enabling the degradation of cresols, it only tolerates cresol in very low concentrations [82]. Additionally, 2-cyclopenten-1-one was reported to be the most toxic compound among the tested ones [50].

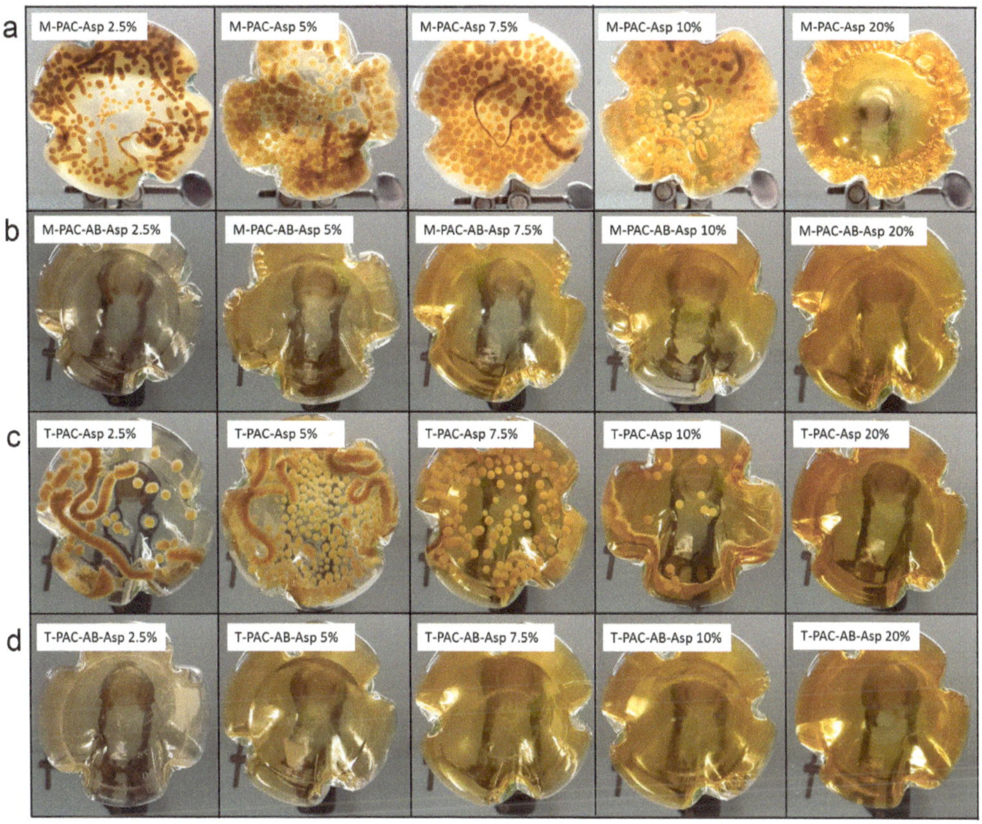

Figure 6. Growth of *A. oryzae* in aerobic flasks containing medium from syngas fermentations and abiotic controls. Rows (**a**) and (**c**) show fungal growth in medium from mesophilic and thermophilic syngas culture fermentations, respectively. Rows (**b**) and (**d**) show the results of fungal growth in medium from the abiotic incubation of PAC and BA medium.

Malate Production from Acetate by *A. oryzae*

The ability of *A. oryzae* to convert glucose and VFAs from various sources into L-malate or biomass has been studied in previous works [49,83–86]. Here, the acetate detected at the start of the *A. oryzae* fermentations derived from different sources: syngas fermentation; acetate originally contained in the PAC; and PAC detoxification.

Complete acetate consumption was recorded in all flasks containing medium from bottle fermentations with up to 10% PAC (Figure 7a,c). L-malate production was detected in all bottles alongside acetate consumption. For both A-M-PAC 20% and A-T-PAC 20%, no acetate consumption nor L-malate production were detected.

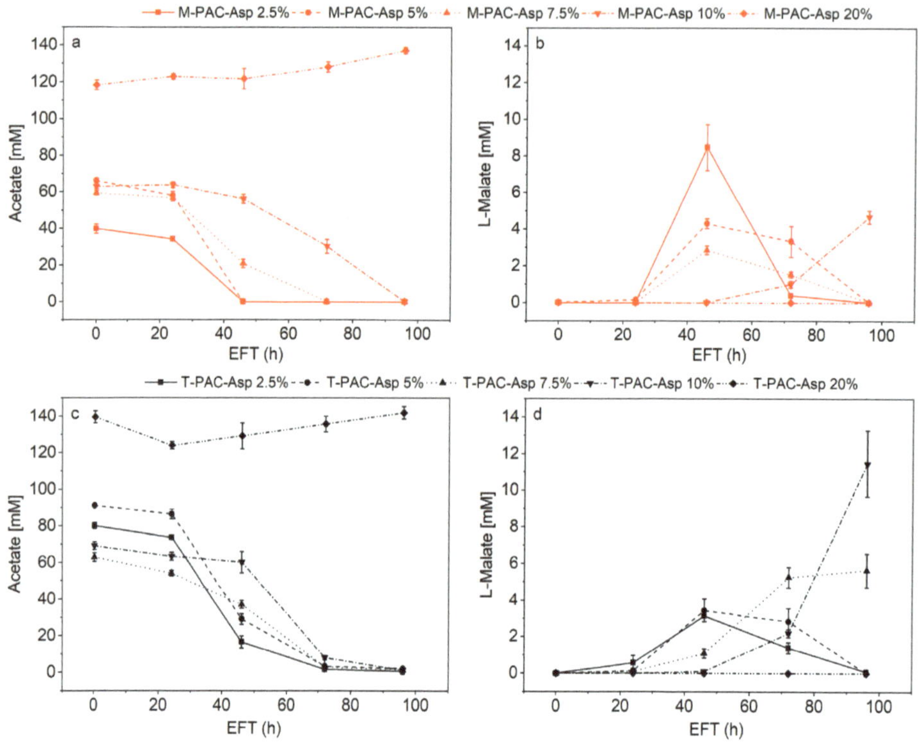

Figure 7. Acetate and L-malate from *A. oryzae* fermentations in the medium from mesophilic syngas fermentations (**a**,**b**) and thermophilic syngas fermentations (**c**,**d**).

For the medium from mesophilic syngas fermentations, the highest amount and yield of malate from acetate of 8.47 ± 0.21 mM and 0.21 mM/mM, respectively, were obtained in M-PAC-Asp 2.5%. Overall, L-malate yields decreased at increasing PAC concentrations for M-PAC-Asp fermentations. On the other hand, when considering the medium from thermophilic syngas fermentations, the highest amount of L-malate produced was detected for T-PAC-Asp 10% at 11.46 ± 0.16 mM with the highest yield of 0.17 mM/mM. Contrarily to M-PAC-Asp fermentations, L-malate yields increased at increasing PAC concentrations. Process optimization for L-malate production exceeded the scope of this work; however, the highest malate yields detected in this study are comparable to the 0.20 g of malic acid per gram of acetate for concentrations of 40 g/L of acetate reported by Kövilein et al. [49]. Kövilein et al. [49] tested acetate concentrations between 10 and 55 g/L for malate production in *A. oryzae* shake flasks cultures. Malate production was reported to be highly dependent on acetate concentration with the highest yield for concentrations of up to 40 g/L [49]. Similarly, Uwineza et al. [84] grew *A. oryzae* on VFAs from the anaerobic digestion of food waste with maximum concentrations of acetate of 9 g/L yielding 0.29 gCDW/gVFAs. Higher concentrations of acetate did not affect the yield. Oswald et al. [83] presented a process concept, in which malate was produced from acetate generated from syngas fermentation by *C. ljungdahlii*. Malate production by *A. oryzae* in the medium from the syngas fermentations with acetate as sole carbon source reached yields of 0.33 g of malate per gram of acetate [83]. The overall conversion of CO and H_2 into malate was calculated to be 0.22 g malate per gram of syngas [83]. The high malate yields achieved in this work, as already hypothesized by Oswald et al. [83], might be linked to the richness in micronutrients of the medium from the previous fermentations.

4. Conclusions

In this study, PAC and syngas were co-fermented by mesophilic and thermophilic mixed cultures and the effects of increasing concentrations of PAC were evaluated. PAC could be used effectively to inhibit methanogenesis and steer microbial metabolism towards other metabolites. Fermenting PAC and syngas in the mesophilic range led to acetate, propionate and *n*-butyrate accumulation in the fermentation broth with net H_2 consumption, whereas fermentations at the thermophilic range produced primarily acetate and H_2. These results show that the mixed cultures performed the dual task of fixing C1 compounds from syngas and detoxifying PAC. Treating PAC together with syngas enabled carboxylates valorization to platform chemicals such as L-malate by *A. oryzae* via a sequential secondary fermentation stage. Mesophilic carboxylate production and thermophilic biohydrogen production via mixed culture syngas fermentations are becoming the center of extensive interest for biochemical or biofuel production. Thus, exploring alternative and effective methods for the inhibition of methanogenesis is still necessary, and inhibitors, such as PAC, are ideal candidates. This work contributes towards a better understanding of the efficient integration of thermochemical processes and mixed culture anaerobic fermentations. Further studies should test the feasibility of this work in continuous bioreactors, aiming to gain a better understanding of the microbial interactions that are contributing to the PAC degradation and syngas metabolism.

Supplementary Materials: The following supporting information can be downloaded at: https://www.mdpi.com/article/10.3390/fermentation8100512/s1. Table S1: Conversion factors for carbon and electron balances; Table S2: Productivities (mM/d) of selected metabolites calculated at 39 days EFT for bottles of the control experiments M-CTRL, T-CTRL, M-BES, T-BES; Table S3: Acetate, propionate and n-butyrate productivities for all bottles of M-PAC and T-PAC experiments; Table S4: Productivities of CO, CH_4, H_2 CO_2 and VFAs in mM/d at increasing PAC concentrations and different temperatures. Negative productivity indicates consumption; Figure S1: Cumulative CO uptake rate in mmol for experiments M-BES, M-CTRL and M-PAC; Figure S2: Cumulative H_2 uptake rate in mmol for experiments M-BES, M-CTRL and M-PAC; Figure S3: Cumulative CO_2 uptake rate in mmol for experiments M-BES, M-CTRL and M-PAC. Negative values mean consumption; Figure S4: Cumulative CH_4 uptake rate in mmol for experiments M-BES, M-CTRL and M-PAC; Figure S5: Cumulative CO uptake rate in mmol for experiments T-BES, T-CTRL and T-PAC; Figure S6: Cumulative H_2 uptake rate in mmol for experiments T-BES, T-CTRL and T-PAC. A negative uptake means production; Figure S7: Cumulative CO_2 uptake rate in mmol for experiments T-BES, T-CTRL and T-PAC. A negative uptake means production; Figure S8: Cumulative CH_4 uptake rate in mmol for experiments T-BES, T-CTRL and T-PAC.

Author Contributions: Conceptualization, A.R., C.K., F.C.F.B., K.O. and A.N.; methodology, A.R., C.K. and F.C.F.B.; formal analysis, A.R.; investigation, A.R., C.W. and C.K.; resources, A.R. and A.N.; data curation, A.R.; writing—original draft preparation, A.R.; writing—review and editing, C.K., F.C.F.B., K.O., A.N.; visualization, A.R.; supervision, A.N.; project administration, A.N.; funding acquisition, A.N. All authors have read and agreed to the published version of the manuscript.

Funding: The authors would like to thank the Helmholtz Research Program "Materials and Technologies for the Energy Transition (MTET), Topic 3: Chemical Energy Carriers" and the support from the KIT-Publication Fund of the Karlsruhe Institute of Technology. Open access funding enabled and organized by Projekt DEAL.

Institutional Review Board Statement: Not applicable.

Acknowledgments: The authors acknowledge Institute of Catalysis Research & Technology, Karlsruhe Institute of Technology, for providing the PAC, Habibu Aliyu for mentoring and the technical staff at Institute of Process Engineering in Life Sciences 2: Technical Biology, Karlsruhe Institute of Technology.

Conflicts of Interest: The authors declare no conflict of interest.

References

1. Cherubini, F.; Ulgiati, S. Crop residues as raw materials for biorefinery systems—A LCA case study. *Appl. Energy* **2010**, *87*, 47–57. [CrossRef]
2. Rathmann, R.; Szklo, A.; Schaeffer, R. Land use competition for production of food and liquid biofuels: An analysis of the arguments in the current debate. *Renew. Energy* **2010**, *35*, 14–22. [CrossRef]
3. Popp, J.; Lakner, Z.; Harangi-Rákos, M.; Fári, M. The effect of bioenergy expansion: Food, energy, and environment. *Renew. Sustain. Energy Rev.* **2014**, *32*, 559–578. [CrossRef]
4. Agler, M.T.; Wrenn, B.A.; Zinder, S.H.; Angenent, L.T. Waste to bioproduct conversion with undefined mixed cultures: The carboxylate platform. *Trends Biotechnol.* **2011**, *29*, 70–78. [CrossRef] [PubMed]
5. Gil, A. Current insights into lignocellulose related waste valorization. *Chem. Eng. J. Adv.* **2021**, *8*, 100186. [CrossRef]
6. Funke, A.; Morgano, M.T.; Dahmen, N.; Leibold, H. Experimental comparison of two bench scale units for fast and intermediate pyrolysis. *J. Anal. Appl. Pyrolysis* **2017**, *124*, 504–514. [CrossRef]
7. Niebel, A.; Funke, A.; Pfitzer, C.; Dahmen, N.; Weih, N.; Richter, D.; Zimmerlin, B. Fast Pyrolysis of Wheat Straw—Improvements of Operational Stability in 10 Years of Bioliq Pilot Plant Operation. *Energy Fuels* **2021**, *35*, 11333–11345. [CrossRef]
8. Leng, L.; Yang, L.; Chen, J.; Hu, Y.; Li, H.; Li, H.; Jiang, S.; Peng, H.; Yuan, X.; Huang, H. Valorization of the aqueous phase produced from wet and dry thermochemical processing biomass: A review. *J. Clean. Prod.* **2021**, *294*, 126238. [CrossRef]
9. Basaglia, M.; Favaro, L.; Torri, C.; Casella, S. Is pyrolysis bio-oil prone to microbial conversion into added-value products? *Renew. Energy* **2021**, *163*, 783–791. [CrossRef]
10. Liang, Y.; Zhao, X.; Chi, Z.; Rover, M.; Johnston, P.; Brown, R.; Jarboe, L.; Wen, Z. Utilization of acetic acid-rich pyrolytic bio-oil by microalga Chlamydomonas reinhardtii: Reducing bio-oil toxicity and enhancing algal toxicity tolerance. *Bioresour. Technol.* **2013**, *133*, 500–506. [CrossRef]
11. Lange, J.; Müller, F.; Bernecker, K.; Dahmen, N.; Takors, R.; Blombach, B. Valorization of pyrolysis water: A biorefinery side stream, for 1,2-propanediol production with engineered Corynebacterium glutamicum. *Biotechnol. Biofuels* **2017**, *10*, 277. [CrossRef]
12. Arnold, S.; Moss, K.; Dahmen, N.; Henkel, M.; Hausmann, R. Pretreatment strategies for microbial valorization of bio-oil fractions produced by fast pyrolysis of ash-rich lignocellulosic biomass. *GCB Bioenergy* **2019**, *11*, 181–190. [CrossRef]
13. Lian, J.; Garcia-Perez, M.; Coates, R.; Wu, H.; Chen, S. Yeast fermentation of carboxylic acids obtained from pyrolytic aqueous phases for lipid production. *Bioresour. Technol.* **2012**, *118*, 177–186. [CrossRef]
14. Lian, J.; Chen, S.; Zhou, S.; Wang, Z.; O'Fallon, J.; Li, C.-Z.; Garcia-Perez, M. Separation, hydrolysis and fermentation of pyrolytic sugars to produce ethanol and lipids. *Bioresour. Technol.* **2010**, *101*, 9688–9699. [CrossRef]
15. Kubisch, C.; Ochsenreither, K. Detoxification of a pyrolytic aqueous condensate from wheat straw for utilization as substrate in Aspergillus oryzae DSM 1863 cultivations. *Biotechnol. Biofuels Bioprod.* **2022**, *15*, 18. [CrossRef]
16. Arnold, S.; Henkel, M.; Wanger, J.; Wittgens, A.; Rosenau, F.; Hausmann, R. Heterologous rhamnolipid biosynthesis by P. putida KT2440 on bio-oil derived small organic acids and fractions. *AMB Express* **2019**, *9*, 80. [CrossRef]
17. Arnold, S.; Tews, T.; Kiefer, M.; Henkel, M.; Hausmann, R. Evaluation of small organic acids present in fast pyrolysis bio-oil from lignocellulose as feedstocks for bacterial bioconversion. *GCB Bioenergy* **2019**, *11*, 1159–1172. [CrossRef]
18. Vítězová, M.; Kohoutová, A.; Vítěz, T.; Hanišáková, N.; Kushkevych, I. Methanogenic microorganisms in industrial wastewater anaerobic treatment. *Processes* **2020**, *8*, 1546. [CrossRef]
19. Anukam, A.; Mohammadi, A.; Naqvi, M.; Granström, K. A Review of the Chemistry of Anaerobic Digestion: Methods of accelerating and optimizing process efficiency. *Processes* **2019**, *7*, 504. [CrossRef]
20. Hashsham, S.A.; Fernandez, A.S.; Dollhopf, S.L.; Dazzo, F.B.; Hickey, R.F.; Tiedje, J.M.; Criddle, C.S. Parallel processing of substrate correlates with greater functional stability in methanogenic bioreactor communities perturbed by glucose. *Appl. Environ. Microbiol.* **2000**, *66*, 4050–4057. [CrossRef] [PubMed]
21. Werner, J.J.; Knights, D.; Garcia, M.L.; Scalfone, N.B.; Smith, S.; Yarasheski, K.; Cummings, T.A.; Beers, A.R.; Knight, R.; Angenent, L.T. Bacterial community structures are unique and resilient in full-scale bioenergy systems. *Proc. Natl. Acad. Sci. USA* **2011**, *108*, 4158–4163. [CrossRef]
22. Hübner, T.; Mumme, J. Integration of pyrolysis and anaerobic digestion—Use of aqueous liquor from digestate pyrolysis for biogas production. *Bioresour. Technol.* **2015**, *183*, 86–92. [CrossRef]
23. Wen, C.; Moreira, C.M.; Rehmann, L.; Berruti, F. Feasibility of anaerobic digestion as a treatment for the aqueous pyrolysis condensate (APC) of birch bark. *Bioresour. Technol.* **2020**, *307*, 123199. [CrossRef] [PubMed]
24. Torri, C.; Fabbri, D. Biochar enables anaerobic digestion of aqueous phase from intermediate pyrolysis of biomass. *Bioresour. Technol.* **2014**, *172*, 335–341. [CrossRef] [PubMed]
25. Zhou, H.; Brown, R.C.; Wen, Z. Anaerobic digestion of aqueous phase from pyrolysis of biomass: Reducing toxicity and improving microbial tolerance. *Bioresour. Technol.* **2019**, *292*, 121976. [CrossRef] [PubMed]
26. Seyedi, S.; Venkiteshwaran, K.; Zitomer, D. Toxicity of various pyrolysis liquids from biosolids on methane production yield'. *Front. Energy Res.* **2019**, *7*, 1–12. [CrossRef]
27. Fabbri, D.; Torri, C. Linking pyrolysis and anaerobic digestion (Py-AD) for the conversion of lignocellulosic biomass. *Curr. Opin. Biotechnol.* **2016**, *38*, 167–173. [CrossRef]

28. Torri, C.; Pambieri, G.; Gualandi, C.; Piraccini, M.; Rombolà, A.G.; Fabbri, D. Evaluation of the potential performance of hyphenated pyrolysis-anaerobic digestion (Py-AD) process for carbon negative fuels from woody biomass. *Renew. Energy* **2020**, *148*, 1190–1199. [CrossRef]
29. Giwa, A.S.; Chang, F.; Xu, H.; Zhang, X.; Huang, B.; Li, Y.; Wu, J.; Wang, B.; Vakili, M.; Wang, K. Pyrolysis of difficult biodegradable fractions and the real syngas bio-methanation performance. *J. Clean. Prod.* **2019**, *233*, 711–719. [CrossRef]
30. Righi, S.; Bandini, V.; Marazza, D.; Baioli, F.; Torri, C.; Contin, A. Life Cycle Assessment of high ligno-cellulosic biomass pyrolysis coupled with anaerobic digestion. *Bioresour. Technol.* **2016**, *212*, 245–253. [CrossRef] [PubMed]
31. Antoniou, N.; Monlau, F.; Sambusiti, C.; Ficara, E.; Barakat, A.; Zabaniotou, A. Contribution to Circular Economy options of mixed agricultural wastes management: Coupling anaerobic digestion with gasification for enhanced energy and material recovery. *J. Clean. Prod.* **2019**, *209*, 505–514. [CrossRef]
32. Funke, A.; Mumme, J.; Koon, M.; Diakité, M. Cascaded production of biogas and hydrochar from wheat straw: Energetic potential and recovery of carbon and plant nutrients. *Biomass Bioenergy* **2013**, *58*, 229–237. [CrossRef]
33. Salman, C.A.; Schwede, S.; Thorin, E.; Yan, J. Enhancing biomethane production by integrating pyrolysis and anaerobic digestion processes. *Appl. Energy* **2017**, *204*, 1074–1083. [CrossRef]
34. Navarro, S.S.; Cimpoia, R.; Bruant, G.; Guiot, S.R. Biomethanation of syngas using anaerobic sludge: Shift in the catabolic routes with the CO partial pressure increase. *Front. Microbiol.* **2016**, *7*, 1188. [CrossRef]
35. Alves, J.I.; Stams, A.J.M.; Plugge, C.M.; Alves, M.M.; Sousa, D.Z. Enrichment of anaerobic syngas-converting bacteria from thermophilic bioreactor sludge. *FEMS Microbiol. Ecol.* **2013**, *86*, 590–597. [CrossRef]
36. Liu, C.; Luo, G.; Wang, W.; He, Y.; Zhang, R.; Liu, G. The effects of pH and temperature on the acetate production and microbial community compositions by syngas fermentation. *Fuel* **2018**, *224*, 537–544. [CrossRef]
37. Angenent, L.T.; Richter, H.; Buckel, W.; Spirito, C.M.; Steinbusch, K.J.J.; Plugge, C.M.; Strik, D.P.B.T.B.; Grootscholten, T.I.M.; Buisman, C.J.N.; Hamelers, H.V.M. Chain Elongation with Reactor Microbiomes: Open-Culture Biotechnology to Produce Biochemicals. *Environ. Sci. Technol.* **2016**, *50*, 2796–2810. [CrossRef]
38. Baleeiro, F.C.F. Syngas-aided anaerobic fermentation for medium-chain carboxylate and alcohol production: The case for microbial communities. *Appl. Microbiol. Biotechnol.* **2019**, *103*, 8689–8709. [CrossRef]
39. Grimalt-Alemany, A.; Łężyk, M.; Kennes-Veiga, D.M.; Skiadas, I.V.; Gavala, H.N. Enrichment of Mesophilic and Thermophilic Mixed Microbial Consortia for Syngas Biomethanation: The Role of Kinetic and Thermodynamic Competition. *Waste Biomass Valorization* **2020**, *11*, 465–481. [CrossRef]
40. Conrad, R.; Wetter, B. Influence of temperature on energetics of hydrogen metabolism in homoacetogenic, methanogenic, and other anaerobic bacteria. *Arch. Microbiol.* **1990**, *155*, 94–98. [CrossRef]
41. Hosseini, S.E.; Wahid, M.A. Hydrogen production from renewable and sustainable energy resources: Promising green energy carrier for clean development. *Renew. Sustain. Energy Rev.* **2016**, *57*, 850–866. [CrossRef]
42. de Groof, V.; Coma, M.; Arnot, T.; Leak, D.J.; Lanham, A.B. Medium chain carboxylic acids from complex organic feedstocks by mixed culture fermentation. *Molecules* **2019**, *24*, 398. [CrossRef]
43. Zhang, F.; Ding, J.; Zhang, Y.; Chen, M.; Ding, Z.-W.; van Loosdrecht, M.C.; Zeng, R.J. Fatty acids production from hydrogen and carbon dioxide by mixed culture in the membrane biofilm reactor. *Water Res.* **2013**, *47*, 6122–6129. [CrossRef] [PubMed]
44. Baleeiro, F.C.F.; Kleinsteuber, S.; Sträuber, H. Recirculation of H_2, CO_2, and Ethylene Improves Carbon Fixation and Carboxylate Yields in Anaerobic Fermentation. *ACS Sustain. Chem. Eng.* **2022**, *10*, 4073–4081. [CrossRef]
45. Gibbs, P.A.; Seviour, R.J.; Schmid, F. Growth of filamentous fungi in submerged culture: Problems and possible solutions. *Crit. Rev. Biotechnol.* **2000**, *20*, 17–48. [CrossRef]
46. Ferreira, J.A.; Mahboubi, A.; Lennartsson, P.R.; Taherzadeh, M.J. Waste biorefineries using filamentous ascomycetes fungi: Present status and future prospects. *Bioresour. Technol.* **2016**, *215*, 334–345. [CrossRef]
47. Uwineza, C.; Sar, T.; Mahboubi, A.; Taherzadeh, M.J. Evaluation of the cultivation of aspergillus oryzae on organic waste-derived vfa effluents and its potential application as alternative sustainable nutrient source for animal feed. *Sustainability* **2021**, *13*, 12409. [CrossRef]
48. Mahboubi, A.; Ferreira, J.A.; Taherzadeh, M.J.; Lennartsson, P.R. Value-added products from dairy waste using edible fungi. *Waste Manag.* **2017**, *59*, 518–525. [CrossRef]
49. Kövilein, A.; Umpfenbach, J.; Ochsenreither, K. Acetate as substrate for l-malic acid production with Aspergillus oryzae DSM 1863. *Biotechnol. Biofuels* **2021**, *14*, 518–525. [CrossRef] [PubMed]
50. Dörsam, S.; Kirchhoff, J.; Bigalke, M.; Dahmen, N.; Syldatk, C.; Ochsenreither, K. Evaluation of pyrolysis oil as carbon source for fungal fermentation. *Front. Microbiol.* **2016**, *7*, 2059. [CrossRef] [PubMed]
51. Telliard, W.A. *Method1684 Total, Fixed, and Volatile Solids in Water, Solids, and Biosolids*; Draft January 2001; U.S. Environmental Protection Agency Office of Water Office of Science and Technology Engineering and Analysis Division (4303)U.S. EPA': Washington, DC, USA, 2001; pp. 1–13.
52. Pfitzer, C.; Dahmen, N.; Tröger, N.; Weirich, F.; Sauer, J.; Günther, A.; Müller-Hagedorn, M. Fast Pyrolysis of Wheat Straw in the Bioliq Pilot Plant. *Energy Fuels* **2016**, *30*, 8047–8054. [CrossRef]
53. Grimalt-Alemany, A.; Łęzyk, M.; Lange, L.; Skiadas, I.V.; Gavala, H.N. Enrichment of syngas-converting mixed microbial consortia for ethanol production and thermodynamics-based design of enrichment strategies. *Biotechnol. Biofuels* **2018**, *11*, 198. [CrossRef] [PubMed]

54. Grimalt-Alemany, A.; Skiadas, I.V.; Gavala, H.N. Syngas biomethanation: State-of-the-art review and perspectives. *Biofuels Bioprod. Biorefining* **2018**, *12*, 139–158. [CrossRef]
55. Mahamkali, V.; Valgepea, K.; Lemgruber, R.D.S.P.; Plan, M.; Tappel, R.; Köpke, M.; Simpson, S.D.; Nielsen, L.K.; Marcellin, E. Redox controls metabolic robustness in the gas-fermenting acetogen Clostridium autoethanogenum. *Proc. Natl. Acad. Sci. USA* **2020**, *117*, 13168–13175. [CrossRef]
56. Wang, S.; Huang, H.; Kahnt, H.H.; Mueller, A.P.; Köpke, M.; Thauer, R.K. NADP-Specific electron-bifurcating [FeFe]-hydrogenase in a functional complex with formate dehydrogenase in clostridium autoethanogenum grown on CO. *J. Bacteriol.* **2013**, *195*, 4373–4386. [CrossRef]
57. Liu, Y.; Wan, J.; Han, S.; Zhang, S.; Luo, G. Selective conversion of carbon monoxide to hydrogen by anaerobic mixed culture. *Bioresour. Technol.* **2016**, *202*, 1–7. [CrossRef] [PubMed]
58. Daniels, L.; Fuchs, G.; Thauer, R.K.; Zeikus, J.G. Carbon monoxide oxidation by methanogenic bacteria. *J. Bacteriol.* **1977**, *132*, 118–126. [CrossRef] [PubMed]
59. Rother, M.; Metcalf, W.W. Anaerobic growth of Methanosarcina acetivorans C2A on carbon monoxide: An unusual way of life for a methanogenic archaeon'. *Proc. Natl. Acad. Sci. USA* **2004**, *101*, 16929–16934. [CrossRef]
60. Sipma, J.; Lens, P.N.L.; Stams, A.J.M.; Lettinga, G. Carbon monoxide conversion by anaerobic bioreactor sludges. *FEMS Microbiol. Ecol.* **2003**, *44*, 271–277. [CrossRef]
61. Slepova, T.V.; Rusanov, I.I.; Sokolova, T.G.; Bonch-Osmolovskaya, E.A.; Pimenov, N.V. Radioisotopic tracing of carbon monoxide conversion by anaerobic thermophilic prokaryotes. *Microbiology* **2007**, *76*, 523–529. [CrossRef]
62. Wang, Y.Q.; Yu, S.J.; Zhang, F.; Xia, X.Y.; Zeng, R.J. Enhancement of acetate productivity in a thermophilic (55 °C) hollow-fiber membrane biofilm reactor with mixed culture syngas (H_2/CO_2) fermentation. *Appl. Microbiol. Biotechnol.* **2017**, *101*, 2619–2627. [CrossRef] [PubMed]
63. Shen, N.; Dai, K.; Xia, X.Y.; Zeng, R.J.; Zhang, F. Conversion of syngas (CO and H_2) to biochemicals by mixed culture fermentation in mesophilic and thermophilic hollow-fiber membrane biofilm reactors. *J. Clean. Prod.* **2018**, *202*, 536–542. [CrossRef]
64. Liu, H.; Wang, J.; Wang, A.; Chen, J. Chemical inhibitors of methanogenesis and putative applications. *Appl. Microbiol. Biotechnol.* **2011**, *89*, 1333–1340. [CrossRef]
65. Thompson, M.A.; Mohajeri, A.; Mirkouei, A. Comparison of pyrolysis and hydrolysis processes for furfural production from sugar beet pulp: A case study in southern Idaho, USA. *J. Clean. Prod.* **2021**, *311*, 127695. [CrossRef]
66. Jönsson, L.J.; Alriksson, B.; Nilvebrant, N.-O. Bioconversion of lignocellulose: Inhibitors and detoxification. *Biotecnol. Biofuels* **2013**, *6*, 16. [CrossRef] [PubMed]
67. Ljungdahl, L.G. Acetate Synthesis in Acetogenic Bacteria. *Annu. Rev. Microbiol.* **1986**, *40*, 415–450. [CrossRef] [PubMed]
68. Bache, R.; Pfennig, N. Selective isolation of Acetobacterium woodii on methoxylated aromatic acids and determination of growth yields. *Arch. Microbiol.* **1981**, *130*, 255–261. [CrossRef]
69. Schink, B.; Pfennig, N. Fermentation of trihydroxybenzenes by Pelobacter acidigallici gen. nov. sp. nov., a new strictly anaerobic, non-sporeforming bacterium. *Arch. Microbiol.* **1982**, *133*, 195–201. [CrossRef]
70. Kato, S.; Chino, K.; Kamimura, N.; Masai, E.; Yumoto, I.; Kamagata, Y. Methanogenic degradation of lignin-derived monoaromatic compounds by microbial enrichments from rice paddy field soil. *Sci. Rep.* **2015**, *5*, 14295. [CrossRef] [PubMed]
71. Chen, Y.; Cheng, J.J.; Creamer, K.S. Inhibition of anaerobic digestion process: A review. *Bioresour. Technol.* **2008**, *99*, 4044–4064. [CrossRef]
72. Fedorak, P.M.; Hrudey, S.E. The effects of phenol and some alkyl phenolics on batch anaerobic methanogenesis. *Water Res.* **1984**, *18*, 361–367. [CrossRef]
73. Battersby, N.S.; Wilson, V. Survey of the anaerobic biodegradation potential of organic chemicals in digesting sludge. *Appl. Environ. Microbiol.* **1989**, *55*, 433–439. [CrossRef] [PubMed]
74. Schink, B.Y.; Philipp, B.; Müller, J. Anaerobic Degradation of Phenolic Compounds. *Naturwissenschaften* **2000**, *87*, 12–23. [CrossRef]
75. Milledge, J.J.; Nielsen, B.V.; Harvey, P.J. The inhibition of anaerobic digestion by model phenolic compounds representative of those from Sargassum muticum. *J. Appl. Phycol.* **2019**, *31*, 779–786. [CrossRef]
76. Kato, K.; Kozaki, S.; Sakuranaga, M. Degradation of ligning compounds by bacteria from termite guts. *Biotechnol. Lett.* **1998**, *20*, 459–462. [CrossRef]
77. Huang, X.-F.; Santhanam, N.; Badri, D.V.; Hunter, W.J.; Manter, D.K.; Decker, S.R.; Vivanco, J.M.; Reardon, K.F. Isolation and characterization of lignin-degrading bacteria from rainforest soils. *Biotechnol. Bioeng.* **2013**, *110*, 1616–1626. [CrossRef]
78. Ozbayram, E.G.; Kleinsteuber, S.; Nikolausz, M.; Ince, B.; Ince, O. Bioaugmentation of anaerobic digesters treating lignocellulosic feedstock by enriched microbial consortia. *Eng. Life Sci.* **2018**, *18*, 440–446. [CrossRef]
79. Tuesorn, S.; Wongwilaiwalin, S.; Champreda, V.; Leethochawalit, M.; Nopharatana, A.; Techkarnjanaruk, S.; Chaiprasert, P. Enhancement of biogas production from swine manure by a lignocellulolytic microbial consortium. *Bioresour. Technol.* **2013**, *144*, 579–586. [CrossRef]
80. Qiu, Y.L.; Hanada, S.; Ohashi, A.; Harada, H.; Kamagata, Y.; Sekiguchi, Y. Syntrophorhabdus aromaticivorans gen. nov., sp. nov., the first cultured anaerobe capable of degrading phenol to acetate in obligate syntrophic associations with a hydrogenotrophic methanogen. *Appl. Environ. Microbiol.* **2008**, *74*, 2051–2058. [CrossRef]
81. Fedorak, P.M.; Hrudey, S.E. Inhibition of anaerobic degradation of phenolics and methanogenesis by coal coking wastewater. *Water Sci. Technol.* **1987**, *19*, 219–228. [CrossRef]

82. Machida, M.; Asai, K.; Sano, M.; Tanaka, T.; Kumagai, T.; Terai, G.; Kusumoto, K.-I.; Arima, T.; Akita, O.; Kashiwagi, Y.; et al. Genome sequencing and analysis of Aspergillus oryzae. *Nature* **2005**, *438*, 1157–1161. [CrossRef]
83. Oswald, F.; Dörsam, S.; Veith, N.; Zwick, M.; Neumann, A.; Ochsenreither, K.; Syldatk, C. Sequential mixed cultures: From syngas to malic acid. *Front. Microbiol.* **2016**, *7*, 891. [CrossRef]
84. Uwineza, C.; Mahboubi, A.; Atmowidjojo, A.; Ramadhani, A.; Wainaina, S.; Millati, R.; Wikandari, R.; Niklasson, C.; Taherzadeh, M.J. Cultivation of edible filamentous fungus Aspergillus oryzae on volatile fatty acids derived from anaerobic digestion of food waste and cow manure. *Bioresour. Technol.* **2021**, *337*, 125410. [CrossRef]
85. Kövilein, A.; Aschmann, V.; Hohmann, S.; Ochsenreither, K. Immobilization of Aspergillus oryzae DSM 1863 for l-Malic Acid Production. *Fermentation* **2022**, *8*, 26. [CrossRef]
86. Schmitt, V.; Derenbach, L.; Ochsenreither, K. Enhanced l-Malic Acid Production by Aspergillus oryzae DSM 1863 Using Repeated-Batch Cultivation. *Front. Bioeng. Biotechnol.* **2022**, *9*, 1–15. [CrossRef]

fermentation

Article

Significance of Intermittent Mixing in Mesophilic Anaerobic Digester

Buta Singh [1,*], Kornél L. Kovács [2], Zoltán Bagi [2], Máté Petrik [1], Gábor L. Szepesi [1], Zoltán Siménfalvi [1] and Zoltán Szamosi [1]

[1] Institute of Energy Engineering and Chemical Machinery, University of Miskolc, 3515 Miskolc, Hungary
[2] Department of Biotechnology, University of Szeged, 6720 Szeged, Hungary
* Correspondence: butasingh.chahal@uni-miskolc.hu

Abstract: The mixing of slurry in an anaerobic digester (AD) is one of many key parameters, which have a significant effect on specific biogas yield (BY) and volatile solid (VS) removal rate. The determination of the optimum mixing regime in a digester is very complex as it depends on a large number of internal and external factors such as microbial community, the rheology of slurry, digester and impeller design, mixing intensity, and mixing intervals. The novelty of this study is the investigation of the optimum mixing regime in a lab-scale digester under semi-continuous mixing regimes by the continuous monitoring of the physicochemical properties of the digestate. In this study, a helical ribbon (HR) impeller was used for the agitation of the slurry operated at 67 rpm for 5 min under various agitation intervals (1 h, 2 h, 3 h, and 4 h). The results showed a 6–12% reduction in BY as the time between mixing operations increased. The highest BY was observed at a mixing frequency of 5 min/h, which produced a total of 54.1 L of biogas as compared to the mixing frequencies of 2 h, 3 h, and 4 h, where the BYs were recorded as 51.2 L, 49.8 L, and 47.3 L, respectively. Volatile fatty acids (VFAs) and FOS/TAC ratio were stabilized at 5–7 Gl^{-1} and 0.3–0.5, respectively. The appropriate mixing intensity was determined to obtain the highest biogas production, which could lead to lower power consumption for mixing operations.

Keywords: biogas; mixing; mechanical; optimization; design

Citation: Singh, B.; Kovács, K.L.; Bagi, Z.; Petrik, M.; Szepesi, G.L.; Siménfalvi, Z.; Szamosi, Z. Significance of Intermittent Mixing in Mesophilic Anaerobic Digester. *Fermentation* **2022**, *8*, 518. https://doi.org/10.3390/fermentation8100518

Academic Editor: Sanjay Nagarajan

Received: 15 September 2022
Accepted: 5 October 2022
Published: 7 October 2022

1. Introduction

Waste-to-energy conversion is one of the most appealing renewable energy technologies in the modern era and recent political/economical/military turbulences. The anaerobic digestion (AD) process is one of them which is receiving more attention day by day due to its low carbon footprints per unit of power production [1]. It has been intensively studied over the past years in order to enhance energy recovery from the organic waste in biogas plants [2–4]. The efficiency of a biogas plant is mainly determined by the amount of methane production and the internal power consumption of the plant. Internal operations in a commercial biogas plant consume power for loading, transportation, the mechanical pretreatment of the substrate, the mixing of slurry, etc. The mixing of slurry is one of the prominent factors that determine efficiency in terms of both biogas production and internal power consumption. Anaerobic digesters are usually configured as continuously stirred reactors in order to provide effective mixing for the homogeneous dispersion of the substrate and the heating energy, reduce solid settling and short-circuiting, and preserve the retention time [5]. The optimum hydrodynamic mixing plays a vital role to achieve maximum biogas recovery and a higher organic removal rate. The adequate mixing of slurry also refers to the uniform distribution of fresh substrate to decrease the reaction time and the release of entrapped biogas, avoiding temperature and pH gradients in the entire volume of an AD slurry and hydraulic shear force without breaking the bacterial/archaea morphology [6].

Inadequate mixing can cause the failure of a biogas plant, the accumulation of solids at the bottom along with the formation of foam, scum, and crust at the top of the surface.

Nearly 44% of large-scale biogas plants experience sedimentation and crust layer formation from time to time due to improper mixing [7]. Moreover, the mixing operation also consumes 54% of the total electricity consumption by the biogas plant. One possibility to increase the efficiency of a biogas plant is to reduce the mixing time. Therefore, optimum mixing in an AD becomes very critical, especially while operating at a high organic loading rate (OLR) and short hydraulic retention times (HRT). Typically, the mixing operation in a biogas plant is undertaken by three modes, i.e., mechanical, slurry recirculation, and biogas recirculation. According to Lemmer et al. [8], mechanical mixing is the most common among all mixing methods. In mechanical mixing, both mixing intensity and duration impact the biogas production in a biogas plant substantially [9].

Numerous studies can be found which include the optimization of the impeller design and the mixing speed of the impeller [10–15]. Singh et al. [6,9] compared the various designs, which were used for the optimization of mixing in the digester. However, documented data on the influence of mixing and ideal intensity are insufficient due to the variation in various parameters in individual experiments such as temperature, total solid (TS) content, and HRT, along with the impeller and digester design. The mixing regime refers to the rotational speed of the impeller and the time gap between the operations of the mixer in a semi-continuous mixing operation [15–17]. While intense mixing has been suggested to boost the biodegradation of volatile solids by raising their solubility and area of contact with bacteria, it has also been reported to inhibit the creation of flocks, where syntrophic microbial interactions can occur, and to cause propionate accumulation [18,19]. Higher mixing intensities can significantly affect the composition of the microbial communities (disruption of bacterial flocs) and hence biomass activity, which may result in the reduction in biogas production rates [20–23]. In the case of mechanical mixing techniques, the different impeller designs and speeds are analyzed [21,24], whereas, for the slurry recirculation, the biogas recirculation, the inlet injection positions and design of nozzles, the recirculation rate, and the mass flow rates are evaluated to optimize mixing in an anaerobic digester [25].

Various studies demonstrated that intermittent mixing operations resulted in the higher efficiency of an anaerobic digester as compared to continuous mixing operations [13,14,26]. In the recent advancements in mixing techniques in an anaerobic digester, conductive additives such as granular activated carbon (GAC) and powdered activated carbon (PAC) are used to evaluate methane productivity and kinetics along with a comprehensive characterization of the quantitative and qualitative features of microbial communities to elucidate the underlying mechanisms of the impacts of different mixing conditions [25]. Tian et al. [15] compared BYs at different agitation intervals (continuous, 2 h, 4 h, 8 h, and 12 h) in a lab scale digester. It was observed that at an agitation interval of 2 h, the highest biogas production rate of 508 $\pm$ 49 biogas mL gTS^{-1} was achieved, which was 14-29% higher than other mixing regimes due to a reduction in floating layers. The floating layers and sedimentation were attributed to the poor contact of the substrate microorganisms. According to Semen et al. [27] the lower the decrease in the agitation interval in a full-scale biogas plant, the higher the specific yield of 11.7% and a decrease in internal energy consumption of 10.4%. Kowalczyk et al. [28] found that 29% of power was saved when the mixing operation time was reduced from 7 h to 2 h day^{-1} without any negative impact on BY.

Previous studies showed how the results of the optimization of mixing in digesters can vary. In addition, the data regarding the rheology of the slurry, the impeller, and the digester design are missing in most studies, which is a very important aspect to evaluate the mixing in any vessel because the mixing operation is a physical process [29]. The present study investigated the effect of increasing the mixing interval time on the efficiency of the AD process and overall BY. Various parameters such as VFAs, pH, the FOS/TAC ratio, and ammonia concentrations were measured throughout the experiment to have a deeper insight into the process. Based on previous experiments [30], the rotational speed of the mixer was adjusted to 67 rpm.

2. Experimental Setup

The experiments were undertaken in three identical, single-stage, continuously fed 5 L lab-scale digesters, with a headspace volume of 1 L, formed from stainless steel by Biospin Ltd. Szeged, Hungary [31]. The digesters were operated in identical operating conditions. The schematic 2D diagram of the setup is shown in Figure 1. The key parameters (temperature, mixing speed, and pH) were automatically controlled by computer software. The digesters were named as F1, F2, and F3 for reference. All the impellers were operated by the same electric motor in order to maintain identical mixing conditions. The temperature in the reactor was controlled with an electric heating jacket around the vessel with an accuracy of ±0.5 °C. Three sets of experiments were conducted in parallel to observe the effect of varying shear rates on biogas production rates and methane content. The experiment lasted for 100 days, including two weeks of a pre-run phase.

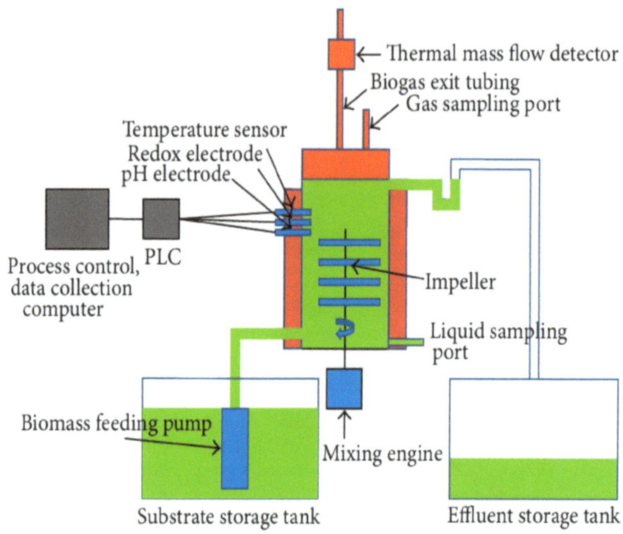

Figure 1. The schematic 2D diagram of the experimental setup.

2.1. Inoculum Feeding, Substrates, and Sampling

The digestate was collected from a commercial biogas plant in Szeged and stored at 4 °C before the start of the experiment. The substrate consisted of a mixture of pig slurry and ensilaged sweet sorghum. In total, 5 L of incubated substrate slurry was added to each of the three digesters. To ensure a steady digestion process and consistent biogas output, the experiment was pre-run for two weeks. At the start of the experiment, ultrapure nitrogen gas was utilized to replace the headspace. Every day, 1 gVS L^{-1} of α-cellulose was injected into the digester. The digesters were run at a mesophilic temperature (37 °C) with a 15-day HRT. Table 1 displays several parameters of the initial sludge. The TS concentration of the slurry remained constant at 4.28 percent. Digestate samples were taken from the bottom of the digester after mixing.

Table 1. Characteristics of the substrate used in the experiment.

Parameter	Value Range
TS (%)	4.28
SS (g L^{-1})	57.8 ± 10.0
Total carbon (%)	46.2
TVS (g L^{-1})	87.6 ± 3.4
COD (g L^{-1})	141 ± 6.4
VFA (g L^{-1})	4.15 ± 1.38
ρ (kg m^{-3})	1068
HRT (d)	18

2.2. *Mixing Mode*

Mechanical mixing using a helical ribbon impeller (Figure 2) was used in this experiment throughout. The intermittent mixing operation was carried out at an agitation rate of 67 rpm for all three digesters. For this purpose, the impellers were turned on for a period of 5 min with turn-off resting intervals of 1 h, 2 h, 3 h, and 4 h.

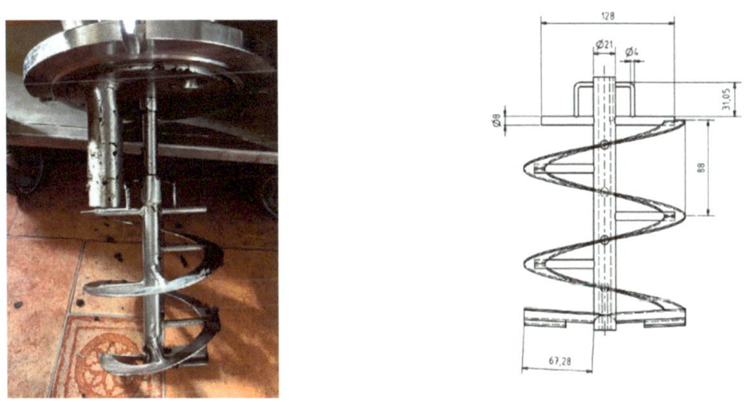

Figure 2. The geometry of the helical ribbon impeller used in this study.

2.3. *Analytical Methods*

2.3.1. Gas Analysis

Gas volume values were recorded every four hours by means of direct mass flow controllers (DMFC, Brooks Instruments) attached to each gas exit port. Biogas composition was analyzed using a gas chromatograph (6890N Net-work GC system, Agilent Technologies). A 250 μL gas sample was collected from the headspace and injected into a gas chromatograph that was equipped with a 5 Å molecular sieve column (length 30 m, I.D. 0.53 megabore, film 23 μm) and a thermal conductivity detector.

2.3.2. Volatile Fatty Acids

HPLC (Hitachi Elite, equipped with an ICSep ICE-COREGEL 64H column and a refractive index detector L2490) was used to determine Volatile acids under the following conditions: a solvent of 0.1 N H_2SO_4, a flow rate of 0.8 mL min^{-1}, a column temperature of 50 °C, and a detector temperature of 41 °C. In total, 10 mL of samples were collected from each digester for analysis. The samples were centrifuged at 13,000 rpm for 10 min to separate the solid and liquid and then filtered through a 0.45 μm membrane. The samples were analyzed every week.

2.3.3. TS and oTS Content

The dry matter content was determined by drying the substrate at 105 °C for 24 h and measuring the residues. The residue was further heated at 550 °C in the oven until its weight did not alter to determine the overall organic solid material.

3. Results and Discussion

3.1. CFD Analysis

In a previous study [30], various mixing speeds (10 rpm, 30 rpm, and 67 rpm) of the impeller in the same AD were compared. Cfd analysis demonstrated the occurrence of substantial unmixed zones at lower mixing speeds, which reached near zero velocities. Slurry homogeneity was achieved at a speed of 67 rpm. Previous experimental results also indicated that increasing the impeller speed to a certain optimal level may increase the mixing performance. Beyond that limit, the power consumption rapidly increases with only a slight improvement in the mixing performance and a decrease in the biogas output. The maximum particle velocity achieved at 67 rpm was noted as 0.24 ms^{-1}, whereas, the average velocity in the entire experiment was calculated as 0.10 ms^{-1}. The dead zone volumes at 67 rpm were reduced to 2%. Following the determination of the optimum mixing regime, the effect of various mixing intervals on biogas yield was further studied.

3.2. Experimental Analysis

3.2.1. Start-Up Phase

All digesters were pre-run until a stable biogas production rate was obtained. The OLRs by α-cellulose substrate were set at 5 g day^{-1}, i.e., 1 g L^{-1}, for the entire experiment, and the reactors were fed once a day. During the start-up phase, process instabilities were observed that might be caused by increased hydrogen concentrations, which result in a better breakdown to propionic acid rather than acetic acid, carbon dioxide, and hydrogen [32]. After the first two weeks of the pre-run, the operation of the digesters became stable with 3.38 ± 0.56–3.56 ± 0.45 L day^{-1} gas production. After the pre-run period of 15 days, the biogas production was constant and the FOS/TAC ratio was recorded as 0.39 ± 0.06, which is considered normal as it indicates that the digestion process was stable. Different mixing intervals were started at day 15.

3.2.2. Effect of Mixing Intervals on Overall Biogas Yield (BY)

According to the results of our previous study [30], the mixing speed of 67 rpm was selected for further investigation of the effect of interval time, i.e., the resting, non-mixing period in between the mixing operations, on biogas production rate. All three digesters were run with identical parameters such as TS content, temperature, and mixing regimes. A mixing interval time of 1 h was selected in the initial stage (after the start-up phase) of the experiment from day 1 to day 20, which further increased to 2 h, 3 h, and 4 h. The use of three parallel digesters is preferred to obtain more precise data on the effects of varying parameters during the whole experiment. Accordingly, similar trends in all reactors were observed at a particular defined mixing regime.

The results from this study indicated that BY was closely related to the mixing interval time. In Figure 3 F represents the digester, and H represents the resting time (in hours) between mixing operations. The daily maximum biogas yields during the resting times of 1, 2, 3, and 4 h were noted as 3.84 L day^{-1}, 3.36 L day^{-1}, 3.12 L day^{-1}, and 2.94 L day^{-1}, respectively (Figure 3a–c). In Figure 3d, the green bars indicate the total biogas production in fermenters 1, 2, and 3 at one hour of resting time. The average daily biogas yields during all the mixing regimes were noted as 3.3, 2.9, 2.8, and 2.5 L day^{-1}, as depicted in Table 2. Similar results were demonstrated by Latha et al. [33] where the mixing regimes were 15 min hr^{-1} and 30 min hr^{-1}. The maximum biogas yield was observed at a mixing rate of 15 min hr^{-1} between 50 rpm–200 rpm. The observed higher biogas yields at the minimum resting time are attributed to have favored a better interaction among methanogenic and acetogenic granules and further enhanced the bacterial contact between the substrate and microbes.

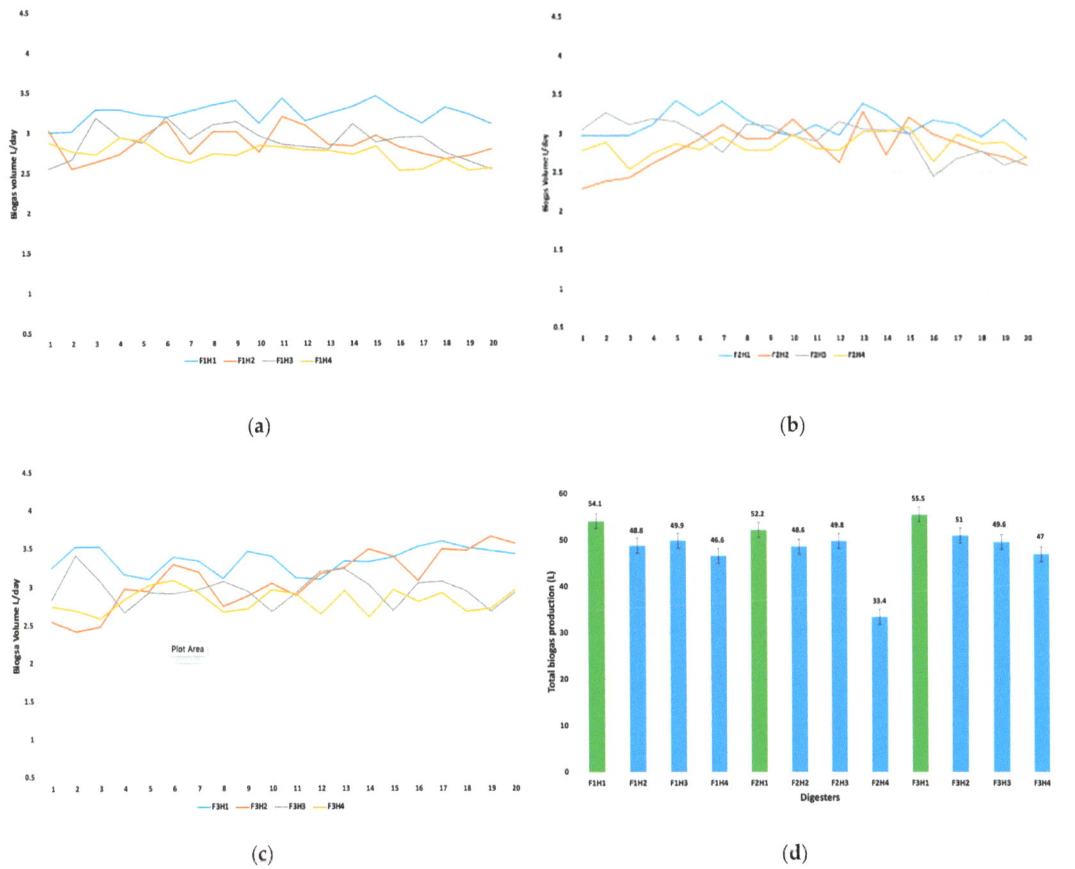

Figure 3. Represents the daily biogas production for continuous 20 days at different mixing intervals. (**a**) for digester 1, (**b**) for digester 2, and (**c**) for digester 3, respectively. (**d**) represents the overall biogas production from all three digesters.

Table 2. Comparison of biogas production from all three digesters under similar working conditions.

	Mixing Regime 1 (1 h Resting Time) (Period 15–35 days)			**Mixing Regime 1 (2 h Resting Time) (Period 35–55 days)**			**Mixing Regime 3 (3 h Resting Time) (Period 55–75 days)**			**Mixing Regime 4 (4 h Resting Time) (Period 75–95 days)**		
Fermenters →	**F1**	**F2**	**F3**	**F1**	**F2**	**F3**	**F1**	**F2**	**F3**	**F1**	**F2**	**F3**
Total Biogas production (L/d)	54.1	52.24	55.5	48.9	48.6	51.2	49.9	49.8	49.6	46.6	33.4	47.3
Maximum BY (daily) (L/d)	3.84	3.24	3.71	3.36	3.20	2.91	2.98	3.12	2.34	2.94	2.82	2.35
Minimum BP (daily) (L/d)	2.50	2.44	2.43	2.36	2.39	2.42	2.34	2.53	2.77	2.31	2.22	2.43
Average (L/d)	2.70	2.61	2.77	2.44	2.43	2.55	2.49	2.82	2.48	2.33	2.51	2.36

Similar trends can be recognized in all three digesters in terms of biogas yields. Figure 3 demonstrates the daily and cumulative biogas production in the digesters. The overall biogas production in fermenter 1 at the resting times of 1 h, 2 h, 3 h, and 4 h was 54.1 L, 48.8 L, 49.9 L, and 46.6 L, respectively. The higher yield of biogas at the resting time of 1 h is assumed to be due to the consequences of better chemical equilibrium along with better buffer action gained during the non-mixing time. The effect of increasing the interval between the mixing periods was observed from day 35 when the resting time was reduced from 1 h to 2 h. The daily biogas production dropped from 0.59 L to 0.41 L in F1, from 0.52 L to 0.41 L in F2, and from 0.61 L to 0.55 L in F3. This variation at different mixing intervals might be attributed to the more frequent mixing, which allowed for more interaction between the substrate and the microorganisms.

The mixing interval of 4 h represented the adverse effect on BY as compared to other intervals. The BY was recorded as 46.6 L, 33.4 L, and 47.3 L in all three digesters, respectively at this particular mixing interval. The main reason for the lower BY was the settling of solid particles at the bottom of the digester during the longer resting times. A BY as low as 14–30% lower was recorded in all digesters in comparison with reduced mixing interval times. The formation of floating layers in the digester is also one of the main factors responsible for deviation in the BY at various mixing intervals [15]. Lowering the mixing interval time possibly led to the prevention of floating layer formation, which was responsible for the smooth discharge of biogas from the slurry. Floating layers are also directly associated with the OLR and TS content [13]. In our case, the OLR was 1 gVS $L^{-1}d^{-1}$, which is believed to fall within the optimum range where the reduction in the pause in the mixing could lead to a reduction in floating layer formation. Table 3 represents the statical analysis of BY at different mixing intervals in all of the digesters. The results demonstrate the values of $p > 0.05$, which means the biogas production at the identical mixing intervals in all the digesters produced was similar in terms of quantity.

Table 3. Statistical analysis of biogas production at various resting times during mixing.

Resting Time (hr)		Data Set		*p* Values
1	F1	F2	F3	0.63712
2	F1	F2	F3	0.092734
3	F1	F2	F3	0.17832
4	F1	F2	F3	0.28374

Figure 4 displays the reduction in the VS content in all the digesters at different mixing-resting intervals. Similar trends can be recognized in the BY. The highest VS reduction rates (64.2%–68.5%) were observed at lower mixing–resting intervals. Nevertheless, at the mixing interval of 4 h, the VS reduction was recorded as 58.3 ± 1.4%, 53.6 ± 2.8%, and 56.7 ± 2.5% for F1, F2, and F3, respectively (Table 4). According to the recent study by Caillet et al. [34], the variation in both the TS and reduction in VS was also found when the different samples were taken from both the top and bottom of a lab-scale digester under various mixing speeds of 30, 40, and 50 rpm. The current study of TS and VS contents showed the effect of mixing on the displacement of solid matter. As a result, biogas production can be enhanced by an appropriate mixer design, mixing speed, and mixing–resting interval times. Furthermore, it is suggested that intermittent mixing is adequate for the anaerobic digestion process. Based on these findings, it can be concluded that biogas output could be increased with the reactor design and that the operating parameters (intermittent mixing mode at lower mixing–resting intervals and OLR) can be favorable for both the substrate and the microorganisms.

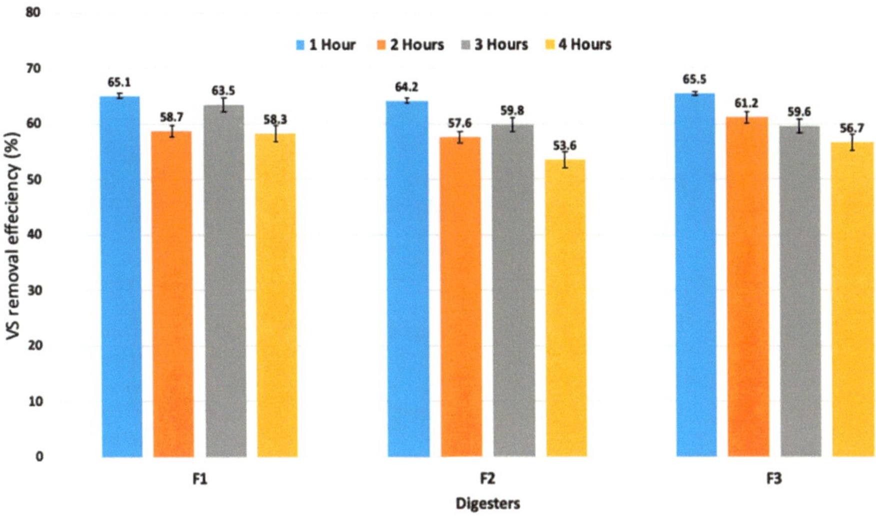

Figure 4. Performance of all three anaerobic digesters for total volatile solids removal with different mixing intervals at 67 rpm.

Table 4. Represents the total VS reduction (%) in all the digesters at various mixing intervals.

Digesters	Mixing Intervals (hours)	VS Removal Efficiency (%)	Total Biogas Production (L)
F1	1	65.1 ± 3.4	54.1 ± 0.24
	2	58.7 ± 2.1	48.9 ± 0.31
	3	63.2 ± 3.6	49.7 ± 0.25
	4	58.3 ± 1.4	46.6 ± 0.14
F2	1	64.2 ± 2.5	52.2 ± 0.32
	2	57.6 ± 3.2	48.6 ± 0.13
	3	58.8 ± 5.2	49.8 ± 0.26
	4	53.6 ± 2.8	33.4 ± 0.27
F3	1	68.5 ± 2.7	55.5 ± 0.17
	2	61.2 ± 3.4	51.2 ± 0.30
	3	59.6 ± 3.2	49.6 ± 0.23
	4	56.7 ± 2.5	47.3 ± 0.21

3.2.3. Impact of Mixing Intervals on VFA Concentration, Alkalinity, pH, and Ammonia

The mixing intensity, mixing mode, and frequency directly influence the AD bioprocess equilibrium and have a major impact on overall biogas production yields [33,35,36]. VFAs such as acetic acid, propionic acid, butyric acid, iso-butyric acid, and valeric acid are produced during the acidogenesis reaction. The rise in VFA concentration has an effect on the efficiency with which substrates are converted to biogas. In this section, the effect of the mixing operation on VFA, pH, the FOS/TAC ratio, and free NH_3 is analyzed (Table 5). All parameters were measured twice a week after the completion of the mixing cycle to obtain a homogeneous sample. The rheology of the substrate is one of the major parameters, which have a significant effect on the performance during mixing. The digesters were operated at 4.2% TS content throughout the experiment; therefore, the rheological parameters of the slurry could remain constant, and more precise results can be obtained.

Table 5. Average performance values of the digesters at various mixing-resting intervals.

	Mixing Regime 1 (1 h Resting Time) (Period 1–20 days)			Mixing Regime 2 (2 h Resting Time) (Period 20–40 days)			Mixing Regime 3 (3 h Resting Time) (Period 40–60 days)			Mixing Regime 4 (4 h Resting Time) (Period 60–80 days)		
VFA's (g/L)	7.3	6.1	5.8	3.1	3.8	4.9	1.1	1.9	2.2	2.4	4.3	3.1
pH	7.2	7.0	7.7	7.1	8.0	7.7	8.5	7.9	8.1	7.6	8.0	7.9
NH_4^+-N (g/L)	0.94	0.89	9.15	0.88	0.82	0.85	0.66	0.75	0.58	0.78	0.82	0.99
FOS/TAC Ratio	0.34	0.49	0.45	0.35	0.44	0.47	0.29	0.24	0.42	0.24	0.35	0.44

The average pH of the reactor content remained between the optimal limits of 6.8–7.5 throughout the experiment. For one hour of resting time, the average pH recorded in all the digesters is 7.2, 7.0, and 7.7 in digesters 1, 2, and 3, respectively. At the maximum resting time of 4 h, the pH values changed to 7.6, 8.0, and 7.9 in digesters 1, 2, and 3, respectively (Figure 5). Higher pH values at higher resting times were presumably due to a lower accumulation of VFAs during that period. The average VFA levels were noted as 5.8 g/L, 4.4 g/L, 5.3 g/L, and 7.7 g/L for the resting time periods of 1 h, 2 h, 3 h, and 4 h, respectively, followed by a concomitant increase in biogas production (Figure 6). According to Caillet et al. [34], the increase in the VFA content had no detrimental impact on biogas generation. In terms of the biogas production and changes in the ammonium and VFA concentrations, no substantial fluctuation in these two concentrations was found to explain the differences in biogas output. Higher VFA concentrations and lower ammonium concentrations resulted in a higher biogas output, whereas, in the current study, the ammonia concentration was found to be in equilibrium during the whole experiment (Figure 7. Similarly, Franke et al. [37] found that the higher VFA levels (8–10 g L^{-1}) and pH values did not destabilize the anaerobic process.

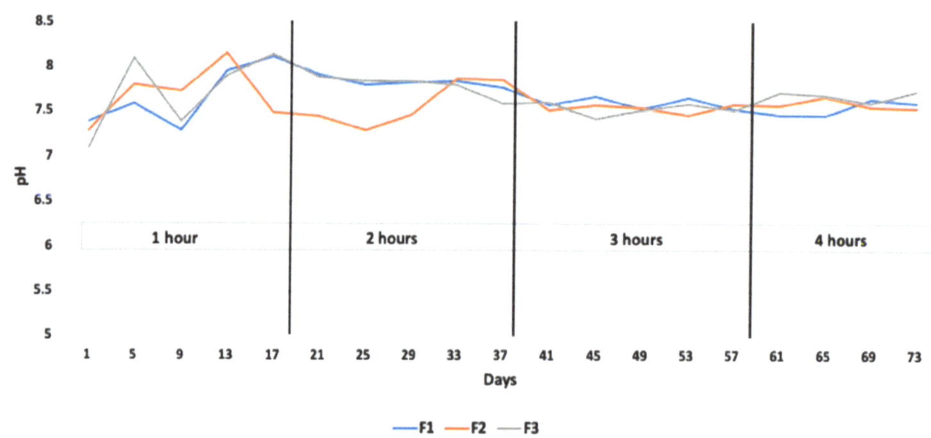

Figure 5. pH levels during the various mixing intervals throughout the experiment.

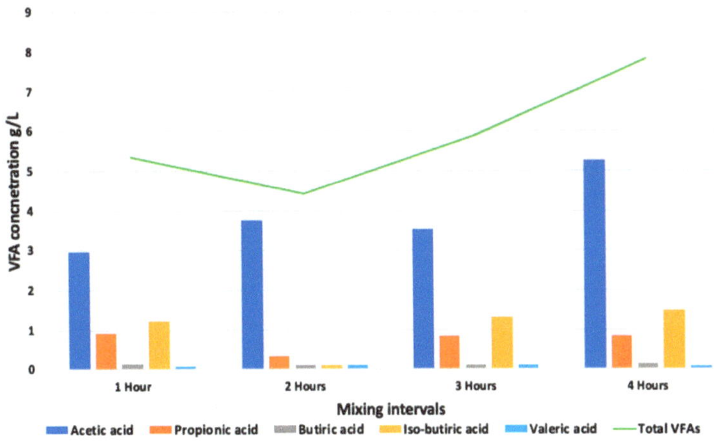

Figure 6. VFA concentration at various mixing regimes.

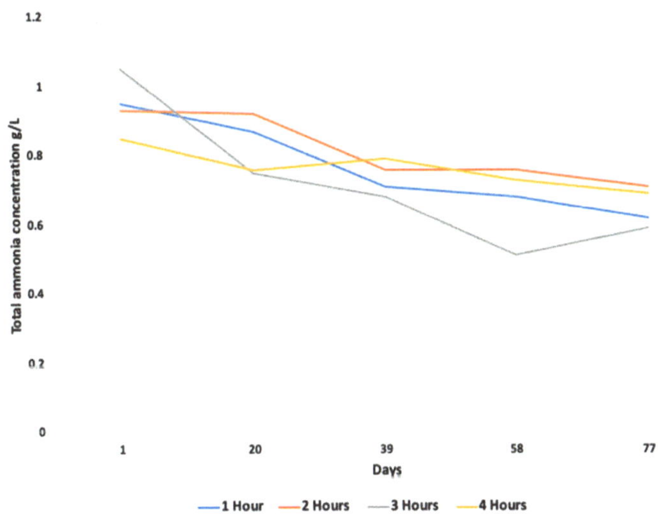

Figure 7. Total ammonia concentrations during the entire experiment.

The concentrations of VFA and alkalinity, as well as the corresponding ratios of VFA-to-alkalinity (FOS/TAC), were used to assess the system's stability (Table 5). The average FOS/TAC was recorded as 0.39, which was reported below the threshold value of 0.5 for a stable process to avoid the failure of digesters during transient conditions [38]. As a result, the startup with stable digesters was deemed successful prior to commencing varied mixing intervals to prevent the impact of shock loading. Greater alkalinity resulted in increased biogas generation. This outcome was predicted since the digestive environment alkalinity was more conducive to the AD process. Furthermore, a rise in the VFA concentration resulted in an increase in the pH. The average pH for the VFA concentration of 5.8 g L^{-1} was 7.6, whereas the average pH for the VFA concentration of 7.79 g L^{-1} was 7.9.

3.2.4. System Mixing Intensity for Semi-Continuous Mixing Mode

The importance of efficient sludge mixing in anaerobic digesters has been recognized as a key design criterion for full-scale anaerobic digesters. For application in the design and operation of systems incorporating mechanical mixing devices, Camp and Stein [39] coined the term velocity gradient:

$$G = [\frac{P}{\mu V}]^{1/2} \quad (1)$$

where G is the average velocity gradient, P is the power dissipation, V is the reactor volume, and μ is the liquid viscosity. For this particular design and construction of setup, the value of G is 10 S^{-1} as a slow mixing value was applied to the system by adjusting the mixing power to achieve this velocity gradient.

Due to the biochemical process of anaerobic digestion, which includes various microbes and the formation of flocs, the velocity gradient is not the only parameter which determines the overall biogas production rates but also the mixing time and the interval between the mixing regimes. Therefore, the parameter of velocity gradient mixing time integral in the case of the semi-continuous mixing mode is calculated by the following equation:

$$mixing\ intensity\ no. = G \times T_m \quad (2)$$

where T_m is the mixing time in seconds. The mixing intensity number can be accurately calculated and can be used to determine the appropriate mixing time of the impeller (Figure 8). In this case, the mixing intensity number of 72,000 is found to be the optimum mixing intensity number, which means mixing the slurry every hour for 5 min at 67 rpm can result in the highest biogas production as compared to other mixing regimes (Table 6).

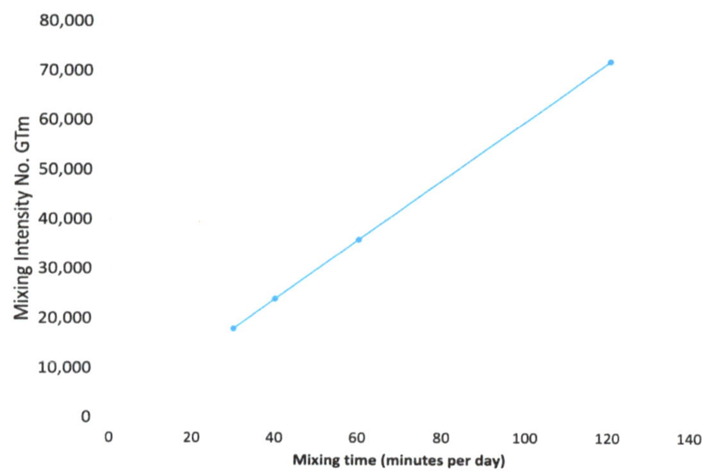

Figure 8. Relationship between mixing intensity number and mixing time.

Table 6. Comparison between total biogas production and the mixing intensity number.

Resting Time (Hours)	Mixing Time (Minutes Per Day)	Mixing Intensity No.	Total Biogas Production (L)
1	120	72,000	54.1
2	60	36,000	51.2
3	40	24,000	49.9
4	30	18,000	46.6

4. Conclusions

Three digesters were operated under identical inoculation and operating parameters with various mixing–resting intervals. It is concluded that the efficiency of the mesophilic digester is directly associated with the mixing–resting time interval. The mixing regime has an effect on the physicochemical properties of the substrate. The digester performance was best under the minimum resting time of 1 h at 67 rpm impeller speed in this system. During the employed mixing regime, the biogas yield was 5–12% higher as compared to longer resting times. The FOS/TAC ratio was below 0.5, and the VS reduction was noted as 66.1 $\pm$ 2.6%. The drop in biogas yield could be due to VFA accumulation to some extent, along with the formation of floating layers and sedimentation at a longer resting time between the mixing operations. The appropriate agitation interval might not only accomplish high biogas generation but also boost the energy efficiency of the process. The findings can be used to run an anaerobic digester in an efficient and cost-effective manner.

Author Contributions: Conceptualization, B.S., K.L.K.; methodology B.S., Z.S. (Zoltán Siménfalvi), Z.S. (Zoltán Szamosi); formal analysis, B.S., K.L.K., Z.B., M.P., G.L.S.; writing—original draft preparation, B.S., K.L.K., supervision, Z.S. (Zoltán Siménfalvi), Z.S. (Zoltán Szamosi). All authors have read and agreed to the published version of the manuscript.

Funding: The support by the Hungarian Research Fund OTKA to ZB (FK123902) is gratefully acknowledged. Additional funding from the National Research, Development and Innovation Office (NKFIH), Hungary was provided through the projects NKFIH-2020-1.1.2-PIACI-KFI-2020-00117 and NKFIH-INNOV-397-13/PALY-2020.

Institutional Review Board Statement: Not applicable.

Informed Consent Statement: Not applicable.

Data Availability Statement: Not applicable.

Conflicts of Interest: The authors declare no conflict of interest.

References

1. Suryawanshi, P.C.; Chaudhari, A.B.; Kothari, R.M. Thermophilic anaerobic digestion: The best option for waste treatment. *Crit. Rev. Biotechnol.* **2010**, *30*, 31–40. [CrossRef] [PubMed]
2. Wirth, R.; Kovács, E.; Maróti, G.; Bagi, Z.; Rákhely, G.; Kovács, K.L. Characterization of a biogas-producing microbial community by short-read next generation DNA sequencing. *Biotechnol. Biofuels* **2012**, *5*, 41. [CrossRef]
3. Bagi, Z.; Ács, N.; Bálint, B.; Horváth, L.; Dobó, K.; Perei, K.R.; Rákhely, G.; Kovács, K.L. Biotechnological intensification of biogas production. *Appl. Microbiol. Biotechnol.* **2007**, *76*, 473–482. [CrossRef]
4. Kovács, E.; Wirth, R.; Maróti, G.; Bagi, Z.; Nagy, K.; Minárovits, J.; Rákhely, G.; Kovács, K.L. Augmented biogas production from protein-rich substrates and associated metagenomic changes. *Bioresour. Technol.* **2015**, *178*, 254–261. [CrossRef]
5. Ghanimeh, S.A.; Al Saniour a, D.N.; Saikaly, P.E.; El-Fadel, M. Correlation between system performance and bacterial composition under varied mixing intensity in thermophilic anaerobic digestion of food waste. *J. Environ. Manag.* **2018**, *206*, 472–481. [CrossRef]
6. Singh, B.; Szamosi, Z.; Siménfalvi, Z. State of the art on mixing in an anaerobic digester: A review. *Renew. Energy* **2019**, *141*, 922–936. [CrossRef]
7. Naegele, H.-J.; Lemmer, A.; Oechsner, H.; Jungbluth, T. Electric Energy consumption of the full scale research biogas plant "Unterer Lindenhof": Results of longterm and full detail measurements. *Energies* **2012**, *5*, 5198–5214. [CrossRef]
8. Lemmer, A.; Naegele, H.-J.; Sondermann, J. How efficient are agitators in biogas digesters? determination of the efficiency of submersible motor mixers and incline agitators by measuring nutrient distribution in full-scale agricultural biogas digesters. *Energies* **2013**, *6*, 6255–6273. [CrossRef]
9. Singh, B.; Szamosi, Z.; Siménfalvi, Z. Impact of mixing intensity and duration on biogas production in an anaerobic digester: A review. *Crit. Rev. Biotechnol.* **2020**, *40*, 508–521. [CrossRef]
10. Karaeva, J.V.; Khalitova, G.R.; Kovalev, D.A.; Trakhunova, I.A. Study of the Process of Hydraulic Mixing in Anaerobic Digester of Biogas Plant. *Chem. Process Eng -Inz. Chem. i Proces.* **2015**, *36*, 101–112. [CrossRef]
11. Monteith, H.; Stephenson, J. Mixing efficiencies in full-scale anaerobic digesters by tracer methods. *Water Pollut. Control Fed.* **1981**, *53*, 78–84. Available online: http://www.jstor.org/stable/25041020 (accessed on 6 August 2022). [CrossRef]
12. Wang, B.; Björn, A.; Strömberg, S.; Nges, I.A.; Nistor, M.; Liu, J. Evaluating the influences of mixing strategies on the Biochemical Methane Potential test. *J. Environ. Manag.* **2017**, *185*, 54–59. [CrossRef] [PubMed]
13. Kaparaju, P.; Buendia, I.; Ellegaard, L.; Angelidakia, I. Effects of mixing on methane production during thermophilic anaerobic digestion of manure: Lab-scale and pilot-scale studies. *Bioresour. Technol.* **2008**, *99*, 4919–4928. [CrossRef] [PubMed]

14. Ong, H.K.; Greenfield, P.F.; Pullammanappallil, P.C. Effect of mixing on biomethanation of cattle-manure slurry. *Environ. Technol.* **2002**, *23*, 1081–1090. [CrossRef]
15. Tian, L.; Zou, D.; Yuan, H.; Wang, L.; Zhang, X.; Li, X. Identifying proper agitation interval to prevent floating layers formation of corn stover and improve biogas production in anaerobic digestion. *Bioresour. Technol.* **2015**, *186*, 1–7. [CrossRef] [PubMed]
16. McLeod, J.D.; Othman, M.Z.; Parthasarathy, R. Process intensification of anaerobic digestion: Influence on mixing and process performance. *Bioresour. Technol.* **2019**, *274*, 533–540. [CrossRef]
17. Subramanian, B.; Pagilla, K.R. Anaerobic digester foaming in full-scale cylindrical digesters – Effects of organic loading rate, feed characteristics, and mixing. *Bioresour. Technol.* **2014**, *159*, 182–192. [CrossRef] [PubMed]
18. Kress, P.; Nägele, H.-J.; Oechsner, H.; Ruile, S. Effect of agitation time on nutrient distribution in full-scale CSTR biogas digesters. *Bioresour. Technol.* **2018**, *247*, 1–6. [CrossRef] [PubMed]
19. Lewis, K.; Hughes, W. Optimisation of Methane Production from Anaerobically Digested Cow Slurry Using Mixing Regime and Hydraulic Retention. Ph.D. Thesis, University of Exeter, Exeter, UK, 2015.
20. Liu, Y.; Tay, J.-H. The essential role of hydrodynamic shear force in the formation of biofilm and granular sludge. *Water Res.* **2002**, *36*, 1653–1665. [CrossRef] [PubMed]
21. Lebranchu, A.; Delaunay, S.; Marchal, P.; Blanchard, F.; Pacaud, S.; Fick, M.; Olmos, E. Impact of shear stress and impeller design on the production of biogas in anaerobic digesters. *Bioresour. Technol.* **2017**, *245*, 1139–1147. [CrossRef]
22. Jiang, J.; Wu, J.; Poncin, S.; Li, H.Z. Effect of hydrodynamic shear on biogas production and granule characteristics in a continuous stirred tank reactor. *Process Biochem.* **2016**, *51*, 345–351. [CrossRef]
23. Hoffmann, R.A.; Garcia, M.L.; Veskivar, M.; Karim, K.; Al-Dahhan, M.H.; Angenent, L.T. Effect of shear on performance and microbial ecology of continuously stirred anaerobic digesters treating animal manure. *Biotechnol. Bioeng.* **2007**, *100*, 38–48. [CrossRef] [PubMed]
24. Amiraftabi, M.; Khiadani, M.; Mohammed, H.A. Performance of a dual helical ribbon impeller in a two-phase (gas-liquid) stirred tank reactor. *Chem. Eng. Process. Process Intensif.* **2020**, *148*, 107811. [CrossRef]
25. Bose, R.S.; Chowdhury, B.; Zakaria, B.S.; Tiwari, M.K.; Dhar, B.R. Significance of different mixing conditions on performance and microbial communities in anaerobic digester amended with granular and powdered activated carbon. *Bioresour. Technol.* **2021**, *341*, 125768. [CrossRef] [PubMed]
26. Lindmark, J.; Thorin, E.; Bel Fdhila, R.; Dahlquist, E. Effects of mixing on the result of anaerobic digestion: Review. *Renew. Sustain. Energy Rev.* **2014**, *40*, 1030–1047. [CrossRef]
27. Önen, S.; Nsair, A.; Kuchta, K. Innovative operational strategies for biogas plant including temperature and stirring management. *Waste Manag. Res.* **2019**, *37*, 237–246. [CrossRef]
28. Kowalczyk, A.; Harnisch, E.; Schwede, S.; Gerber, M.; Span, R. Different mixing modes for biogas plants using energy crops. *Appl. Energy* **2013**, *112*, 465–472. [CrossRef]
29. Singh, B.; Singh, N.; Čonka, Z.; Kolcun, M.; Siménfalvi, Z.; Péter, Z.; Szamosi, Z. Critical analysis of methods adopted for evaluation of mixing efficiency in an anaerobic digester. *Sustainability* **2021**, *13*, 6668. [CrossRef]
30. Singh, B.; Kovács, K.L.; Bagi, Z.; Nyári, J.; Szepesi, G.L.; Petrik, M.; Siménfalvi, Z.; Szamosi, Z. Enhancing efficiency of anaerobic digestion by optimization of mixing regimes using helical ribbon impeller. *Fermentation* **2021**, *7*, 251. [CrossRef]
31. Kovács, K.L.; Ács, N.; Kovács, E.; Wirth, R.; Rákhely, G.; Strang, O.; Herbel, Z.; Bagi, Z. Improvement of biogas production by bioaugmentation. *BioMed Res. Int.* **2013**, *2013*, 482653. [CrossRef]
32. Bryant, M.P. Microbial methane production—Theoretical aspects. *J. Anim. Sci.* **1979**, *48*, 193–201. [CrossRef]
33. Latha, K.; Velraj, R.; Shanmugam, P.; Sivanesan, S. Mixing strategies of high solids anaerobic co-digestion using food waste with sewage sludge for enhanced biogas production. *J. Clean. Prod.* **2019**, *210*, 388–400. [CrossRef]
34. Caillet, H.; Madyira, D.M.; Adelard, L. Study of the performances of a vinasse mesophilic anaerobic digester behavior submitted to intermittent mixing: Monitoring of the physicochemical properties of the digestate and local samples of the digester. *Bioresour. Technol. Rep.* **2021**, *16*, 100837. [CrossRef]
35. Nandi, R.; Saha, C.K.; Alam, M.M. Effect of mixing on biogas production from cowdung. *Eco-friendly Agril. J.* **2017**, *10*, 7–13.
36. Wang, Y.; Dentel, S. The effect of high speed mixing and polymer dosing rates on the geometric and rheological characteristics of conditioned anaerobic digested sludge (ADS). *Water Res.* **2010**, *44*, 6041–6052. [CrossRef] [PubMed]
37. Franke-Whittle, I.H.; Walter, A.; Ebner, C.; Insam, H. Investigation into the effect of high concentrations of volatile fatty acids in anaerobic digestion on methanogenic communities. *Waste Manag.* **2014**, *34*, 2080–2089. [CrossRef]
38. Ferrer, I.; Vázquez, F.; Font, X. Long term operation of a thermophilic anaerobic reactor: Process stability and efficiency at decreasing sludge retention time. *Bioresour. Technol.* **2010**, *101*, 2972–2980. [CrossRef] [PubMed]
39. Camp, T.R. Velocity gradients and internal work in fluid motion. *J. Bost. Soc. Civ. Eng.* **1943**, *30*, 219–230.

fermentation

MDPI

Article

Degradation Kinetics of Lignocellulolytic Enzymes in a Biogas Reactor Using Quantitative Mass Spectrometry

Jan Küchler [1,†], Katharina Willenbücher [2,3,†], Elisabeth Reiß [1], Lea Nuß [1], Marius Conrady [4,5], Patrice Ramm [4], Ulrike Schimpf [6], Udo Reichl [1,7], Ulrich Szewzyk [3] and Dirk Benndorf [1,7,8,*]

1 Bioprocess Engineering, Max Planck Institute for Dynamics of Complex Technical Systems, 39106 Magdeburg, Germany
2 System Microbiology, Department Bioengineering, Leibniz Institute for Agricultural Engineering and Bioeconomy (ATB), Max-Eyth-Allee 100, 14469 Potsdam, Germany
3 Environmental Microbiology, Institute of Environmental Technology, Faculty of Process Sciences, Technische Universität Berlin, Ernst-Reuter-Platz 1, 10587 Berlin, Germany
4 Institute of Agricultural and Urban Ecological Projects, Berlin Humboldt University (IASP), Philippstr. 13, 10115 Berlin, Germany
5 Biosystems Engineering Division, Humboldt Universität zu Berlin, Albrecht-Thaer-Weg 3, 14195 Berlin, Germany
6 Division of Glycoscience, Department of Chemistry, School of Engineering Science in Chemistry, Biotechnology and Health (CBH), Alba Nova University Center, Royal Institute of Technology (KTH), Roslagstullsbacken 21, 10691 Stockholm, Sweden
7 Bioprocess Engineering, Otto von Guericke University Magdeburg, Universitätsplatz 2, 39106 Magdeburg, Germany
8 Applied Biosciences and Process Engineering, Anhalt University of Applied Sciences, Bernburger Straße 55, 06366 Köthen, Germany
* Correspondence: benndorf@mpi-magdeburg.mpg.de
† These authors contributed equally to this work.

Citation: Küchler, J.; Willenbücher, K.; Reiß, E.; Nuß, L.; Conrady, M.; Ramm, P.; Schimpf, U.; Reichl, U.; Szewzyk, U.; Benndorf, D. Degradation Kinetics of Lignocellulolytic Enzymes in a Biogas Reactor Using Quantitative Mass Spectrometry. *Fermentation* **2023**, *9*, 67. https://doi.org/10.3390/fermentation9010067

Academic Editor: Sanjay Nagarajan

Received: 14 December 2022
Revised: 5 January 2023
Accepted: 5 January 2023
Published: 12 January 2023

Abstract: The supplementation of lignocellulose-degrading enzymes can be used to enhance the performance of biogas production in industrial biogas plants. Since the structural stability of these enzyme preparations is essential for efficient application, reliable methods for the assessment of enzyme stability are crucial. Here, a mass-spectrometric-based assay was established to monitor the structural stability of enzymes, i.e., the structural integrity of these proteins, in anaerobic digestion (AD). The analysis of extracts of *Lentinula edodes* revealed the rapid degradation of lignocellulose-degrading enzymes, with an approximate half-life of 1.5 h. The observed low structural stability of lignocellulose-degrading enzymes in AD corresponded with previous results obtained for biogas content. The established workflow can be easily adapted for the monitoring of other enzyme formulations and provides a platform for evaluating the effects of enzyme additions in AD, together with a characterization of the biochemical methane potential used in order to determine the biodegradability of organic substrates.

Keywords: mass spectrometry; biogas production; fungal enzymes; lignocellulose conversion; *Lentinula edodes*

1. Introduction

Biogas reactors produce renewable energy by degrading agricultural crops, animal waste, livestock residues, and other industrial by-products. Biogas production should preferably be carried out with by-products and waste products in order to reduce the use of energy crops and increase the sustainability of the process. Through anaerobic digestion (AD), organic matter is converted into biogas, composed mainly of methane and carbon dioxide. The AD process consists of distinct steps and is catalyzed by complex microbial communities. In the first step, biomass is broken down by bacteria and archaea into amino acids, sugars, and shorter-chain carbohydrates. This hydrolysis is hampered by the recalcitrant structure of lignocellulosic substrates, composed of cellulose, hemicellulose, pectin, lignin, extractives, and various inorganic materials [1–3]. Therefore, long retention times are required for the lignocellulose-rich substrates in the anaerobic digester [3]. Although, bacteria are capable of synthesizing hydrolytic enzymes, such as cellulases and hemicellulases. However, even with a longer retention time, lignin cannot be degraded under anaerobic conditions [2,4]. Physical, chemical, and biological pre-treatments, such as acid hydrolysis and steam explosion [3], as well as enzymatic, fungal, and bacterial pre-treatment to enhance substrate conversion, are the focus of numerous investigations [5]. Treatment or pre-treatment with enzymes is especially advantageous as this accelerates the process [6,7].

A promising strategy to improve the degradation of lignocellulosic biomass is the addition of enzymes derived from cellulose-consuming fungi. The application of fungal enzymes is advantageous because they can be extracted from the by-products of industrial mushroom cultivation, resulting in sustainable processes with low costs [8]. However, Binner et al. [9] have shown that externally added enzymes are often structurally unstable under the conditions required for AD.

The effect of enzyme supplementation is often quantified by measuring the biochemical methane potential (BMP). Here, for a more comprehensive evaluation of the effects of external enzymes on AD, the structural stability of fungal enzymes in AD was investigated using quantitative mass spectrometry. Therefore, a preparation from the edible fungus *L. edodes*, containing lignocellulolytic enzymes, such as cellulase, xylanase, and pectinase, was investigated.

2. Materials and Methods

2.1. Bioreactor Set-Up and Sampling

Two identical continuous stirred tank reactors (CSTRs) were operated to evaluate the structural stability of the supplemented enzymes in the biogas reactor medium. The CSTRs had a total volume of 24 L, with a working volume of approximately 16.5 L. Operation of the CSTRs was described in a previous study on the semi-continuous AD of whole-crop cereal silage (wheat: rye [1:1], DAH Energie, Oberkrämer, Germany) at an organic loading rate of 3.5 g of organic dry mass (ODM) L^{-1} d^{-1} under mesophilic conditions at 38 °C [10]. During this experiment, 75 μL of fungal enzyme preparation was added to the biogas reactor per L reactor volume, daily. The inoculum for the reactor start-up was obtained from an agricultural biogas plant (Kaim Agrar-Energie, Nauen, Germany). The collection of reactor content for subsequent experiments was conducted six months after inoculation under steady-state conditions and 21 h after the addition of substrate. The mean biogas yields for both CSTRs were 0.63 L g^{-1} ODM and 0.60 L g^{-1} ODM, respectively, with a methane percentage of 54%. The reactor content was frozen in liquid nitrogen immediately after sampling and stored at −20 °C until further analysis.

2.2. Enzyme Preparation

Partially purified lignocellulolytic enzymes obtained from the by-products of industrial mushroom cultivation were prepared as previously described [8] with minor modifications. Briefly, a spent mushroom substrate from the cultivation of *L. edodes* was chopped after the fruiting bodies were harvested. An enzyme extract was generated by watering, pressing, and filtering the chopped residues. The resulting liquid (pressed juice) was centrifuged at 4 °C for 5 min at 20,000× *g* (Sigma Zentrifugen, Osterode, Germany). The supernatant was concentrated ten-fold using tangential flow filtration (Lab scale TFF System, Merck, Darmstadt, Germany) with a 10 kDa MWCO membrane (Biomax, Merck, Darmstadt, Germany). The concentrated enzyme extract was supplemented with maltodextrin and sodium benzoate to final concentrations of 4% and 0.5% for conservation, respectively, and stored at −20 °C.

2.3. *Addition of Fungal Enzymes*

To assess the structural stability of the fungal enzymes in the preparation, various dilutions were added to 2 mL of reactor content. The volume of fungal enzymes was set to 0.015 µL (factor 1), 0.15 µL (factor 10), 1.5 µL (factor 100), and 15 µL (factor 1000). Factor 1 corresponds to the amount of enzyme added to the reactors during the experiment as described above, whereas the multiplied factors (10-, 100-, and 1000-fold) were added in order to determine the sensitivity of the quantitative mass spectrometry and to provide stronger signals for measuring the structural stability of the fungal enzymes. A control sample of the biogas reactor medium (no treatment), was served as a blank. After adding the fungal enzyme preparation, the bioreactor medium was incubated at 37 °C and sampled after 0, 4, 10, and 24 h.

2.4. *Sample Preparation*

2.4.1. Protein Precipitation

The phenol extraction was performed according to the protocol of Heyer et al. [11]. In brief, a 2 mL sample was added into a 15 mL tube with 5 g of silica beads (Ø = 0.5 mm), 2 mL of sucrose solution, and 3.5 mL of phenol solution (10 g of phenol dissolved in 1 mL of ultrapure water). Then, the proteins were extracted by adding an equal volume of trichloroacetic acid (20%, *w*/*v*, Sigma Aldrich) to the samples. After incubation for 1 h at 4 °C, the samples were centrifuged (10 min, 16,400× *g*, 4 °C) (Micro Star 17R, VWR, Darmstadt, Deutschland) and precipitated twice with 1.5 mL of ice-cold acetone (80% *v*/*v*, 99.8%) and ethanol (70%, *v*/*v*, 99.8%) for 15 min, respectively. The dried pellets were resuspended in 200 µL of urea buffer (8 M of urea (Applichem) in 0.1 M of Tris-HCl, pH 8.5).

2.4.2. Protein Quantification

The protein concentrations were determined in triplicates using a modified amido black assay as reported previously [11]. Briefly, 300 µL of amido black solution was added to 50 µL of the sample and mixed. After a centrifugation step (5 min, RT, 16,400× *g*), the supernatant was discarded, and the pellets were washed twice with 10% acetic acid in methanol in order to remove the unbound staining solution. The pellets were then dissolved in 0.1 M of sodium hydroxide and the absorbance was measured at 615 nm using a spectrophotometer (Genesys 10S UV-Vis spectrophotometer, Thermo Scientific, Waltham, MA, USA).

2.5. *Tryptic Digestion*

Tryptic digestion of the proteins was performed according to a modified filter-assisted sample preparation protocol described elsewhere [12]. For this purpose, a sample volume equivalent to 100 µg of protein was diluted with 8 M of urea to a total volume of 200 µL. The samples were then loaded onto a 10 kDa filter (Pall Nanosep, VWR, 516-8492) in a 1.5 mL reaction tube. After washing with 200 µL of urea buffer, the samples were treated with 40 mM of dithiothreitol (20 min, 300 rpm, 56 °C) and 0.55 M of iodoacetamide (20 min, 300 rpm, room temperature (RT), in the dark) and centrifuged (10,000× *g*, 10 min, RT, also for subsequent steps). The filter was washed once with 100 µL of urea buffer and three times with 50 mM of ammonium bicarbonate. For subsequent tryptic digestion, 200 µL of trypsin solution (2.5 µg mL^{-1} trypsin in ammonium bicarbonate buffer; Trypsin mass spectrometric (MS)-approved, Serva) was added to the filter. The samples were then incubated at 37 °C and 300 rpm for 2 h and centrifuged. The samples were washed with 50 µL of extraction buffer (50 mM of ammonium bicarbonate +5% LC-MS grade acetonitrile) and 50 µL of water (LC-MS grade). The flow-through from the trypsin incubation and the two subsequent washing steps was collected and 30 µL of each was transferred to separate tubes for acidification. Three microliters of trifluoroacetic acid (99.99%, *w/v*, Sigma Aldrich) were added and the samples were centrifuged (10 min, 4 °C, 10,000× *g*) (Micro Star 17R) before being transferred to vials for mass spectrometric measurements.

2.6. Chromatography and Mass Spectrometry

Analysis of the samples was performed using a liquid chromatography (LC)-coupled mass spectrometry system. Three microliters of each sample were injected and separated by an UltiMate® 3000 nano splitless reversed phase nanoHPLC (Thermo Fisher Scientific), equipped with a reversed-phase trap column (nano trap cartridge, 300 μm i.d. × 5 mm, packed with Acclaim PepMap100 C18, 5 μm, 100 Å, nanoViper) and a reversed-phase separation column (Acclaim PepMap RSLC, C18, 2 μm, 75 μm, 50 cm). The gradient was set from 5 to 35% mobile phase B (LC-MS grade acetonitrile, 0.1% formic acid, (99% *v/v*)) at a flow rate of 0.4 μL min^{-1}. The LC was coupled with a timsTOF Pro mass spectrometer (Bruker Daltonik GmbH), equipped with a captive spray ionization source operated in a positive ion mode with a capillary voltage of 1400 V and a capillary temperature of 200 °C. In total, three different modes were used for the MS measurements: Parallel Accumulation–Serial Fragmentation (PASEF) with a 120 min gradient, a Sequential Window Acquisition of all Theoretical Mass Spectra (SWATH-MS) [13], and multiple reaction monitoring mode (MRM) with a 30 min LC gradient. The spectra acquisition rate was set to 7.5 Hz for SWATH-MS and 4 Hz for MRM. For SWATH-MS, the scan range for MS1 was set to 400–1000 m/z with 24 isolation windows of 26 m/z (1 m/z overlap). For all three modes, the resulting total cycle time for a complete scan of the mass range was 3.2 s. The collision energy ranged from 27 to 48 eV (slope = 0.042; intercept = 9.45). The scan range for MS2 was set from 150 to 2200 m/z.

2.7. Data Analysis

The PASEF data were processed using the spectrometry-based pro software CompassData Analysis (vs. 5.3, Bruker Daltonik). The mgf files were searched with MASCOT against a database containing the Uniprot genome of the fungus *L. edodes*. For the initial analysis, the open-source software Skyline (vs. 19.1) [14] was used to analyze the raw files from the MRM and SWATH-MS measurements. In order to validate the spectra, the dat files from the MASCOT searches were used to set-up a spectral library containing peptide spectra of *L. edodes*. The parameters of the software were set to "enzyme = trypsin", "maximal missed cleavages = 0", and "structural modifications = carbamidomethyl (cysteine)". The peptide areas were exported as CSV files and processed into pivot tables with excel.

In order to compare the MRM and SWATH-MS results and to determine the half-life constant of kinetic degradation for ten selected peptides of the fungal enzymes, a limit of detection (LoD) and a limit of quantification (LoQ) was introduced. The calculations were based on the peak area of blank samples (factor 0, time point 0) for every peptide for the measured duplicates. The LoD was determined as the sum of the average peak area of the two replicates (xmean) and the standard deviation (ysd) times 3.29. In order to determine the LoQ, the standard deviation was multiplied by ten, as described previously [15,16].

$$\text{LoD} = \text{xmean} + 3.29 \cdot \text{ysd} \quad (1)$$

$$\text{LoQ} = \text{xmean} + 10 \cdot \text{ysd} \quad (2)$$

In order to quantify the degradation of the enzymes, the peak area values from the MS measurements were fitted by first-order kinetics. From this, the values for the half-life (t1/2) of each selected peptide were calculated using the formula for exponential degradation and the decay constant of the regression fit (b):

$$t1/2 = \ln(2)/b \quad (3)$$

3. Results and Discussion

3.1. Detection of Enzyme Proteins

The addition of lignocellulose-degrading enzymes for the improvement of biogas production processes has been described previously in the literature and was tested in various experimental studies [9,17,18]. Since the structural stability of enzyme preparations is crucial for efficient application, reliable methods for assessing the structural stability of these proteins are needed. In contrast to studies that have been conducted in this field that used mainly semi-quantitative techniques [9,19], this study aimed to monitor the degradation kinetics of the lignocellulolytic enzymes of fungal origin using an MS-based approach. Consequently, the first step was to develop a quantitative MS-based assay that would allow the detection of enzymes in AD samples.

In order to investigate the sensitivity of our workflow, the fungal enzymes were added to samples of the biogas reactor content at different concentrations. For the first assessment, standard PASEF measurements and subsequent database searches against the genome of *L. edodes* were carried out. Although the detection of fungal enzymes with cellulolytic activity failed in the samples representing working concentrations in the bioreactor (factor 1, factor 10), higher concentrations, in particular factor 1000, enabled the detection of proteins from *L. edodes* (Table 1 and Figure S1). In contrast, hydrolytic enzymes from bacteria (glucanase, xylanase) were detected with high abundance in the biogas reactors [20] using the applied workflow [11] combining phenol extraction and mechanical disruption in the bead mill. The low abundance of fungal proteins in highly sensitive PASEF measurements, as well as the limitation of PASEF in quantification using spectral counts, required the setup of SWATH-MS and MRM techniques as more sensitive and quantitative MS methods.

Table 1. Detected proteins of the fungal enzyme preparation. MASCOT search results are shown as representatives for the factor 1000 sample at 0 h after the addition of the enzyme preparation against the UniProt_Lentinula_Edodes database.

Acc	Description	Score	Mass	Matches	Sequences
A0A1Q3EJ89	A0A1Q3EJ89_LENED Glucanase OS = *Lentinula edodes* OX = 5353 GN = LENED_009211 PE = 3 SV = 1	1028	55,356	33	4
A0A1Q3EU06	A0A1Q3EU06_LENED Glucanase OS = *Lentinula edodes* OX = 5353 GN = LENED_012880 PE = 3 SV = 1	636	62,086	21	6
A0A1Q3E856	A0A1Q3E856_LENED Subtilisin-like protein OS = *Lentinula edodes* OX = 5353 GN = LENED_004953 PE = 4 SV = 1	620	63,032	26	6
A0A1Q3ENW5	A0A1Q3ENW5_LENED Beta-xylanase OS = *Lentinula edodes* OX = 5353 GN = LENED_010993 PE = 3 SV = 1	389	120,358	13	4
A0A1Q3EGI8	A0A1Q3EGI8_LENED Glycoside hydrolase family 55 protein OS = *Lentinula edodes* OX = 5353 GN = LENED_008227 PE = 4 SV = 1	385	67,565	9	4
Q9C1R4	Q9C1R4_LENED Glucanase OS = *Lentinula edodes* OX = 5353 GN = cbhII-1 PE = 2 SV = 1	324	46,832	13	3
A0A1Q3EBY2	A0A1Q3EBY2_LENED Glycoside hydrolase family 5 protein OS = *Lentinula edodes* OX = 5353 GN = LENED_006487 PE = 3 SV = 1	318	43,559	8	2
A0A0A1I5X1	A0A0A1I5X1_LENED Glycoside hydrolase family 5 endoglucanase (Fragment) OS = *Lentinula edodes* OX = 5353 GN = glu PE = 3 SV = 1	231	8509	5	1
A0A1Q3DWF6	A0A1Q3DWF6_LENED Cu-oxidase-domain containing protein OS = *Lentinula edodes* OX = 5353 GN = LENED_000509 PE = 3 SV = 1	273	57,232	14	3
A0A1Q3EAX5	A0A1Q3EAX5_LENED Glycoside hydrolase family 12 protein OS = *Lentinula edodes* OX = 5353 GN = LENED_006131 PE = 3 SV = 1	237	26,422	7	3

3.2. Selection of Signature Peptides and Validation of the Assay

Based on the detected fungal enzymes in the AD samples using PASEF at factor of 1000, the discovered proteins were filtered for peptides specific to the lignocellulolytic enzymes. We focused on the technical criteria: detectability, mainly intensity and signal-to-noise ratio. In addition, we aimed for a wide coverage of the different families of hydrolytic enzymes. Finally, ten peptides were selected for further investigation of the degradation dynamics based on the initial SWATH-MS measurements (Table 2).

Table 2. Selected peptides of the given enzyme preparation in this study. The enzyme preparation (from the fungus *L. edodes*) was added to the biogas reactor content at different concentrations and incubated at 37 °C for 24 h. The resulting mass spectra were matched with the proteome of *L. edodes*.

Protein	ID	Peptide Sequence	Abbreviation	*m*/*z*	Collision Energy [eV]	LoD MRM	LoQ MRM	LoD SWATH-MS	LoQ SWATH-MS
Glycoside Hydrolase Family 16 Protein	A0A1Q3DXQ1	ADFTTILDPNGPGR	ADF	737.3703	42	9265.24	20,818.55	13,190.26	32,121.56
Carbohydrate Esterase Family 15 Protein	A0A1Q3EF14	IALTIPQESGSGGDAGWR	IAL	907.9552	48	4319.99	10,469.10	10,078.47	26,025.36
Glycoside Hydrolase Family 5 endoglucanase (EC 3.2.1.4) (fragment)	A0A0A1I5X1	LADATSWLQSTGIK	LAD	745.8961	42	11,100.85	29,059.49	25,864.80	183,437.42
Glucanase (EC 3.2.1.-)	Q96VU3 Q9C1R6	MGDTSFYGPGLTVDTTSK VANIPTFIWLDQVAK	MGD VAN	938.9353 857.9800	48 45	5227.55 7526.52	11,224.83 19,345.53	12,511.12 19,870.98	32,676.04 49,686.59
Glycoside Hydrolase Family 5 Protein	A0A1Q3EBY2	LPFLLER	LPF	444.2711	27	37,104.11	92,768.77	39,710.62	105,827.83
Carbohydrate Esterase Family 16 Protein	A0A1Q3DZ58	SFLVVDVYGR	SFL	577.8139	31	24,580.91	60,929.95	83,059.27	221,305.20
Family S53 Protease (Kinesin-like Protein)	A0A1Q3DXE6	TDISSATTFTLQTLDGGSDPQAA	TDI	1227.093	48	3451.58	8628.03	25,740.59	65,377.70
Cupper Radical Oxidase	A0A1Q3E003	VQFLNPPFLSR	VQL	659.3693	39	17,981.82	37,553.65	30,637.34	80,912.03
Beta-Mannosidase (EC 3.2.1.25)	A0A1Q3E4F1	GSNLVPFDPFYSR	GSN	749.8699	42	9253.63	22,135.45	48,250.42	127,927.94

To facilitate the detection of the selected peptides, SWATH-MS and MRM were tested in order to maximize the signal-to-noise ratio and intensity. The signal-to-noise ratio and intensity were 5-fold higher in the MRM mode than in SWATH-MS, resulting in an overall higher sensitivity compared to SWATH-MS (Figure 1, Figures S1 and S2). Furthermore, the linearity of the peak area ranged from a factor of 1 to a factor of 1000 in the MRM mode, whereas in the SWATH-MS mode, the linearity ranged from factor 10 to 1000 only (Figure S1). These results are in accordance with previous studies showing a higher sensitivity and a better signal-to-noise ratio for MRM, especially for complex samples [21,22]. Overall, this suggested that the MRM mode was better suited to detect a small number of the selected peptides due to its precise selection of precursor ions. In contrast, SWATH-MS with widely defined mass windows for precursor selection seems to be better suited to obtain an overview of the proteins and peptides present in samples. In conclusion, SWATH-MS could support the selection of signature peptides, whereas the MRM provides maximum sensitivity.

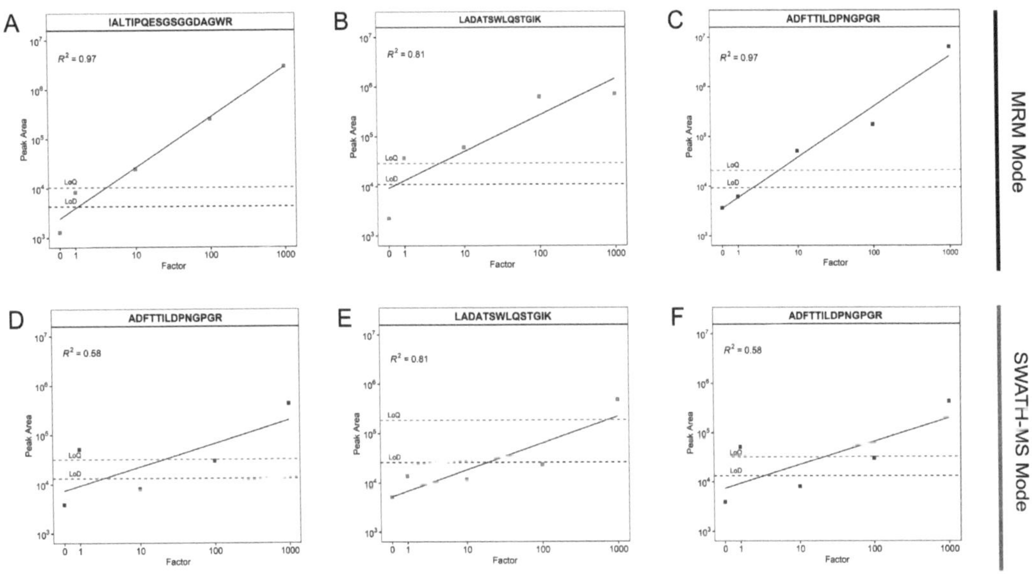

Figure 1. Correlation of MRM and SWATH-MS measurements for different dilution factors of the fungal enzyme preparation added to the biogas reactor content and incubated for 24 h. (**A–C**), correlation between the measured peak area of three selected peptides for the MRM mode; (**D–F**), SWATH-MS mode. The titles of the subplots display the peptide sequences. The correlation coefficients are shown in each graph.

3.3. *Dynamics of Enzyme Degradation*

After selecting the characteristic peptides, enzyme degradation was further investigated by quantifying these peptides over an incubation period of 24 h. The peak areas were measured for all the selected peptides in order to evaluate the structural stability of the enzyme preparation (presented for the peptide IAL in Figure 2). Immediately after the addition of the enzyme preparation (0 h), all four concentrations of peptide IAL were detectable above the LoD. After 4 h of incubation, the concentration of peptide IAL in factor 1 and factor 10 decreased below the LoD. At 24 h, the peptide IAL was only detectable with an initial concentration of the enzyme preparation that was 1000-fold higher than that applied in the bioreactor (Figure 2).

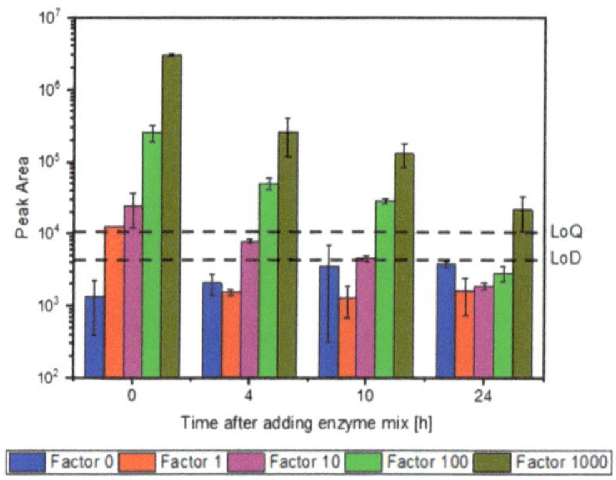

Figure 2. Degradation kinetics of the peptide IAL. The enzyme preparation was added in different dilutions to the biogas content and incubated over 24 h. IAL represents the ten selected peptides.

In order to estimate the degradation kinetics of the fungal enzymes, the peak areas of all the selected peptides were quantified in factor 1000 samples and fitted to first-order kinetics in order to calculate the half-lives (Figure 3). A high correlation coefficient for all the peptides confirmed the assumption of first-order degradation kinetics. The corresponding half-lives of the selected peptides ranged from 0.87 h to 25.66 h, with an average of 1.5 h (Table 3). These results are consistent with previous studies that investigated the structural stability of fungal enzyme preparations in a biogas reactor medium using SDS-PAGE. In particular, most of the fungal enzymes with cellulolytic activity were degraded within a few hours after their addition [9,17]. In contrast to SDS-PAGE, the mass spectrometric quantification of the fungal enzymes was superior since it provided the half-life constants for distinct enzymes with good reproducibility, as evidenced by comparable values of the two replicates (Table 3). Minor differences could be attributed to the inhomogeneity of the samples from AD even if they originate from the same reactor [23].

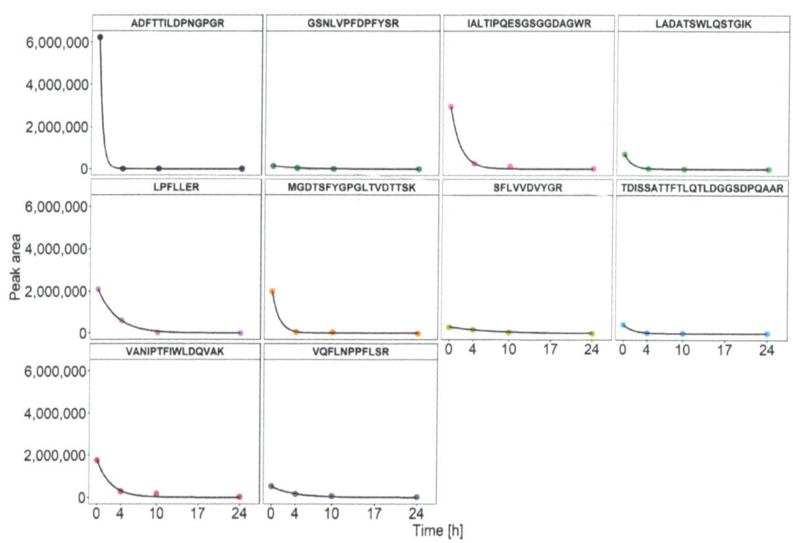

Figure 3. Degradation kinetics of the peptide IAL for factor 1000. The fungal enzyme preparation was added in different dilutions to the biogas reactor medium and incubated for 24 h. Measurements were carried out in the MRM mode. The exponential fit was performed with R.

Table 3. Half-life constants of selected peptides representing the enzyme preparation used in this study. The three peptides (LAD, VAN, and GSN) were excluded from calculations of the average half-life constants as their values significantly exceeded the range of the other peptides.

Protein	Peptide	Half-Life [h]		
		Replicate A	Replicate B	Average
Carbohydrate Esterase Family 15 Protein	IAL	1.17	1.07	1.12
Glycoside Hydrolase Family 16 Protein	ADF	0.44	0.44	0.44
Glycoside Hydrolase Family 5 endoglucanase (EC 3.2.1.4) (fragment)	LAD	22.84	22.51	22.68
Glucanase (EC 3.2.1.)	MGD	0.82	1.50	1.16
	VAN	20.62	25.66	23.14
Glycoside Hydrolase Family 5 Protein	LPF	1.69	3.10	2.40
Carbohydrate Esterase Family 16 Protein	SFL	1.01	1.57	1.29
Family S53 Protease (Kinesin-like Protein)	TDI	1.28	1.54	1.41
Copper Radical Oxidase	VQL	1.38	1.79	1.59
Beta-Mannosidase (EC 3.2.1.25)	GSN	5.27	7.27	6.27
	Average	1.17	1.78	1.48

Compared to the other peptides investigated, the VAN and LAD peptides had significantly prolonged half-live constants exceeding 20 h. LAD belongs to the glycoside hydrolase family 5, some members of which have high thermal stability due to their molecular structure [24]. Accordingly, the use of enzymes from this family could have some positive long-term effects on cellulose degradation in biogas reactors.

The selection of peptides from multiple enzymes enabled a comprehensive overview about the structural stability of many possible fungal enzymes in a complex sample. However, calculated half-life values may vary when measuring multiple peptides of a single protein [25]. Therefore, the selection of precursors should be adapted to the research question, focussing either on a small number of proteins for the calculation of precise half-life constants or on a more comprehensive set of proteins for the raw estimation of half-life values. This limitation of MRM can be circumvented by using a more recent instrument configuration and measurement mode. In particular, using the dia-PASEF mode available for the timsTOF Pro II mass spectrometer enables the combination of high sensitivity and selectivity for a large number of selected precursors.

The enzyme extract of the white-rot fungus *L. edodes* [8] has been taken as an example for evaluating the potential for degradation of lignocellulosic biomass in AD. The observed low structural stability of lignocellulose-degrading enzymes corresponded with previous results obtained for other enzyme preparations [9,17]. Moreover, it correlated with the rather poor overall effect of the addition of fungal enzymes on the sampled bioreactors, which showed comparable biogas yields. Fast degradation appears to be a more appropriate reason for the minor effects of enzyme additions than suboptimal reaction conditions [26]. Accordingly, endogenous proteases have been identified in anaerobic digesters [27] and their activity has been confirmed [17].

Nevertheless, this new workflow can easily be adapted to monitor other enzyme preparations by exchanging the selected peptides in the bioinformatic workflow. Thus, it provides a general platform for evaluating the effect of the addition of enzyme preparations on AD in addition to the BMP that is conventionally used. Furthermore, this workflow could also be applied for the quantification of enzymes of the archaeal or bacterial microbial community involved in AD, e.g., methyl coenzyme-M reductase (McrA) as a key enzyme for methanogenesis. If necessary, absolute quantification of the selected enzymes would be possible by adding peptides labelled with stable isotopes as internal standards [28].

4. Conclusions

A mass spectrometric assay was established in order to investigate the structural stability of various enzymes of *L. edodes* added to AD over a period of 24 h. Monitoring of the selected peptides using MS-based approaches clearly suggested a rapid degradation of most of the lignocellulose-degrading enzymes, with an approximate half-life of 1.5 h.

The observed low structural stability of the investigated lignocellulose-degrading enzymes in AD coincided with previous results obtained from SDS-PAGE measurements. The workflow that was established can easily be adapted for the monitoring of other enzymes and provides a platform for evaluating their effect on AD in addition to BMP. Furthermore, the platform enables the targeted absolute quantification of other key enzymes of AD, aiming for individually customized enzyme preparations.

Supplementary Materials: The following supporting information can be downloaded at: https://www.mdpi.com/article/10.3390/fermentation9010067/s1, Figure S1: Correlation of MRM measurements over different dilutions of the enzyme preparation logarithmically plotted. Given fungal enzyme preparation was added in different dilutions to biogas reactor content and incubated for 24 hours. Peak areas for all measured factors of seven peptides (A–G) are shown with their limits of detection (LoD) and limit of quantification (LoQ). Correlation coefficients are shown in each graph, respectively; Figure S2: Correlation of SWATH-MS measurements over different dilutions of the enzyme preparation logarithmically plotted. Given fungal enzyme preparation was added in different dilutions to biogas reactor content and incubated for 24 hours. Peak areas for all measured factors of seven peptides (A–G) are shown with their limits of detection (LoD) and limit of quantification (LoQ). Correlation coefficients are shown in each graph, respectively.

Author Contributions: Conceptualization, D.B.; methodology, D.B.; formal analysis, E.R., L.N., and J.K.; investigation, K.W., J.K., and D.B.; resources, U.R. and K.W.; data curation, J.K.; writing—original draft preparation, J.K. and K.W.; writing—review and editing, D.B., P.R., M.C., U.R., and U.S. (Ulrich Szewzyk); visualization, J.K. and K.W.; supervision, D.B., U.R., and U.S. (Ulrich Szewezyk); project administration, D.B.; funding acquisition, U.S. (Ulrike Schimpf) and D.B. All authors have read and agreed to the published version of the manuscript.

Funding: The operation of the sampled biogas reactor was realized within the subproject 'Application and effect of the biocatalysts' of the joint research project 'Biocatalysts in bioreactors: monitoring, control and multi-criteria optimization of biogas processes (BIOKAT)' funded by the BMEL (project carrier: FNR, grant nos. 22402516 and 22403816). Open Access funding provided by the Max Planck Society.

Institutional Review Board Statement: Not applicable.

Informed Consent Statement: Not applicable.

Data Availability Statement: https://panoramaweb.org/QidMT3.url.

Acknowledgments: The authors thank Anne Ballmann for providing excellent technical assistance regarding protein sample preparation and Marten Suhr for the graphical abstract.

Conflicts of Interest: The authors declare no conflict of interest.

References

1. Taherzadeh, M.J.; Karimi, K. Pretreatment of lignocellulosic wastes to improve ethanol and bi-ogas production: A review. *Int. J. Mol. Sci.* **2008**, *9*, 1621–1651. [CrossRef] [PubMed]
2. Parawira, W. Enzyme research and applications in biotechnological intensification of biogas production. *Crit. Rev. Biotechnol.* **2012**, *32*, 172–186. [CrossRef] [PubMed]
3. Speda, J.; Johansson, M.A.; Odnell, A.; Karlsson, M. Enhanced biomethane production rate and yield from lignocellulosic ensiled forage ley by in situ anaerobic digestion treatment with endogenous cellulolytic enzymes. *Biotechnol. Biofuels* **2017**, *10*, 129. [CrossRef] [PubMed]

4. Fernandes, T.; Bos, G.K.; Zeeman, G.; Sanders, J.; van Lier, J. Effects of thermo-chemical pre-treatment on anaerobic biodegradability and hydrolysis of lignocellulosic biomass. *Bioresour. Technol.* **2009**, *100*, 2575–2579. [CrossRef] [PubMed]
5. Ferdeș, M.; Dincă, M.N.; Moiceanu, G.; Zăbavă, B.Ș.; Paraschiv, G. Microorganisms and Enzymes Used in the Biological Pretreatment of the Substrate to Enhance Biogas Production: A Review. *Sustainability* **2020**, *12*, 7205. [CrossRef]
Ometto, F.; Quiroga, G.; Pšenička, P.; Whitton, R.; Jefferson, B.; Villa, R. Impacts of microalgae pre-treatments for improved anaerobic digestion: Thermal treatment, thermal hydrolysis, ultrasound and en-zymatic hydrolysis. *Water Res.* **2014**, *65*, 350–361. [CrossRef]
7. Szűcs, C.; Kovács, E.; Bagi, Z.; Rákhely, G.; Kovács, K.L. Enhancing biogas production from agroindustrial waste pre-treated with filamentous fungi. *Biol. Futur.* **2021**, *72*, 341–346. [CrossRef]
8. Schimpf, U.; Schulz, R. Industrial by-products from white-rot fungi production. Part I: Generation of enzyme preparations and chemical, protein biochemical and molecular biological characterization. *Process Biochem.* **2016**, *51*, 2034–2046. [CrossRef]
9. Binner, R.; Menath, V.; Huber, H.; Thomm, M.; Bischof, F.; Schmack, D.; Reuter, M. Comparative study of stability and half-life of enzymes and enzyme aggregates implemented in anaerobic biogas processes. *Biomass Conv. Bioref.* **2011**, *1*, 1–8. [CrossRef]
10. Willenbücher, K.; Wibberg, D.; Huang, L.; Conrady, M. Phage Genome Diversity in a Biogas-Producing Microbiome Analyzed by Illumina and Nanopore GridION Sequencing. *Microorganisms* **2022**, *10*, 368. [CrossRef]
11. Heyer, R.; Schallert, K.; Büdel, A.; Zoun, R.; Dorl, S.; Behne, A.; Kohrs, F.; Püttker, S.; Siewert, C.; Muth, T.; et al. A Robust and Universal Metaproteomics Workflow for Research Studies and Routine Diagnostics Within 24 h Using Phenol Extraction, FASP Digest, and the MetaProteomeAnalyzer. *Front. Microbiol.* **2019**, *10*, 1883. [CrossRef]
12. Wiśniewski, J.R.; Zougman, A.; Nagaraj, N.; Mann, M. Universal sample preparation method for proteome analysis. *Nat. Methods* **2009**, *6*, 359–362. [CrossRef]
13. Gillet, L.C.; Navarro, P.; Tate, S.; Röst, H.; Selevsek, N.; Reiter, L.; Bonner, R.; Aebersold, R. Targeted data extraction of the MS/MS spectra generated by data-independent acquisition: A new concept for consistent and ac-curate proteome analysis. *Mol. Cell. Proteom. MCP* **2012**, *11*, O111.016717. [CrossRef]
14. MacLean, B.; Tomazela, D.M.; Shulman, N.; Chambers, M.; Finney, G.L.; Frewen, B.; Kern, R.; Tabb, D.L.; Liebler, D.C.; MacCoss, M.J. Skyline: An open source document editor for creating and analyzing targeted proteomics experiments. *Bioinformatics* **2010**, *26*, 966–968. [CrossRef]
15. Currie, L.A. Limits for qualitative detection and quantitative determination. In *Application to Radiochemistry*; Analytical Chemistry Division, National Bureau of Standards: Washington, DC, USA, 1968.
16. Mani, D.R.; Abbatiello, S.E.; Carr, S.A. Statistical characterization of multiple-reaction monitoring mass spectrometry (MRM-MS) assays for quantitative proteomics. *BMC Bioinform.* **2012**, *13* (Suppl. 16), S9. [CrossRef]
17. Odnell, A.; Recktenwald, M.; Stensén, K.; Jonsson, B.-H.; Karlsson, M. Activity, life time and effect of hydrolytic enzymes for enhanced biogas production from sludge anaerobic digestion. *Water Res.* **2016**, *103*, 462–471. [CrossRef]
18. Davidsson, Å.; Wawrzynczyk, J.; Norrlöw, O.; Jansen, J.L.C. Strategies for Enzyme Dosing to Enhance Anaerobic Digestion of Sewage Sludge. *J. Residuals Sci. Technol.* **2007**, *4*, 1–7.
19. Garcia, N.H.; Mattioli, A.; Gil, A.; Frison, N.; Battista, F.; Bolzonella, D. David Evaluation of the methane potential of different agricultural and food processing substrates for improved biogas produc-tion in rural areas. *Renew. Sustain. Energy Rev.* **2019**, *112*, 1–10. [CrossRef]
20. Heyer, R.; Schallert, K.; Siewert, C.; Kohrs, F.; Greve, J.; Maus, I.; Klang, J.; Klocke, M.; Heiermann, M.; Hoffmann, M.; et al. Metaproteome analysis reveals that syntrophy, competition, and phage-host interaction shape microbial communities in biogas plants. *Microbiome* **2019**, *7*, 69. [CrossRef]
21. Colgrave, M.L.; Byrne, K.; Blundell, M.; Heidelberger, S.; Lane, C.S.; Tanner, G.J.; Howitt, C.A. Comparing Multiple Reaction Monitoring and Sequential Window Acquisition of All Theoret-ical Mass Spectra for the Relative Quantification of Barley Gluten in Selectively Bred Barley Lines. *Anal. Chem.* **2016**, *88*, 9127–9135. [CrossRef]
22. Montemurro, N.; Orfanioti, A.; Manasfi, R.; Thomaidis, N.S.; Pérez, S. Comparison of high resolution mrm and sequential window acquisition of all theoretical fragment-ion acquisition modes for the quantitation of 48 wastewater-borne pollutants in lettuce. *J. Chromatogr. A* **2020**, *1631*, 461566. [CrossRef] [PubMed]
23. Schievano, A.; Scaglia, B.; D'Imporzano, G.; Malagutti, L.; Gozzi, A.; Adani, F. Prediction of biogas potentials using quick laboratory analyses: Upgrading previous models for application to heterogeneous organic matrices. *Bioresour. Technol.* **2009**, *100*, 5777–5782. [CrossRef] [PubMed]
24. Badieyan, S.; Bevan, D.R.; Zhang, C. Study and design of stability in GH5 cellulases. *Biotechnol. Bioeng.* **2012**, *109*, 31–44. [CrossRef] [PubMed]
25. Trötschel, C.; Poetsch, A. Current approaches and challenges in targeted absolute quantification of membrane proteins. *Proteomics* **2015**, *15*, 915–929. [CrossRef]
26. Liew, Y.X.; Chan, Y.J.; Manickam, S.; Chong, M.F.; Chong, S.; Tiong, T.J.; Lim, J.W.; Pan, G.-T. Enzymatic pretreatment to enhance anaerobic bioconversion of high strength wastewater to biogas: A review. *Sci. Total Environ.* **2020**, *713*, 136373. [CrossRef]

27. Aghtaei, H.K.; Püttker, S.; Maus, I.; Heyer, R.; Huang, L.; Sczyrba, A.; Reichl, U.; Benndorf, D. Adaptation of a microbial community to demand-oriented biological methanation. *Biotechnol. Biofuels Bioprod.* **2022**, *15*, 125. [CrossRef]
28. Küchler, J.; Püttker, S.; Lahmann, P.; Genzel, Y.; Kupke, S.; Benndorf, D.; Reichl, U. Absolute quantification of viral proteins during single-round replication of MDCK suspension cells. *J. Proteom.* **2022**, *259*, 104544. [CrossRef]

MDPI AG
Grosspeteranlage 5
4052 Basel
Switzerland
Tel.: +41 61 683 77 34

Fermentation Editorial Office
E-mail: fermentation@mdpi.com
www.mdpi.com/journal/fermentation

MDPI